NONLINEAR SCIENCE THEORY AND APPLICATIONS

Numerical experiments over the last thirty years have revealed that simple nonlinear system can have surprising and complicated behaviours. Nonlinear phenomena include waves that behave as particles, deterministic equations having irregular, unpredictable solutions, and the formation of spatial structures from an isotropic medium.

The applied mathematics of nonlinear phenomena has provided metaphors and models for a variety of physical process: solitons have been described in biological macromolecules as well as in hydrodynamic systems; irregular activity that has been identified with chaos has been observed in continuously stirred chemical flow reactors as well as in convecting fluids; nonlinear reaction diffusion systems have been used to account for the formation of spatial patterns in homogeneous chemical systems as well as biological morphogenesis; and discrete-time and discrete-space nonlinear systems (cellular automata) provide metaphors for processes ranging from the microworld of particle physics to patterned activity in computing neural and self-replication genetic systems.

Nonlinear Science: Theory and Applications will deal with all areas of nonlinear science – its mathematics, methods and applications in the biological, chemical, engineering and physical sciences.

Nonlinear science: theory and applications

Series editor: Arun V. Holden, *Reader in General Physiology, Centre for Nonlinear Studies, The University, Leeds LS2 9NQ, UK*
Editors: S. I. Amari (Tokyo), P. L. Christiansen (Lyngby), D. G. Crighton (Cambridge), R. H. G. Helleman (Houston), D. Rand (Warwick), J. C. Roux (Bordeaux)

Chaos A. V. Holden (*Editors*)

Control and optimization J. E. Rubio

Automata networks in computer science F. Fogelman Soulié, Y. Robert and M. Tchuente (*Editors*)

Simulation of wave processes in excitable media V. S. Zykov

Introduction to the algebraic theory of invariants of differential equations K. S. Sibirsky

Mathematical models of chemical reactions P. Erdi and J. Tóth

Almost periodic operators and related nonlinear integrable systems V. A. Chulaevsky

Oscillatory evolution processes I. Gumowski

Soliton theory: a survey of results A. P. Fordy (*Editor*)

Fractals in the physical sciences H. Takayasu

Stochastic cellular systems: ergocidity, memory, morphogenesis R. L. Dobrushin, V. I. Kryukov, A. L. Toom (*Editors*)

Chaotic oscillations in mechanical systems T. Kapitaniak

Stability of critical equilibrium states L. G. Khazin and E. E. Shnol

Nonlinear random waves and turbulence in nondispersive media: waves, rays, particles S. N. Gurbatov, A. N. Malakhov and A. I. Saichev

Nonlinear chemical waves Peter J. Ortoleva

Other volumes are in preparation

Nonlinear random waves and turbulence in nondispersive media: waves, rays, particles

S. N. GURBATOV, A. N. MALAKHOV
and A. I. SAICHEV

Translation edited by D. G. Crighton

Manchester University Press
Manchester and New York
Distributed exclusively in the USA and Canada by St. Martin's Press

Published by Manchester University Press
Oxford Road, Manchester M13 9PL, UK
and Room 400, 175 Fifth Avenue,
New York, NY 10010, USA

Distributed exclusively in the USA and Canada by
St. Martin's Press, Inc.,
175 Fifth Avenue, New York, NY 10010, USA

British Library cataloguing in publication data
Gurbatov, S. N.
Nonlinear random waves and turbulence in nondispersive media.
1. Nonlinear waves
I. Title II. Malakhov, A. N. III. Saichev A. I. IV. Series
531.33

Library of Congress cataloging in publication data
Gurbatov, S. N.
Nonlinear random waves and turbulence in nondispersive media: waves, rays, particles / S. N. Gurbatov, A. N. Malakhov and A. I. Saichev.
p. cm. – (Nonlinear science : theory and application)
Translated from the Russian.
Includes bibliographical references and index.
ISBN 0-7190-3275-X
1. Nonlinear waves. 2. Turbulence. I. Malakhov, A. N. II. Saichev, A. I. III. Title. IV. Series: Nonlinear science.
QA927.G87 1991
501'.174—dc20 91-18538
CIP

ISBN 0 7190 3275 X *hardback*

Set in 10/12 pt Times by Graphicraft Typesetters Ltd., Hong Kong
Printed in Great Britain by Biddles Ltd., Guildford & King's Lynn.

Contents

Translation editor's preface

This volume adds to the growing number of books available in English translation which report important Soviet work in nonlinear wave theory and in particular in nonlinear acoustics, a subject badly under-represented in Western science. It deals with stochastic problems for the scalar and vector Burgers equations (and occasionally with one-dimensional nondispersive waves with arbitrary nonlinearity), and in fact for the most part deals with the vanishing dissipation limit which yields the Riemann equation $v_t + vv_x = 0$ together with the weak shock prescription for discontinuities.

The main matter at issue is the relationship between the statistical descriptors of waves in the Eulerian representation (e.g. the correlation function $\langle v(x, t)\, v(x + r, t)\rangle$ and its Fourier transform, the power spectral density of $\langle v^2\rangle$), which is the description usually required in practice, and the statistics in the Lagrangian representation, for which the evolution is extremely simple, $\mathrm{d}v/\mathrm{d}t = 0$. In deterministic problems, analysis usually exploits particular features in particular problems to unravel the Eulerian–Lagrangian transformation; but a complete solution of the Eulerian statistical evolution problem requires one to address the general problem of Eulerian–Lagrangian statistics head-on, and accordingly that topic occupies a significant part of this book. It is material not readily to hand elsewhere, and should be useful in many other contexts.

The discussion generally identifies the field $v(x, t)$ with the velocity field of gasdynamics, but there is frequent mention of the ray optical analogue, in which $v = S_x$, the action S is the wave phase function, and the Hamilton–Jacobi equation $S_t + \frac{1}{2}S_x^2 = 0$ is simply the potential form of the Riemann equation, caustic formation being the optical analogue of shock formation. There is also frequent interpretation, naturally, in terms of Lagrangian flows of noninteracting particles, and of 'sticking' particles in the dissipative (Burgers) model.

In the final chapters, a most unusual topic of long-standing interest is analysed – that of the agglomeration of gravitating matter on the large scale in the Universe, and the formation of a cellular structure with galaxies in the nodes. Here the model put forward (a development of one due originally to Zel'dovich) comprises the vector Burgers equation for velocity $\mathbf{v}(\mathbf{x}, t)$, a constraint that guarantees the irrotationality of $\mathbf{v}(\mathbf{x}, t)$ (and thereby permits the Hopf–Cole linearising tranformation $\mathbf{v} = -2\mu\nabla \ln \Phi$) and a continuity equation for density. From unstructured initial perturbations, a field of (non-parallel) planar shocks is formed, defining a cellular structure of planes, ribs and nodes. Normal velocities jump at the shocks, tangential velocities remain continuous, and the velocity $\mathbf{v}$ is linear in $\mathbf{x}$ between shocks. Therefore the matter is transferred to the shocks, then along them to the ribs, and finally to the nodes, in processes for which this book gives an exact statistical description. Comparisons with recent direct numerical simulations of the motion of a large number of gravitating point masses confirm the essence and accuracy of this description. It will come as a matter of great satisfaction to workers in theoretical nonlinear acoustics (dispersionless nonlinear waves) that the most basic model in that field provides, in its vector form, a remarkably complete and explicit analytical solution of the problem of the large-scale evolution of the matter distribution in the Universe.

I hope that this book will make the work of the authors and of their colleagues in the USSR accessible in the West, and that it will find readers in mechanics, wave theory and astrophysics. I hope also that the scientific message has been correctly and clearly rendered, without undue deviation from the style of the Russian original, in this translation.

D. G. Crighton
Cambridge, March 1990

Authors' preface

This book is devoted to a theoretical study of the evolution laws of random nonlinear waves of diverse physical nature propagating in nondispersive media.

Medium nonlinearity with no dispersion leads to a pronounced enrichment of the wave spectral composition due to intensive generation of interacting harmonics, as a wave propagates. Because of this one cannot analyse the waves under consideration using traditional techniques, widely employed, for example, in nonlinear optics. Additional difficulties are imposed by the wave discontinuities arising in nondispersive media with time.

In view of the above, new ways should be outlined to investigate random strongly nonlinear waves in nondispersive media. In this book we offer the experience gained when developing an approach mainly relying on a body of Lagrangian and Eulerian statistics of random fields. The method developed allows one to embrace in a unified manner a wide variety of random strongly nonlinear waves in nondispersive media, to reveal their specific features and fundamental evolution laws, and, finally, to apply the results to the solution of some particular physical problems.

This book is based on lectures delivered by the authors at the Radiophysical Department of the Gorky State University in different years and is mainly supported by original works of the authors in the statistics of nonlinear random waves of hydrodynamic type. Although there are many mathematical calculations in the book, it has been written in physical language and at a 'physical level' of rigour and, moreover, many analytically intractable problems have been tackled by only making qualitative use of estimated results. References completing the book mostly cover only the contributions of the last one and a half decades and are far from comprehensive.

The authors are grateful to A. L. Melott and S. F. Shandarin, V. I. Arnold, Yu. M. Baryshnikov and I. A. Bogayevsky who kindly agreed to write supplements to the book. Also deserving acknowledgement is the considerable service rendered by G. I. Barenblatt, N. G. Denisov, Ya. B. Zel'dovich, V. I.

Klyatskin, Yu. A. Kravtsov, L. A. Ostrovsky, E. N. Pelinovsky, O. V. Rudenko, V. I. Tatarsky, V. E. Fridman, A. S. Chirkin, S. F. Shandarin and I. G. Yakushkin in multiple discussions of the problems considered in this book. We would also like to acknowledge Blackwell Scientific Publications Ltd for permission to reproduce Figures 6.11–6.17 [167].

University of Nizhni Novgorod
(Gorky State University)

S. N. Gurbatov
A. N. Malakhov
A. I. Saichev

January 1990

Introduction

Evolution of waves of any physical nature is determined by the medium in which a wave propagates. If a medium is characterised by its own spatial and temporal scales, one can speak about waves in dispersive media. If, besides, the medium response is changed with the wave amplitude variation, then a wave of sufficiently large intensity is referred to as nonlinear. Nonlinearity and the dispersive properties of a medium, when combined with geometrical characteristics of a wave and dissipation processes, give rise, in total, to the whole variety of wave processes encountered in nature. Among them we manage to separate classes of related waves possessing similar specific features. The latter are mainly specified by a principal mechanism of formation of the waves belonging to a given class – for example, dispersion and nonlinearity.

This book is concerned with random nonlinear waves in nondispersive media. As a characteristic feature of these waves we consider an avalanche-type growth of the wave harmonic number; the wave components interact efficiently leading, in particular, to the small-scale structure emergence in a wave. Such waves include, for instance, nonlinear acoustic waves whose profile distorts nonlinearly during propagation, to eventually give way to the formation of discontinuities, i.e. creation of shock pressure fronts.

In understanding the peculiarities of the nonlinear wave evolution in nondispersive media, a vital role is played by comparatively simple model equations allowing for the main mechanisms of the wave self-action [I.1–I.3]. First of all, it is the Riemann equation

$$\frac{\partial v}{\partial t} + v\frac{\partial v}{\partial x} = 0,$$

that describes the wave profile nonlinear distortions due to the wave propagation velocity dependence on the wave magnitude, and then the Burgers equation

$$\frac{\partial v}{\partial t} + v\frac{\partial v}{\partial x} = \mu\frac{\partial^2 v}{\partial x^2},$$

the right-hand side of which takes into account the wave small-scale component dissipation responsible for the shock front stabilisation.

The solution of the Riemann equation can have an obvious physical interpretation: it describes the velocity field of noninteracting particle flow. This field can be described in two ways: either we observe a wave profile at a fixed point and at a fixed time instant and have the so-called Eulerian description, or we take interest in the separate particle behaviour and come to the Lagrangian description [I.4]. Whereas under the Lagrangian description the movement of an individual particle proves trivial, that is to say it travels with constant velocity with all nonlinear effects clearly absent, when one observes the velocity profile of the particle flow, such nonlinear effects as profile steepening and change of the spatial spectral composition can be distinguished. As was stated in reference [I.5], dedicated to a laboratory study of caustics, optical modelling of the particle motion, and cosmology: 'consideration of a continuous medium, i.e. not a single particle but a whole ensemble of them, leads to interesting and nontrivial results even in this simple case'.

In this connection it should be emphasised that nonlinear effects may equally well appear in linear hyperbolic systems [I.1]. Thus, if one is interested in the complex phase of the field described by the linear wave equation, the necessity arises to analyse nonlinear equations. Namely, if we confine ourselves to the geometrical optics approximation, then evolution of the arrival angles of a wave front (phase gradient) in a two-dimensional homogeneous medium is reduced to the Riemann equation, and the equation for intensity coincides with that of continuity with respect to the density of the noninteracting particle flows. Consequently, when recording arrival angles and the intensity of a wave, one can observe such typically nonlinear effects as steepening of the slope of the angle profile, including the appearance of non-single-valuedness that physically corresponds to multi-stream regimes of propagation, and the emergence of localised regions of increased intensity.

In Figure I.1 we present a photograph taken from a study by Martin and Flatte [I.6] depicting the intensity fluctuations of an optical wave that has crossed a layer of a randomly inhomogeneous medium. This picture offers the results of numerical simulation. Regions of increased intensity are shown in lighter colours. It is implied here that these regions are substantially anisotropic and form a connected cellular structure. A very similar photograph is given on the fly-leaf of a book by Rytov et al. [I.7] which illustrates experimental results obtained in recording the optical wave intensity fluctuations in a turbulent atmosphere.

Such structures also arise during the development of a cold gas gravitational instability. Figure I.2 suggests the results obtained by Melott and

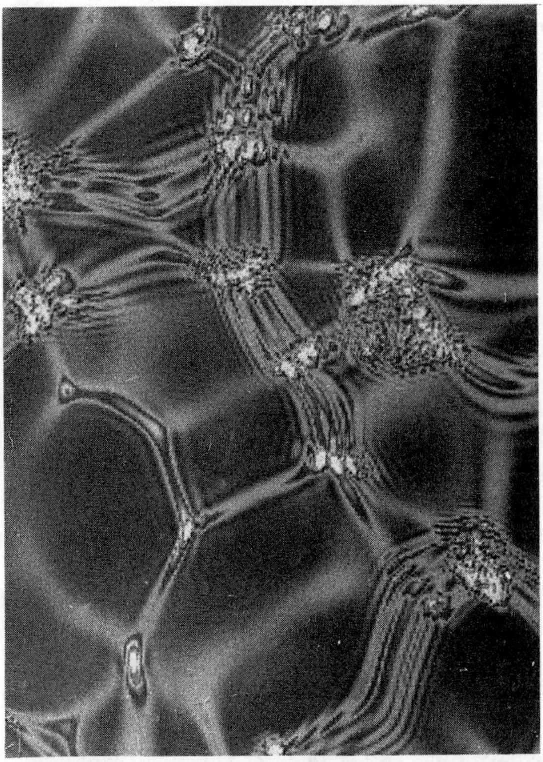

Fig. I.1

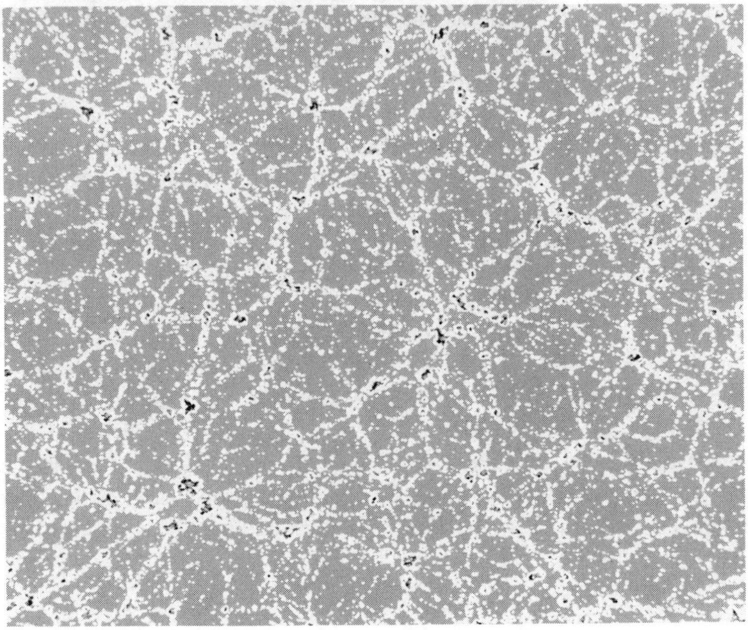

Fig. I.2

Shandarin in numerical simulation of two-dimensional flow of a gas of gravitationally interacting particles [I.8]. Regions of increased density, emphasised by using lighter colours, prove sharply anisotropic in this case as well. Such a gas is a typical example of a nonlinear medium without dispersion. In the initial stage all scales are equally amplified due to gravitational instability since there is no pressure in the medium, and the formation of anisotropic structures is eventually attributed to kinematic nonlinearity. Here also the transition from the Eulerian description to the Lagrangian one seems fruitful. We suggest in this book that using a model of 'sticking particles' based on the three-dimensional Burgers equation enables one to outline the process involved in the formation of a cellular structure for the matter distribution in the Universe at a nonlinear stage of the gravitational instability development.

The above-mentioned equations and their analogues not only adequately embody nonlinear acoustic waves, but successfully simulate wave phenomena of absolutely different physical origins: nonlinear waves in long transmission lines, kinematic waves, optical waves in the geometrical optics approximation, hydrodynamic flows of particles, and so on. Therefore, the variety of applications and heuristic value of the Riemann and Burgers equations render them standard when analysing the behaviour of nonlinear waves in nondispersive media.

Still more important, as compared with the investigation of such nonlinear dynamics effects, appears to be the discussion of the *statistical property evolution* for solutions of the corresponding equations with *random initial conditions*. The fact is that the Burgers equation, for example, serves as the simplest model equation of strong hydrodynamic turbulence that allows for the joint action of the two mechanisms which are of paramount importance in establishing the properties of real hydrodynamic turbulence – inertial nonlinearity and viscosity. For this reason we pay much attention in this book to studying statistical characteristics of Burgers turbulence such as the energy spectra, correlation functions, probability distributions, shock front statistical properties, distances between the shocks, and so on. It is worthwhile to underline that the problem of model Burgers turbulence is intimately related to that of intensive acoustic noise propagation, important from the point of view of application. The main difficulty, when theoretically analysing strongly nonlinear turbulent motions of liquid and gas, is that concerned with the necessity for solution of an infinite hierarchy of coupled equations for increasingly more complicated statistical properties of turbulence. This difficulty has usually been overcome with the help of a particular approximation or physical hypothesis. For random nonlinear waves in nondispersive media, however, an alternative approach to investigating the field statistics can be thought of, which suggests a practically complete statistical description of the most interesting strongly nonlinear regimes.

The book consists of six chapters. The first chapter contains a brief review of physical examples of nonlinear waves in nondispersive media. The key notions are discussed which characterise such waves. Through the example of Burgers turbulence we have illuminated several aspects of the traditional hypotheses and approximations employed to analyse a strongly nonlinear turbulence.

The second chapter offers the exact solution of the Burgers equation for arbitrary inititial conditions. Behaviour of the solution at large Reynolds numbers is considered in detail. An evident geometrical interpretation of the solution at infinite Reynolds number is demonstrated. Evolution of various types of disturbance followed which, as separate bricks, compose more complex fields ending up with the Burgers turbulence. An analogy is revealed between the Burgers equation solution, noninteracting particle hydrodynamics, and the field of an optical wave in the geometrical optics approximation.

The third chapter develops an efficient method to analyse the statistics of nonlinear fields in nondispersive media. It is based on establishing connections between the statistical properties of the fields of hydrodynamic type in Lagrangian and Eulerian coordinate systems. At the beginning of this chapter we give a list of the basic notions and techniques to analyse statistical characteristics of random fields required for better understanding of the ensuing chapters of the book.

The fourth chapter deals with the calculation and physical interpretation of the probability distributions, spectra and correlation functions of the non-interacting particle flows in a homogeneous medium, and of acoustic noise waves at a stage before discontinuity formation. Attention is also drawn to the problem of nonlinear interaction between regular waves and noise. The intensity fluctuations due to caustic singularities of optical waves are invesitgated. An extension to the particle motion in a field of random external forces is carried out.

The fifth chapter develops a sufficiently comprehensive theory of Burgers turbulence which explicitly takes into account the initiation and further multiple merging of the turbulent field discontinuities. A detailed description is given, as regards the quasi-ordered dissipative structure of the Burgers turbulence realisations at the stage of fully developed discontinuities. It is found that at this stage the statistical properties of the Burgers turbulence have a self-preserving character. A concluding stage of the Burgers turbulence linear decay is studied with due attention.

In the sixth chapter we treat the dynamical and statistical properties concerning the evolution of the so-called potential turbulence described by a three-dimensional Burgers equation, which simulates the nonlinear stage of the gravitationally interacting particle gas instability. Within the framework of the given model we follow the formation and transformation of a cellular large-scale structure of matter distribution in the Universe. Model results are compared with numerical ones allowing for gravitational interaction of cold particles.

There is an Appendix in the book which incorporates a set of formulae connected with the use of a body of delta-function properties in statistical problems, and there are two Supplements. Supplement 1, written by A. L. Melott and S. F. Shandarin, gives the results of numerical simulation of a two-dimensional gas of gravitationally interacting particles. The process of formation and evolution of a cellular structure of matter distribution is shown, as well as a very complicated inner structure of singularities. Supplement 2, contributed by V. I. Arnold, Yu. M. Baryshnikov and I. A. Bogayevsky, suggests a rigorous mathematical classification of singularities for the two- and three-dimensional Burgers equation with vanishing viscosity, and outlines the ways of reconstructing these singularities.

Within each chapter of the book formulae are numbered decimally by section (e.g. in Chapter 2, the sixth formula in Section 2.4 is (4.6)). When a formula is referenced in other chapters, its chapter number is given in addition (so (4.6) becomes (2.4.6)). Figures are numbered decimally by chapter (e.g. Chapter 3 includes Figs 3.1–4).

References for Introduction

[I.1] Whitham, G. B., *Linear and Nonlinear Waves*. New York, London: Wiley-Interscience, 1974.
[I.2] Burgers, J. M. *The Nonlinear Diffusion Equation*. Dordrecht: Reidel, 1974.
[I.3] Sachdev, P. L. *Nonlinear Diffusive Waves*. Cambridge: Cambridge University Press, 1987.
[I.4] Zel'dovich, Ya. B. & Myshkis, A. D. *Elements of Mathematical Physics. A Medium of Noninteracting Particles*. Moscow: Nauka, 1973.
[I.5] Zel'dovich, Ya. B., Mamaev, A. V. & Shandarin, S. F. *Usp. Fiz Nauk* **139**, 1 (1983), 153.
[I.6] Martin, J. M. & Flatte, S. M. *Applied Optics*, **27**, (1988), 2111.
[I.7] Rytov, S. M., Kravtsov, Yu. A. & Tatarsky, V. I. *Introduction to Statistical Radiophysics. Part II. Random Fields*. Moscow: Nauka, 1978.
[I.8] Melott, A. L. & Shandarin, S. F. Gravitational instability with high resolution. Preprint 1989, University of Kansas; *Astrophys J.* **343** (1989), 26.

1 One-dimensional waves in nonlinear nondispersive media

The investigation methods for nonlinear waves in nondispersive media are discussed in this chapter. The Riemann and Burgers equations are presented as basic equations for the waves under consideration. The problems of random waves in nonlinear nondispersive media and of one-dimensional Burgers turbulence as a hydrodynamical turbulence model are formulated. The main physical circumstances in which nondispersive nonlinear waves arise are considered. The statistical aspects of nonlinear acoustics are treated in more details.

1.1 The basic equations and statistical problems of the theory of random waves in nondispersive media

1. Among the great variety of wave motions, one can single out the classes of dispersive and hyperbolic waves [1–6]. Dispersive waves exist in media with inherent temporal and spatial scales and are characterised by the wave-frequency-dependent propagation velocity. Hyperbolic waves arise in media without intrinsic scales. Such media are considered to be nondispersive, and all weak waves propagating in them have the same frequency-independent velocity.

The simplest equation describing weak hyperbolic waves has the form

$$\frac{\partial v}{\partial t} + c_0 \frac{\partial v}{\partial x} = 0, \qquad c_0 = \text{const.} \tag{1.1}$$

The solution $v(x, t)$ of this equation is

$$v(x, t) = v_0(x - c_0 t), \tag{1.2}$$

and describes a nondistorted initial field, propagating with velocity c_0.

The nonlinear analogue of equation (1.1) may be expressed in the following form:

$$\frac{\partial v}{\partial t} + c(v)\frac{\partial v}{\partial x} = 0, \tag{1.3}$$

where $c(v)$ is the propagation velocity as a function of the local value of the field v.

According to Whitham [1]: 'The study of this deceptively simple-looking equation will provide all the main concepts for nonlinear hyperbolic waves.' There are still more reasons to apply these words to the statistical problems under consideration, where we need to reconstruct the field statistics at an arbitrary time on the basis of our knowledge about a random initial field $v_0(x)$. All the difficulties and peculiarities inherent in the statistical theory of nonlinear random waves in nondispersive media are distinctly manifested in this simple example.

The equations

$$\frac{\mathrm{d}V}{\mathrm{d}t} = 0, \qquad \frac{\mathrm{d}X}{\mathrm{d}t} = c(V) \tag{1.4}$$

are the system of characteristic equations for the partial differential equation (1.3). Under the initial conditions

$$X(t = 0) = y, \qquad V(t = 0) = v_0(y) \tag{1.5}$$

the solution of (1.4) is given by the relations

$$\begin{gathered} X(y, t) = y + F(y)t, \qquad F(y) = c(v_0(y)), \\ V(y, t) = v_0(y). \end{gathered} \tag{1.6}$$

Thus, the field $v(x, t)$ may be described in two different ways: on the one hand we may consider it at a fixed point of space, at time t. This is the so-called Eulerian description. On the other hand, we may follow the wave evolution along the characteristic $X(y, t)$, defined by the initial coordinate y. This description is a Lagrangian one, with y the Lagrangian coordinate.

To pass from the Lagrangian approach to the Eulerian one it is necessary to solve equation (1.6) with respect to y, i.e. to find an initial Lagrangian coordinate $y(x, t)$ for the wave profile particle which arrives at a given point x at time t.

This procedure is rather difficult for the deterministic cases, and is moreover non-trivial for the random ones. That is why an essential part of this book has been devoted to the peculiarities of the relationship between the Lagrangian and Eulerian statistical characteristics of the field.

Relation (1.6) may be rewritten as

$$v = v_0(x - c(v)t). \tag{1.7}$$

This expression implicitly defines the values of v in point x at time t. It is clear from (1.7) that each initial profile point moves with velocity $c(v)$, being dependent on v, and, consequently, a distortion of the field initial profile takes place during the motion.

2. The solution of (1.7) shows that for $c'_v \neq 0$ any initial profile will topple over, i.e. become multi-valued after a certain time. It is convenient to follow this process in the (x, t)-plane. For the two characteristics, starting from the neighbouring points y and $y + \mathrm{d}y$, we have, from (1.6),

$$X(y + \mathrm{d}y, t) - X(y, t) = (1 + F'(y)t)\,\mathrm{d}y.$$

Thus, if $F'_y < 0$, these characteristics will intersect sooner or later. If we consider the whole totality of characteristics we shall find the solution of (1.3) to be single-valued only within the finite time interval $t \in (0, t_*)$, where

$$t_* = -1/\min_{F'<0} F'(y). \tag{1.8}$$

All the waves described by an equation of the type (1.3) have the same behaviour before the toppling (i.e. for $t < t_*$). After toppling, waves of different nature may be separated into two contrasting types: waves with permitted multi-stream motion (e.g. particle streams), and waves that are single-stream by their nature and for which $v(x, t)$ is an essentially single-valued function of x. Pressure waves in gases may be considered as a typical example of the latter. For such waves we need to take into account the non-local interaction of the sharply varying field with the medium, which prevents the wave toppling over and, along with the nonlinearity, leads to the appearance of steep drops in the wave profile, i.e. results in so-called discontinuities or shock fronts.

There exist two approaches to the analysis of nonlinear waves in non-dispersive media, allowing for shock fronts. The first method (see, for example, references [1, 4, 5]), which almost ignores the process of shock front formation itself, assumes that the fronts are infinitely thin, and singles out from the set of mathematically permitted discontinuous solutions of equation (1.3) the physically correct ones, satisfying some appropriate integral conservation laws. Such discontinuous solutions are named 'generalised solutions'.

The second approach explicitly takes into consideration the non-local nature of the wave interaction with the medium, and yields the more complicated (compared with (1.3)) equations of nonlinear diffusion. For example, it is possible to supplement equation (1.3) as follows [4, 7]:

$$\frac{\partial v}{\partial t} + c(v)\frac{\partial v}{\partial x} = \mu\frac{\partial^2 v}{\partial x^2}. \tag{1.9}$$

As is shown in [4] the solution of (1.9) coincides with the corresponding generalised solution of (1.3) at $\mu \to 0$.

3. The basic equation for waves in nondispersive media is the simple wave or Riemann equation (RE):

$$\frac{\partial v}{\partial t} + v\frac{\partial v}{\partial x} = 0. \tag{1.10}$$

An ordinary physical example which gives a clear interpretation of Riemann waves is the hydrodynamic flow of noninteracting particles, moving along the x-axis. An initial condition (1.5) for equation (1.10) represents, meanwhile, the initial field of the particle velocity. Density $\rho(x, t)$, satisfying the equation of continuity

$$\frac{\partial \rho}{\partial t} + \frac{\partial}{\partial x}(v\rho) = 0, \tag{1.11}$$

is an important field parameter for such a flow, as well as the velocity. Under random initial conditions (1.5), the velocity fluctuations will result in formation and evolution of medium density fluctuations. The solution of (1.10) is single-valued only over the limited time interval $t \in (0, t_*)$ where, according to (1.8),

$$t_* = -1/\min(v_0'(x)). \tag{1.12}$$

From the point of view of particles, t_* is the time when two adjacent particles overtake each other. When $t = t_*$ the gradient catastrophe occurs and v_x' becomes infinite, while for $t > t_*$ the solution of RE becomes a multi-valued one, and the particle motion becomes a many-stream one.

4. Different diffusion-type equation may correspond to the same equation (1.10). In a medium with quadratic nonlinearity and linear dissipation, such an equation has the form

$$\frac{\partial v}{\partial t} + v\frac{\partial v}{\partial x} = \mu\frac{\partial^2 v}{\partial x^2}. \tag{1.13}$$

Although equation (1.13) first appeared apparently, in physics in 1915 in work [8], this equation now rightfully carries the name of Burgers, since it was he who proposed it as a model equation for hydrodynamical turbulence [7, 9, 10]. The Burgers equation (BE) describes two principle effects, inherent in any turbulence. The first one is the nonlinear redistribution of the energy over the spectrum, while the second is the action of viscosity in a small-scale region. Since external forces are not present in (1.13), BE describes the decay of the turbulence as well, i.e. the nonlinear transformation and damping of the initial field $v_0(x)$.

Generally it is customary to consider both weak and strong turbulence. The weak turbulence regime exists for weakly nonlinear waves in strongly dispersive media. For such media one can apply spatial or temporal expansion of nonlinear wave fields in the linear theory modes. This approach reduces the wave problem to the problem of coupled oscillators and appears to be quite fruitful, e.g. in statistical nonlinear optics [11, 12], where to analyse the interaction of a few harmonics the slowly-varying-amplitude method is used. In common cases, however, the applicability of such a procedure is restricted to a small number of oscillators interacting with the one under consideration, and to a weak binding energy with respect to the oscillator's own energy. Some waves, if they satisfy the weak binding condition, may live for quite a long time and may be treated as quasi-modes of the system. It is this state that we call a 'weakly turbulent' one. The assumption of phase randomisation of certain harmonics is correct in this case. Further, the energy in such a turbulence is transferred by stages over the spectrum between non-coherent harmonics. The use of kinetic equations to describe this process made it possible to obtain important results in the theory of weak plasma turbulence, weak acoustic turbulence in dispersive media, and in the theory of surface waves, and so on [13, 14, 115, 116].

Strongly interacting harmonics become strongly coupled ones. In this case it is customary to talk about strong turbulence. To analyse such turbulence it is frequently more convenient to introduce the coordinate representation and new degrees of freedom which correspond to some regular solutions of the dynamical equations, i.e. to nonlinear modes. As an example one can consider hydrodynamical turbulence, where we are leading with the description of random fluid motions as a system of interacting eddies [15, 16]. The parameter of interaction (though between *nonlinear* modes) is rather significant here as well. For instance, the mode interaction in a gas of solitons (formations steadily separated in space) can be treated as a weak and non-coherent interaction [17].

Another situation applies for Burgers turbulence, which appearing in a nondispersive, weakly-dissipative medium. Here, due to coherent interaction of all harmonics, sawtooth shock waves are formed, which may be treated as a particle gas with strong local interaction. While solitons penetrate each other essentially without any interaction, shock front collision actually results in their coalescence, which is equivalent to an inelastic particle collision. As will be shown in the following discussion, the description of such a system, being an example of strong turbulence, is essentially based upon interaction locality, and may be effected both in terms of kinetic equations for quasi-particles and in terms of a rigorous solution of the dynamical problem.

5. Now let us formulate statistical problems of random waves in nondispersive media. As was already noted, the problem of Burgers turbulence (BT) description is reduced to an analysis of the statistical characteristics of the solution to equation (1.13) with random initial conditions (1.5). First of all it is important to know the time dependence of the correlation function:

$$B_v(s, t) = \langle v(x, t)\, v(x + s, t)\rangle, \qquad (1.14)$$

the power spectrum

$$g_v(k, t) = \frac{1}{2\pi} \int_{-\infty}^{\infty} B_v(s, t)\, \mathrm{e}^{iks}\, \mathrm{d}s \tag{1.15}$$

and, in the general case, the many-point probability distributions of the field $v(x, t)$. For strongly nonlinear turbulence, represented for example by BT, very important information is contained in the statistical characteristics of the turbulence quasi-modes, i.e. in our case in the probability distributions of the shock discontinuity amplitudes, of their propagation velocities, and so on.

Now we briefly draw attention to the main concepts of approximate techniques, applied to statistical examination of nonlinear random waves in nondispersive media, including Burgers turbulence.

It is reasonable to point out the following groups of approximate methods.

From BE (1.13) for the correlation function of statistically homogeneous turbulence, one can find the following equation, which is an analogue of the Kármán–Howarth equation in the theory of hydrodynamic turbulence [18]:

$$\frac{\partial B_v(s, t)}{\partial t} + \frac{1}{2}\frac{\partial}{\partial s}[B_{21}(s, t) + B_{12}(s, t)] = 2\mu \frac{\partial^2 B_v(s, t)}{\partial s^2}. \tag{1.16}$$

This equation is not closed, since it includes the third moment functions, having, for example, the form

$$B_{21}(s, t) = \langle v^2(x + s, t)\, v(x, t)\rangle .$$

The equation for B_{21} contains, in turn, the fourth moment functions, and so on. The idea of the first group of methods is the truncation of an infinite chain of equations by means of setting to zero the higher moments or cumulants. These methods are valid for small Reynolds numbers, or when the probability distributions are nearly Gaussian. In particular, the reduction of the third moment by means of Gaussian approximation formulae yields $B_{21} \equiv 0$. Consequently, the first non-trivial approach is achieved, setting the fourth cumulant to zero, i.e. making use of Millionshchikov's hypothesis. However, model probability distributions with a finite (more than two) set of cumulants are not positive everywhere [19]. For this reason, this hypothesis leads for large Reynolds numbers to the occurrence of negative values for the energy spectrum [20].

A detailed review of the closure techniques for turbulence in a space with various numbers of dimensions, BT included, is presented in [21], while various aspects of the method are depicted in [22, 23].

The essence of the second approach is to replace a non-closed equated (1.16) by a model equation of the type

$$\frac{\partial B_v(s, t)}{\partial t} = -\alpha \frac{\partial^2}{\partial s^2}[B_v(s, t) - B_v(0, t)]^2 + 2\mu \frac{\partial^2 B_v(s, t)}{\partial s^2}. \tag{1.17}$$

This is the approximation of Kraichnan's 'direct interactions' [25, 26] and the method of 'Markov stochastic coupled models' [27–29]. Here, depending on the closure procedure, $\alpha = t$ or $\alpha = \tau_0$, where τ_0 is a characteristic temporal scale. A great number of publications have dealt with investigation of equations of type (1.17) (see, for example, [30]). It may be concluded from (1.17), that as $\mu \to 0$, analyticity of the correlation function

$$B_v(s, t) = E(t) - D(t)s^2 + \dots,$$

where $E(t) = \langle v^2(x, t) \rangle$ and $D(t) = \langle (v_x)^2 \rangle$ is violated within a finite time interval t_*. For $t < t_*$ wave energy is conserved, but for $t > t_*$ the correlation function may be represented as

$$B_v(s, t) = E(t) - D(t) \, |s| + \dots,$$

which results in wave energy damping, as follows from (1.17). The failure of the analyticity of the correlation functions is connected, together with the presence of wave energy dissipation even in the limit $\nu \to 0$, with the formation of discontinuities. However, equation (1.17) does not permit us to provide a comprehensive investigation of turbulent behaviour at large times, when the wave evolution begins to depend on the properties of initial perturbations in a large-scale region, i.e. on the correlation function *integral* features.

To analyse BT, the Wiener–Cameron–Martin expansion technique [31–34] has been widely used, in which the random field $v(x, t)$ is expanded in a functional series in stochastic orthogonal functions $M_n(x_1, x_2, \dots, x_n)$:

$$v(x, t) = \int K_1(x - x_1, t) \, M_1(x_1) \, dx_1 + \iint K_2(x - x_1, x - x_2) \, M_2(x_1, x_2) \, dx_1 \, dx_2 + \dots.$$

Here the first term in the series allows for the Gaussian component, while others describe non-Gaussian effects. Because of the nonlinearity of BE (1.13), an equation chain for the kernels K_n will be infinite and its truncation would not make it possible to analyse BT for the practically interesting case of large Reynolds numbers. It is worth mentioning reference [35, 36] where a field expansion in the linear theory modes was used to construct a cascade model of turbulence.

The most general approach to the problem of turbulence is to analyse the Hopf equation for the characteristic functional. The one-dimensionality of BE makes it possible to advance along this line further than in the eddy turbulence problem. However, the absence of stardard methods for the solution of functional equations impedes clear definition of the range of applicability of the results obtained.

In this book we are going to propose another procedure, which is based on BT representation as an ensemble of stable strongly nonlinear modes. But BT is not considered to the only representative in the class of problems related to random waves in nondispersive media. RE and BE may be applied also to the wide range of physical systems with rather weak nonlinearity and without interaction between oppositely travelling waves, and where the dispersion is negligibly small as compared with damping. As illustrations of it, one may point to electromagnetic waves in transmission lines [40–44], nonlinear waves in thermoelastic media [45], large-intensity longitudinal acoustic waves [46–48], and turbulent ionisation waves in a luminous discharge column [50]. Moreover, similar equations also appear when one consider short-wave radiation in dispersive media, where, in the geometrical optics approximation, equations describing the wave propagation coincide with equations of the type (1.3).

1.2 Physical examples of nonlinear waves

1. It may be stated that the most significant example of nonlinear waves in nondispersive media is represented by nonlinear acoustic waves. In this case BE results from the equations of hydrodynamics of a viscous heat-conducting fluid, assuming slow wave-profile variation due to nonlinearity and dissipation [51–53]. A more general problem, of the propagation of nonlinear sound beams with narrow angular spectrum, may be described in the framework of the Zabolotskaya–Khokhlov–Kuznetsov equation: [54, 55]:

$$\frac{\partial}{\partial\tau}\left(\frac{\partial p}{\partial z}-\frac{\varepsilon}{c_0^3\rho_0}p\frac{\partial p}{\partial\tau}-\frac{b}{2c_0^3\rho_0}\frac{\partial^2 p}{\partial\tau^2}\right)=\frac{c_0}{2}\Delta_\perp p, \tag{2.1}$$

where z is distance along the beam axis, $\tau = t - z/c_0$ is the coordinate running with the wave, c_0 is the linear sound velocity, b is the dissipation coefficient, linearly dependent on the medium viscosity and heat conduction, ε is the dimensionless parameter of nonlinearity (for gases in particular $\varepsilon = (\gamma + 1)/2$, where γ is the adiabatic index), $\Delta_\perp$ is the transverse Laplacian. Questions of the dynamical behaviour contained in the deterministic solutions of (2.1), and of many other equations of nonlinear acoustics, are discussed in the forthcoming book by Ostrovsky and Naugol'nykh [160].

Equation (2.1) served as the basis for systematic development of the theory of parametric sound generators [57] and was used in numerical calculations of nonlinear beam diffraction [56]. Various approximate methods play an important role in nonlinear diffraction theory. The possibility of application of approximate solutions of (2.1) to describe nonlinear beam diffraction with space–time statistics is quite thoroughly presented in the review [58]. Here we confine ourselves to a case of regular-field spatial structure, and to a case when the nonlinear beam diffraction problem is reduced to the one-dimensional one, one way or another.

When analysing sound beam propagation we may single out three main physical processes which define its evolution. They are linear damping, linear diffraction and the wave profile nonlinear distortion. Each one of these processes is considered to have a characteristic length for which those events take place. Let the characteristic wavelength be equal to λ and the field amplitude to p_0, beam aperture size being a. Then a characteristic diffraction length is $l_d \sim a^2/\lambda$, a damping length is $l_r \sim \lambda^2 c_0 \rho_0 / b$, and the length scale for nonlinear distortion is $l_n \sim \lambda c_0^2 \rho_0 / p_0$. The ratios of these lengths define two independent parameters: the Reynolds number $R = l_r/l_n$ describing the nonlinear contribution relative to dissipative effects, and the parameter $N = l_n/l_d$, showing the relative influence of diffraction and nonlinear effects. A more rigorous approach requires us to consider local scale ratios $R(z)$, $N(z)$ since, due to diffraction and damping, the relative contribution of nonlinear effects may vary as the wave propagates.

The so-called 'stage-by-stage' approach appears to be rather fruitful to analyse the diffraction of nonlinear sound beams. The method assumes that the whole wave evolution range is separated into several intervals, each with a predominant physical effect. Namely, depending on the $N(z)$ value one may single out such regions as a nonlinearity-dominated region where $N \gg 1$, a diffraction-dominated region where $N \ll 1$, and an intermediate region of solution 'matching', where $N \sim 1$. The stage-by-stage approach may be applied if the characteristic size of the intermediate region is much less than both the nonlinear and diffraction intervals.

If diffraction is small in the initial interval ($N(0) \ll 1$), (2.1) leads to BE

$$\frac{\partial p}{\partial z} - \beta p \frac{\partial p}{\partial \tau} = \mu \frac{\partial^2 p}{\partial \tau^2}, \qquad p(\tau, z = 0) = p_0(\tau), \tag{2.2}$$

where $\beta = \varepsilon / c_0^3 \rho_0$, $\mu = b / 2 c_0^3 \rho_0$. If, in addition, the Reynolds number is large, equation (2.2) is converted into RE:

$$\frac{\partial p}{\partial z} - \beta p \frac{\partial p}{\partial \tau} = 0, \tag{2.3}$$

the solution of which, however, requires the introduction of discontinuities. If, further, beginning with some z_* diffraction effects become dominant, then for $z > z_*$ the field may be described with the help of linear diffraction theory. When $N(0) \gg 1$ diffraction prevails over nonlinearity at the initial stage, and therefore the field at the first stage can be calculated within the linear theory concepts in terms of a non-stationary Kirchhoff integral [59]. Further study of the field distortion is possible within the framework of the nonlinear geometrical acoustics of spherically divergent waves.

In the nonlinear geometrical acoustics approximation (the beam diffraction is ignored) the wave propagation is described along the beam trajectory. In particular, in the problem of spherical wave diffraction the wave evolution is assumed to coincide with the spherical wave evolution. The equation for one-dimensional non-plane waves differs from (2.2) by an additional term allowing for the wave geometric divergence [49, 51–53]:

$$\frac{\partial P}{\partial r} + \frac{n}{2r} P - \beta P \frac{\partial P}{\partial \tau} = \mu \frac{\partial^2 P}{\partial \tau^2}. \tag{2.4}$$

Here r is a distance, measured from a certain radius r_0 where a wave is specified; $n = 2$ and $n = 1$ serve for spherical and cylindrical waves, respectively. It is supposed, in the geometrical acoustics approximation, that on the sphere $r = r_0$ the field P_0 is given, which in the stage-by-stage approach is defined from the Kirchhoff integral. Introducing new variables for spherical waves according to

$$p = Pr/r_0, \quad z = r_0 \ln (r/r_0), \tag{2.5}$$

we have instead of (2.4)

$$\frac{\partial p}{\partial z} - \beta p \frac{\partial p}{\partial \tau} = \mu e^{z/r_0} \frac{\partial^2 p}{\partial \tau^2}. \tag{2.6}$$

As $\mu \to 0$, i.e. at infinite Reynolds numbers, this equation is converted into RE (2.3). Moreover at infinite Reynolds numbers the equation of nonlinear geometrical acoustics, describing intense beam propagation in smoothly inhomogeneous media [41, 60] is also reduced to the uniform RE. Due to this fact, the results for plane waves may be extended to a wide range of nonlinear acoustics problems.

The effect of medium inhomogeneities is reduced to the peculiarities of the passage to real physical only, i.e. to the change of the wave amplitude and real distance over which the nonlinear distortions occur.

BE, in the form (2.2), is generally used in nonlinear acoustics. In particular, in aeronautical acoustics the problem of depicting the intense noise spectrum transformation is considered to be of great importance [58, 113, 114]. Here the problem of describing the intense wave propagation is to find a field $p(\tau, z)$ and its statistics at an arbitrary station z, from an initial noise perturbation $p_0(z)$ specified at $z = 0$. The wave spherical divergence may be taken into consideration by means of replacing the variables according to (2.5). Equation (2.2) is to be solved under the initial condition $p(\tau, z = 0) = p_0(\tau)$. To pass from BE (2.2) to the classic form of BE (1.13), we formally replace the variables with

$$v = -p, \quad t = \beta z, \quad \tau = x, \quad \mu = \mu/\beta. \tag{2.7}$$

It seems to be obvious that all results obtained for the Burgers turbulence may be automatically extended to problems of intense one-dimensional acoustic noise propagation. For such noise the power spectrum is known to be one of the most important features,

$$S_p(\omega, z) = \frac{1}{2\pi} \int_{-\infty}^{\infty} \langle p(\tau, z)\, p(\tau + \kappa, z)\rangle e^{i\omega\kappa}\, d\kappa. \tag{2.8}$$

Analysing its evolution, we are able to state the nature of the noise energy redistribution over frequencies ω in the course of the noise propagation. If the law of the Burgers turbulence spectrum evolution $g_v(k, t)$ (1.15) is known, then under the same initial conditions $v(x, t = 0) = -p(\tau, z = 0)$, $\tau = x$, $\mu = \mu/\beta$, the spectra (2.8) and (1.15) are related by

$$S_p(\omega, z) = g_v(\omega, \beta z). \tag{2.9}$$

2. As an important example leading to the equations of Riemann and Burgers type, one may consider the so-called kinematic wave [1] defining the continuous distribution of a substance or the state of a medium. In the one-dimensional case we may introduce the density of these waves, $v(x, t)$ per unit length, and the flux $q(x, t)$ through a cross-section per unit time. Supposing that the rate of change of the amount of the substance within $x_1 < x < x_2$ is completely compensated by the total flux through the sections x_1 and x_2, we find that

$$\frac{d}{dt}\int_{x_1}^{x_2} v(x, t)\,\mathrm{d}x + q(x_2, t) - q(x_1, t) = 0. \tag{2.10}$$

If we assume that $v(x, t)$ has continuous derivatives and the flux through the cross-section is a function of density $q = Q(v)$, it is easy to demonstrate, passing to the limit $x_2 \to x_1$, that (2.10) reduces to an equation of type (1.3)

$$\frac{\partial v}{\partial t} + c(v)\frac{\partial v}{\partial x} = 0, \quad c(v) = Q'(v). \tag{2.11}$$

There are many physical problems that lead to equations of type (2.11). For example, for high water (flood) waves in rivers, v is the cross-sectional area of the stream and $Q(v)$ measures the variation of the stream rate through the given cross-section with the change of area. Similar equations appear in the problems of gas chromatography when describing such exotic examples (for a physicist) as transport streams and glacier movement.

A more rigorous description results if we replace the functional connection between q and v by a differential one,

$$q = Q(v) - \mu v_x,$$

i.e. assuming that the flux depends not only on the density but on its gradient v_x as well. In this case (2.10) will transform into a diffusion-type equation, similar to BE.

3. The most important example leading to nonlinear hydrodynamic-type equations is manifested in the equations of geometrical optics. They may be equally interpreted as kinematic equations representing the conservation of the wave crest number during the wave propagation.

In the geometrical optics approximation the fields are sought in the form

$$f(\mathbf{r}, t) = A(\mathbf{r}, t) \exp\,(i\varphi(\mathbf{r}, t)), \tag{2.12}$$

where φ is the phase and A is the amplitude, slowly varying in comparison with $\exp\,(i\varphi)$. Let the spatial–temporal inhomogeneities of the medium be characterised by a parameter $n(\mathbf{r}, t)$. Then the local frequency $\omega(\mathbf{r}, t)$ and the local wave vector $\mathbf{k}(\mathbf{r}, t)$:

$$\omega = \frac{\partial \varphi}{\partial t}, \qquad \mathbf{k} = -\nabla\varphi, \tag{2.13}$$

satisfy the eikonal equation

$$\omega = \omega_d(\mathbf{k}, n(\mathbf{r}, t)). \tag{2.14}$$

Here $\omega = \omega_d(\mathbf{k}, n)$ is the dispersion equation in a homogeneous medium. From (2.13), (2.14) for the local frequency and the local wave vector we have the following set of equation [61–63]

$$\begin{aligned} &\frac{\partial \omega}{\partial t} + (\mathbf{u}.\nabla)\,\omega = F_\omega\,(\omega, \mathbf{k}, n), \\ &\frac{\partial \mathbf{k}}{\partial t} + (\mathbf{u}.\nabla)\,\mathbf{k} = F_\mathbf{k}\,(\omega, \mathbf{k}, n), \end{aligned} \tag{2.15}$$

where $\mathbf{u}$ is the group velocity

$$\mathbf{u} = \frac{\partial \omega_d}{\partial \mathbf{k}}, \qquad F_\omega = \frac{\partial \omega_d}{\partial t}, \qquad F_k = \frac{\partial \omega_d}{\partial \mathbf{k}}.$$

To fully describe the wave field one needs to supplement (2.15) with the equation for the slowly varying field amplitudes.

For a one-dimensional medium without fluctuations the equation for the wave vector $k(x, t)$ has the form

$$\frac{\partial k}{\partial t} + v(k)\frac{\partial k}{\partial x} = 0, \tag{2.16}$$

where $v(k)$ is the group velocity, depending on k in a dispersive medium. It is evident that (2.16) coincides with (1.3), which gives the propagation of the finite-amplitude waves in a nondispersive medium.

However, there are important distinctions between these two classes of waves. While the multi-valued solution (1.3) cannot be realised physically, the multi-valued solution of the geometrical optics equations corresponds to the *actual* multi-beam pattern, i.e. when waves with different frequencies and wave numbers arrive at the same point.

A more detailed study of the aforementioned distinction will be presented later, taking the stationary geometrical optics as an example.

4. In the instances discussed above, the evolution of the wave fields in the Lagrangian representation was reduced either to a free motion of non-interacting particles (1.6) and (1.10), or to particle (beam) motion under a given potential. Nevertheless there are such situations when the Lagrangian approach becomes efficient in the statistical analysis of nonlinear wave

motion and for interacting particle flows. Thus in Chapters 2 and 5 it will be shown that the asymptotic solution of BE as $\mu \to 0$ describes particle flows with absolutely inelastic contact collisions.

Our second example refers to nonlinear Langmuir oscillations of a cold plasma in which electrons interact via electric field. It is known that such oscillations are represented by the equations of hydrodynamic type [129]:

$$\frac{\partial v}{\partial t} + v\frac{\partial v}{\partial x} = -\frac{e}{m} E,$$

$$\frac{\partial E}{\partial x} = 4\pi e(\rho_0 - \rho), \tag{2.17}$$

$$\frac{\partial \rho}{\partial t} + \frac{\partial}{\partial x}(v\rho) = 0,$$

where v, ρ are electron velocity and density, E is an electric field, e, m are the electron charge and mass, and ρ_0 is a motionless ion density. Assuming that as $x \to -\infty$ the flux $v\rho$ equals zero, the equation for $E(x, t)$ may be written as

$$\frac{\partial E}{\partial t} + v\frac{\partial E}{\partial x} = 4\pi e\rho_0 v. \tag{2.18}$$

As a result, the system of characteristic equations for the electron coordinate X, its velocity V and the electric field acting on the electron has the following expression:

$$\frac{\mathrm{d}X}{\mathrm{d}t} = V, \quad \frac{\mathrm{d}V}{\mathrm{d}t} = -\frac{e}{m}E, \quad \frac{\mathrm{d}E}{\mathrm{d}t} = 4\pi e\rho_0 V. \tag{2.19}$$

This system of linear equations describes the finite-amplitude electron oscillations, and on passing to the Eulerian fields $v(x, t)$, $E(x, t)$, $\rho(x, t)$ one may observe such typically nonlinear effects as, for example, generation of spatial harmonics under sinusoidal initial perturbation. It is rather easy to carry out statistical analysis of the behaviour for an individual electron under random initial conditions. However, to investigate the statistics of the Eulerian fields it is necessary to know the relations between the two statistical descriptions: the Lagrangian and the Eulerian. An analysis of the single-point Eulerian characteristics of the nonlinear random Langmuir oscillations of a cold plasma has been accomplished in reference [117]. In the following we shall omit this example, while in Chapter 5 we are going to consider the case of gravitationally interacting particles, which may be reduced to (2.17) if, formally, one assumes that $\rho_0 < 0$.

Let us mention finally the nonlinear random waves in a Hookean elastic body [118]. In the framework of the Eulerian approach, the velocity field $v(x, t)$ and strain field $j(x, t)$ satisfy the following nonlinear equations:

$$\frac{\partial v}{\partial t} + v\frac{\partial v}{\partial x} = c^2(j+1)\frac{\partial j}{\partial x},$$
$$\frac{\partial j}{\partial t} + v\frac{\partial j}{\partial x} = (j+1)\,\frac{\partial v}{\partial x}. \tag{2.20}$$

In the Lagrangian representation, the Lagrangian coordinate $X(y, t)$ of a fixed body particle with an initial coordinate y satisfies the linear equation

$$\frac{\partial^2 X}{\partial t^2} = c^2\frac{\partial^2 X}{\partial y^2}, \tag{2.21}$$

which may be substantially easier to subject to a statistical analysis than the initial system (2.20).

2 One-dimensional wave dynamics

This chapter deals with the general solution of the Burgers equation; its behaviour at large Reynolds numbers and its relation to the Riemann equation solution are discussed. It is shown that the BE solution at large Reynolds numbers permits obvious interpretation in terms of the hydrodynamic flows of noninteracting particles as well as of beams in geometrical optics. The analogy between BE and quasioptics parabolic equation is stated. Evolution of the main types of perturbation is considered.

2.1 Exact solution of the Burgers equation; Reynolds number

1. Let us study the general solution of the Burgers equation (BE)

$$\frac{\partial v}{\partial t} + v\frac{\partial v}{\partial x} = \mu\frac{\partial^2 v}{\partial x^2} \tag{1.1}$$

for the field $v(x, t)$ and discuss its relation with the solution of the Riemann equation (RE)

$$\frac{\partial v}{\partial t} + v\frac{\partial v}{\partial x} = 0, \tag{1.2}$$

assuming the initial field profile to be given as

$$v(x, t = 0) = v_0(x). \tag{1.3}$$

By means of nonlinear change of the variables, suggested by Hopf [64] and Cole [65], BE may be reduced to the linear diffusion equation. To demonstrate this, we represent the field $v(x, t)$ as the gradient of a function $s(x, t)$ (and call it an action):

$$v(x, t) = \frac{\partial}{\partial x}s(x, t). \tag{1.4}$$

After integrating (1.1) over x we obtain for $s(x, t)$

$$\frac{\partial s}{\partial t}+\frac{1}{2}\left(\frac{\partial s}{\partial x}\right)^2=\mu\frac{\partial^2 s}{\partial x^2}. \tag{1.5}$$

By the change of variables

$$s(x, t) = -2\mu \ln \varphi(x, t) \tag{1.6}$$

nonlinear equation (1.5) is reduced to the linear diffusion equation for $\varphi(x, t)$:

$$\frac{\partial \varphi}{\partial t}=\mu\frac{\partial^2 \varphi}{\partial x^2}. \tag{1.7}$$

The desired field $v(x, t)$ can be expressed through the auxiliary field $\varphi(x, t)$ in the following way:

$$v(x, t) = -2\mu\frac{\partial}{\partial x}\ln \varphi(x, t) = -2\mu\frac{\varphi_x}{\varphi}. \tag{1.8}$$

Initial conditions for (1.5), (1.7) are found in accordance with (1.4), (1.6) as

$$s(x, 0) = S_0(x) = \int_{x_0}^{x} v_0(y)\,dy, \tag{1.9}$$

$$\varphi(x, 0) = \varphi_0(x) = \exp\left(-\frac{1}{2\mu}S_0(x)\right). \tag{1.10}$$

The solution of the linear diffusion equation (1.7) has the form

$$\varphi(x, t) = \frac{1}{\sqrt{4\pi\mu t}}\int_{-\infty}^{\infty}\varphi_0(y)\exp\left[-\frac{(x-y)^2}{4\mu t}\right]dy, \tag{1.11}$$

or, allowing for (1.10),

$$\varphi(x, t) = \frac{1}{\sqrt{4\pi\mu t}}\int_{-\infty}^{\infty}\exp\left[-\frac{1}{2\mu}S_0(y)-\frac{(x-y)^2}{4\mu t}\right]dy, \tag{1.12}$$

and, therefore, according to (1.8) the general solution of BE (1.1) under the initial condition (1.3) is written as

$$v(x, t) = \frac{\displaystyle\int_{-\infty}^{\infty}\left(\frac{x-y}{t}\right)\exp\left[-\frac{1}{2\mu}\Phi(x, y, t)\right]dy}{\displaystyle\int_{-\infty}^{\infty}\exp\left[-\frac{1}{2\mu}\Phi(x, y, t)\right]dy}, \tag{1.13}$$

where

$$\Phi(x, y, t) = S_0(y) + \frac{(x-y)^2}{2t}. \tag{1.14}$$

The wide collection of particular solutions of BE is systematised in review [66]. Later we discuss the dependence of the behaviour of general solution on the relationship between nonlinear and dissipative effects.

2. Suppose that the initial perturbation $v_0(x)$ may be characterised by a single amplitude σ_0 and a single spatial scale l_0. Then, from (1.1) we have the estimates of a nonlinear term, $vv_x \sim \sigma_0^2/l_0$ and of the term describing dissipation, $-\mu v_{xx} \sim \mu\sigma_0/l_0^2$. Comparing them with $v_t \sim \sigma_0/t$, one may introduce characteristic times of nonlinear (t_n) and dissipative (t_d) effect manifestation:

$$t_n \approx l_0/\sigma_0, \quad t_d \approx l_0^2/\mu. \tag{1.15}$$

The ratio of these times shows the relative contribution of nonlinear distortions and damping in the process of transformation of the initial field. $v_0(x)$; it is

$$R_0 \approx t_d/t_n \approx \sigma_0 l_0/\mu, \tag{1.16}$$

and is called the acoustic Reynolds number. Since the field is changed with time its characteristic scales are also changed: $\sigma = \sigma(t)$, $l = l(t)$. This leads to the fact that the relative influence of nonlinearity and dissipation, on the field $v(x, t)$ profile distortion, changes in time and the Reynolds number depends on time, $R(t) = \sigma(t)l(t)/\mu$. In the first approximation it may be thought that for $R \ll 1$ the dissipative effects are dominant, while the nonlinearity ones are weak, but for $R \gg 1$ nonlinear effects play a cardinal role in the wave profile distortion.

Let us show that the Reynolds number, naturally, appears in the analysis of the BE exact solution as well. Since the auxiliary field $\varphi(x, t)$ satisfies the linear equation of diffusion (1.7), BE nonlinearity manifests itself in the nonlinear relation (1.8) between the fields φ and v. Consider a group of perturbation with the bounded initial action: $|S_0(x)| < \infty$. For such perturbations $\varphi(x, t)$ is also bounded, and, besides, owing to (1.10), $\varphi \geq 0$. If in $\varphi(x, t)$ we separate a constant component $\overline{\varphi}$ which, according to (1.7), does not depend on time

$$\varphi(x, t) = \overline{\varphi} + \tilde{\varphi}(x, t),$$

we have from (1.8)

$$v(x, t) = -2\mu \frac{\partial}{\partial x} \ln\left(1 + \frac{\tilde{\varphi}(x, t)}{\overline{\varphi}}\right). \tag{1.17}$$

Thus the degree to which nonlinear effects are manifested may be described by the relative changes in $\varphi(x, t)$, and the Reynolds number may be introduced as $R(t) \approx |\Delta\tilde{\varphi}|/\tilde{\varphi}$, where $\Delta\tilde{\varphi}$ is the characteristic change in amplitude of $\varphi(x, t)$. At the initial stage, according to (1.10), relative changes are characterised by ΔS_0, which is the amplitude of initial action variation. Consequently

$$R_0 \approx \left| \frac{\Delta\tilde{\varphi}_0}{\bar{\varphi}} \right| \approx \left| \Delta \ln \varphi_0 (x) \right| \approx \left| \frac{\Delta S_0}{2\mu} \right|. \tag{1.18}$$

It is clear that in order of magitude ΔS_0 is nearly equal to the difference between the maximum and minimum of $S_0(x)$. If, allowing for (1.10), we estimate the initial action increment as $\Delta S_0 \approx \sigma_0 l_0$, (1.18) will coincide with the Reynolds number definition previously introduced.

Finally, for statistically homogeneous random fields the Reynolds number may be expressed through the statistical moments of the field and its derivative as

$$R(t) = \frac{\langle v^2 \rangle}{\mu\sqrt{\langle (v_x)^2 \rangle}}.$$

All these definitions of Reynolds number yield, as a rule, similar-order values. However, it is to be noted that the Reynolds number gives only a rough estimate of the relative influence of nonlinear and dissipative effects, and does always properly describe the evolution stages of sufficiently complex fields.

2.2 Burgers equation solution at large Reynolds numbers

1. As is evident from (1.16), the Reynolds number R_0 varies inversely with the viscosity factor μ. Let us first consider a limit case $R_0 \to \infty$, which corresponds to $\mu \to 0$. Rigorous mathematical analysis of the exact BE solution as $\mu \to 0$ is quite thoroughly described in the classic work by Hopf [64] and monograph [4]. Now we are going to discuss the limiting solution of BE as $R_0 \to \infty$, as well as its relation with BE solution.

Estimating the initial action $S_0(x)$ as $S_0 \sim \sigma_0 l_0$ one can see that for $\sigma_0 l_0 \gg \mu$ a large parameter $R_0 \approx \sigma_0 l_0/\mu$, equal to the Reynolds number, appears in the exponentials of the BE exact solution (1.13). When $R_0 \gg 1$ the main contribution to the integrals in (1.13) comes from the neighbourhood of local minimum points of the function $\Phi(x, y, t)$ where

$$\Phi_y = v_0(y) + \frac{x - y}{t} = 0, \quad \Phi_{yy} > 0. \tag{2.1}$$

In the general case equation (2.1) may have several roots, denoted as $\xi_m(x, t)$. Then, using the method of steepest descent [68], we obtain the following asymptotic expression for the BE solution at large R_0:

$$v(x, t) = \frac{\sum_m \frac{(x-\xi_m)}{t} \mid \Phi''_m \mid^{-\frac{1}{2}} \exp\left(-\frac{1}{2\mu}\Phi_m\right)}{\sum_m \mid \Phi''_m \mid^{-\frac{1}{2}} \exp\left(-\frac{1}{2\mu}\Phi_m\right)}, \tag{2.2}$$

$$\Phi_m = \Phi(x, \xi_m, t), \Phi''_m = \Phi_{yy}(x, \xi_m, t) = v'_0(\xi_m) + \frac{1}{t}. \tag{2.3}$$

Let us discuss now the form of (2.2) at infinite Reynolds number. As $\mu \to 0$ in (2.2), the explession on the right-hand side will be dominated by the summand that has the overall absolute minimum Φ_m and, therefore, the BE solution takes the form

$$v(x, t) = \frac{x - y(x, t)}{t}. \tag{2.4}$$

Here $y(x, t)$ is the y-coordinate of the absolute minimum of the function $\Phi(x, y, t)$ of (1.14),

$$\Phi(x, y, t) = S_0(y) + \frac{(x-y)^2}{2t}, \quad S_0(y) = \int_{y_0}^{y} v_0(x)\,\mathrm{d}x, \tag{2.5}$$

considered at fixed values of x and t.

The limiting solution (2.4) allows us to carry out a very convenient and simple graphical construction for finding the field $v(x, t)$ [7, 69]. Let us view the curve $\Phi(x, y, t)$ and a straight line $\tilde{\alpha}$ with ordinate H, (Fig. 2.1a). It is obvious that the y-coordinate of the absolute minimum appears to be at the same time the abscissa of the first point of tangency of the curve Φ and of the straight line as H increases from $-\infty$. In most cases, however, it is more convenient to use another graphical presentation of this procedure. Introduce a parabola:

$$\alpha(x, y, t) = H - \frac{(x-y)^2}{2t}. \tag{2.6}$$

Then condition (2.1) is identical to the equality $\alpha_y = S'_0$, i.e. to the condition of the parabola α touching the initial action $S_0(y)$. In this case, moreover, the y-coordinate of the absolute minimum of $\Phi(x, y, t)$ coincides with the abscissa of the first point of tangency when the parabola α moves towards the initial action $S_0(y)$ as H increases from $-\infty$ (Fig. 2.1b). If we substitute the resulting coordinate $y(x, t)$ into (2.4), we find the value of the field $v(x, t)$ at point x and at time t. To develop the solution for all x, it is necessary to move the α-parabola centre coordinate, which is equal to x, and search for the abscissa of the first touching point. If we need to find the wave $v(x, t)$ at some later time t, we change the curvature $1/t$ of the parabola α, and carry out the same procedure.

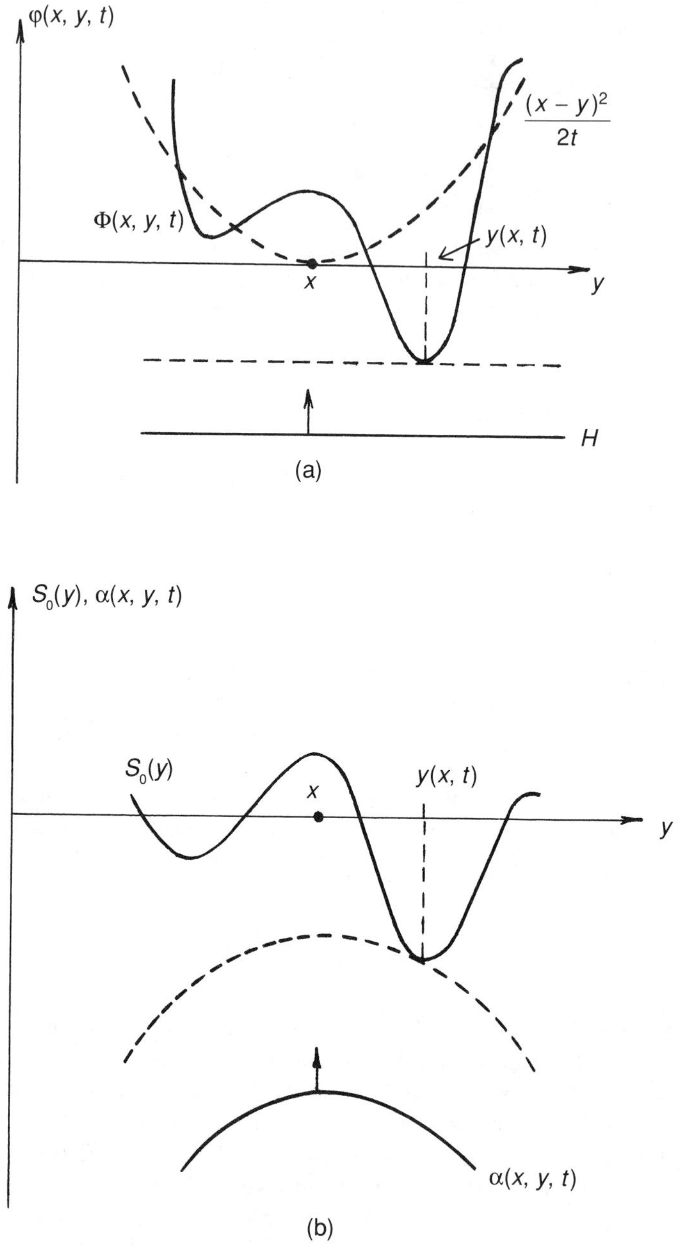

Fig. 2.1 Graphical procedure to seek the absolute minimum coordinates

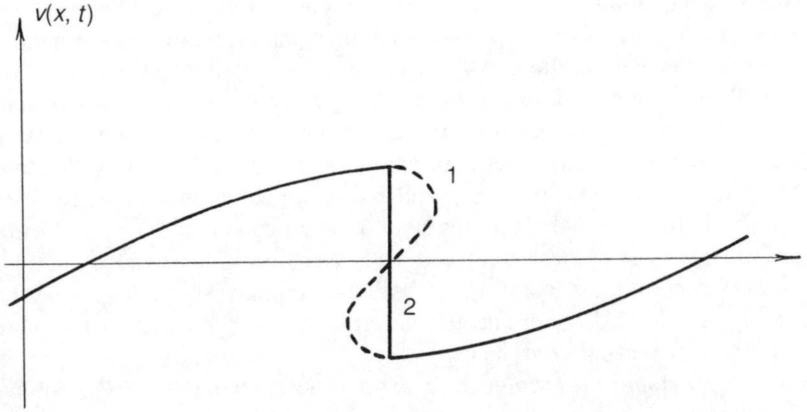

Fig. 2.2 Multi-valued profile of a Riemann wave, 1; an appropriate solution of BE as $\mu \to 0$, 2

2. Let us discuss now the relation of BE asymptotic solution as $\mu \to 0$ with RE (1.2) solution, formally obtained from BE for $\mu = 0$.

If at all y the inequality

$$\Phi_{yy} = v_0'(y) + \frac{1}{t} > 0, \quad \text{that is,} \quad t < -1/v_0'(y) \tag{2.7}$$

is satisfied, then Φ_y is a monotonically increasing function of y, and (2.1) has a single root. Substituting it into (2.4), we come to

$$v(x, t) = v_0(y(x, t)), \tag{2.8}$$

where $y(x, t)$ is the solution of (2.1). If we replace y by the field v with the help of (2.4), we obtain

$$v(x, t) = v_0(x - v(x, t)t), \tag{2.9}$$

an implicit expression for RE ((1.2) and (1.1.7)) solution. If $v_0'^*$ is a gradient minimum value of the initial field $v_0(x)$, inequality (2.7) will be satisfied when $t < -1/v_0'^*$. Therefore, only over a restricted time interval will the BE solution as $\mu \to 0$ coincide with the RE solution.

Estimating $|v_0'^*|$ as σ_0/l_0, one can see that the initial-stage duration, over which the field is described by RE, is equal to $t_n \approx l_0/\sigma_0$. The condition $t < t_n$ means that wave displacement due to nonlinearity, proportional to $\sigma_0 t$, is less than initial field characteristic scale l_0. When $t > t_n$, equation (2.1) already has several roots $\xi_m (x, t)$. Substituting them into expression (2.4) we again acquire the Riemann solution (2.8), which, however, becomes a multi-valued one. Making use of the aforementioned rule for absolute minimum choice, we select, at any x, from the set of formal RE solution branches, only that branch corresponding to the BE asymptotic solution as $\mu \to 0$ (Fig. 2.2).

The passage from an initial evolution stage to a discontinuous one becomes particularly evident when using the graphical procedure for construction of the BE solution. As has been already mentioned, the function $y(x, t)$ which is present in the solution (2.4), coincides with the abscissa of the lowest point of the parabola α (2.6) which touches the initial action $S_0(y)$. The form of tangency between these two curves is defined by the ratio of the parabola curvature $\alpha_{yy} = 1/t$, and the initial action curvature, equal to $v_0'(y) \sim \sigma_0/l_0$. In the initial stage, for small t, when condition (2.7) is fulfilled, the parabola has large curvature, and with variation of the parabola centre coordinate x, the targert point $y(x, t)$ will be continuously sliding over the profile $S_0(y)$ (Fig. 2.3). It immediately follows from (2.4) that in this case the field $v(x, t)$ is itself continuous.

For $t > t_n$, along with segments in which the parabola α slides continuously over initial action $S_0(y)$, jumps of the target point exist when $y(x, t)$ jumps from y_k^+ to y_{k+1}^- during a small variation of x in the neighbourhood of a point x_k (Fig. 2.3). From (2.4), it is evident that at the same point x_k the field $v(x, t)$ has a discontinuity. The localisation of discontinuities x_k may be recovered from the condition of simultaneous touching of the initial action S_0 at two points y_k^+ and y_{k+1}^- by parabola α. This is equivalent to the following equality:

$$\Phi(x_k, y_k^+, t) = \Phi(x_k, y_{k+1}^-, t) \tag{2.10}$$

while in solution of (2.2) the formation of a discontinuity corresponds to the transition of the absolute minimum of Φ from local minimum $\Phi(x, \xi_k, t)$ to local minimum $\Phi(x, \xi_m, t)$ $(m > k)$.

From (2.5) and (2.10) we have for the discontinuity coordinates:

$$x_k = \tfrac{1}{2}(y_{k+1}^- + y_k^+) + V_k t, \tag{2.11}$$

$$V_k = \frac{S_0(y_{k+1}^-) - S_0(y_k^+)}{y_{k+1}^- - y_k^+} = \frac{\int_{y_k^+}^{y_{k+1}^-} v_0(x)\,\mathrm{d}x}{y_{k+1}^- - y_k^+}. \tag{2.12}$$

For the field on the left side of the discontinuity $(x = x_k - 0)$ and on the right side of it $(x = x_k + 0)$ we obtain, due to (2.4) and (2.8) respectively,

$$\begin{aligned} v(x_k - 0, t) &= \frac{1}{t}(x - y_k^+) = v_0(y_k^+), \\ v(x_k + 0, t) &= \frac{1}{t}(x - y_{k+1}^-) = v_0(y_{k+1}^-). \end{aligned} \tag{2.13}$$

The field jump amplitude is represented by

$$\Delta v = \frac{1}{2}[v(x_k - 0, t) - v(x_k + 0, t)] = \frac{1}{2}[v_0(y_k^+ - v_0(y_{k+1}^-)] = \frac{y_{k+1}^- - y_k^+}{2t}. \tag{2.14}$$

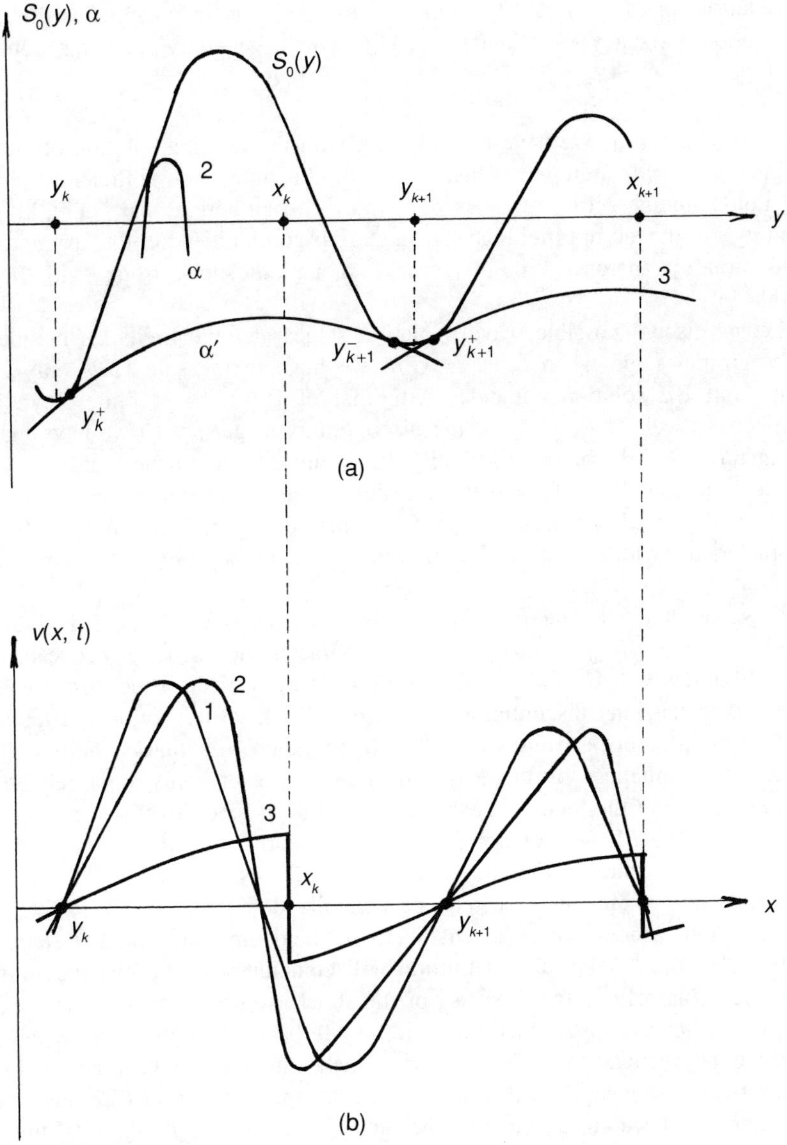

Fig. 2.3 Graphical procedure to seek the absolute minimum coordinate; (a) field profile; (b) at various times: $t = 0$, 1; $t < t_n$, 2; $t > t_n$, 3

Relationships (2.11)–(2.14) define the discontinuity characteristics through the double tangency coordinates y_k^+, y_{k+1}^-, which, generally speaking, depend on time.

3. Up to this time we have been dealing with the limiting solution of BE at infinite Reynolds number, when $\mu \to 0$. At finite but sufficiently large Reynolds numbers ($\mu \neq 0$), viscosity plays an important part only in the vicinity of the abrupt field changes, and discontinuities of the asymptotic solution are transformed in this case into smooth shock fronts with finite width.

Let us discuss possible wave evolution stages at finite Reynolds numbers. At an initial stage when $t < t_n \approx \sigma_0/l_0$ we have in the sum (2.2) only one term, and BE solution coincides with that of RE (2.9) at finite Reynolds numbers as well, if $R_0 \gg 1$. At the discontinuity stage ($t > t_n$) we have many summands in (2.2), but at practically all x only one dominates–the one with the least action. If as $\mu \to 0$ the transition from one local minimum to the other in (2.2) with variation of x had the form of a jump, then for $\mu \neq 0$ the characteristic width of a transition region δ may be estimated from (2.2) as $\delta \approx \mu t/(y_{k+1}^- - y_k^+)$.

If we assume that the field at the discontinuous stage is described by a single scale $l(t) \sim (y_{k+1}^- - y_k^+)$, then the characteristic distance between the discontinuities will be also equal to l, and the ratio of the shock front width $\delta \approx \mu t/l$ to the inter-discontinuity distance l will have the form of $\mu t/l^2$. For $\mu t/l^2 \ll 1$ the shock front width δ will be much less than l, and in this regime the finiteness of the Reynolds number results only in a relatively narrow region in the vicinity of the shock front. Outside the shock fronts, the structure of the BE field at finite though sufficiently large Reynolds numbers coincides with the asymptotic solution of BE as $\mu \to 0$. The current Reynolds number in this case may be naturally defined as $R(t) \approx l(t)/\delta(t)$. If we take into account that the characteristic field amplitude at this stage is equal to $\sigma(t) \approx l/t$, then this definition will coincide with the one previously introduced, namely (1.16). Because of the shock front widening in the course of time, it may happen that this width will become comparable with the distance between discontinuities, and in such situation the very notion of a shock front disappears. At this stage there are many terms in (2.2) and they are found to be comparable. For the auxiliary field $\varphi(x, t)$ of (1.2), this is equivalent to the fact that relative changes $\tilde{\varphi}(x, t)$ become much less with respect to an average level $\tilde{\varphi}$. Accordingly, (1.17) may be linearised and, consequently, this stage is characterised by a linear damping of the field. Later on all these stages will be treated in more detail.

4. Now we discuss the structure of the shock front using the solution (2.2). Let $\xi_n(x, t)$ and $\xi_m(x, t)$ be the coordinates of two competing local minima in (2.2); namely, as $\mu \to 0$ at the discontinuity point $x = x_k$, local minima are

equal, and coordinates ξ_n and ξ_m coincide at $x = x_k$ with the corresponding coordinates of the two tangency points: $\xi_n(x_k, t) = y_k^+$, $\xi_m(x_k, t) = y_{k+1}^-$. While for the local action $\Phi(x, y, t)$, equality (2.10) is fulfilled. In the competition zone of the two local minima, the solution of (2.2) may be written as

$$v(x, t) = \frac{1}{t}\left(x - \frac{\xi_m + \xi_n}{2}\right) - \left(\frac{\xi_m - \xi_n}{2t}\right)\tanh\left[\frac{1}{4\mu t}(y_{k+1}^- - y_k^+)(x - x_k)\right]. \tag{2.15}$$

From this it follows that the shock front width is inversely proportional to the value of the field jump Δv (see (2.14)),

$$\delta = \frac{4\mu t}{y_{k+1}^- - y_k^+} = \frac{2\mu}{\Delta v}. \tag{2.16}$$

Deviating to the left and to the right from the jump by the value δ, one can see that solution (2.15) is turned into the asymptotic solution (2.4) of BE.

If within the width of the shock front δ we can neglect the variation of $\xi_m(x, t)$ and $\xi_n(x, t)$ in (2.15), we see that the shape of the shock front has a universal structure, and its parameters are defined by the field values to the left and to the right of the jump,

$$v(x, t) = \frac{1}{t}\left(x - \frac{y_{k+1}^- + y_k^+}{2}\right) - \left(\frac{y_{k+1}^- + y_k^+}{2t}\right)\tanh\left[\frac{1}{4\mu t}(y_{k+1}^- - y_k^+)(x - x_k)\right], \tag{2.17}$$

$$v(x, t) = \frac{1}{2}\left[v(x_k - 0, t) + v(x_k + 0, t)\right] - (\Delta v)\tanh\left[\frac{\Delta v(x - x_k)}{2\mu}\right]. \tag{2.18}$$

It is easy to prove that (2.18) coincides with the stationary solution of BE $v(x, t) = v(x - v_+ t)$, describing a stationary shock front. If we write $v_1 = v(-\infty)$, $v_2 = v(+\infty)$, then the stationary solution of BE takes the form

$$v(x - v_+ t) = v_+ - v_- \tanh\left[\frac{v_-}{2\mu}(x - v_+ t)\right], \tag{2.19}$$

with

$$v_+ = \frac{v_1 + v_2}{2}, \quad v_- = \frac{v_1 - v_2}{2}, \quad \delta = \frac{2\mu}{v_-}\ (v_2 > v_1), \tag{2.20}$$

and, hence, the solution of BE at finite R_0 may be constructed by ‘gluing’ the discontinuous solution ($\mu \to 0$) to the stationary one (2.19) depicting the shock front structure.

2.3 Evolution of the basic types of disturbance

Let us consider some BE solutions, which are important for investigations of the evolution of various fields including random fields. Analysing the asymptotic behaviour ($\mu \to 0$) of the solution $v(x, t)$, we make use of the aforementioned graphical procedure, according to which the field $v(x, t)$ is given by (2.4), where $y(x, t)$ is the first coordinate for touching of parabola α (2.6) and initial action $S_0(y)$.

1. In this subsection we shall consider the unipolar pulse. Let the initial field be represented as the unipolar (i.e. one-signed) pulse with length l_0 localised in the neighbourhood of $x = 0$. Since BE has the divergence form the field $v(x, t)$ involves an invariant

$$A = \int_{-\infty}^{\infty} v(x, t)\,\mathrm{d}x = \int_{-\infty}^{\infty} v_0(x)\,\mathrm{d}x, \tag{3.1}$$

which implies pulse area conservation. Take, for definiteness $v_0(x) \geq 0$ and $A > 0$. Because of (1.10), the initial action $S_0(x)$ is constant outside the pulse localisation region, equals $S_0(-\infty) = 0$ for $x < 0$ and equals the pulse area $S_0(\infty) = A$ for $x > l_0$ (Fig. 2.4).

Now we present a discussion of the pulse behaviour as ($\mu \to 0$) for sufficiently large times. At times $t \ll l_0^2/A = t_n$, when the curvature of parabola $\alpha_{xx} = 1/t$, is much less than the initial action curvature $S_0'' = A/l_0^2$, the pulse profile assumes a universal form. Indeed, at these times, the critical parabola α_{cr} having two tangency points, is determined from the following set of equations (see Fig. 2.4):

$$\alpha(x_d, x_d, t) = H = A,$$

$$\alpha\,(x_d, 0, t) = H - \frac{x_d^2}{2t} = 0.$$

Here we assume that the left tangency point practically coincides with the pulse origin $x = 0$. Then it is evident that in $x < 0$, $y(x, t) = x$. In $0 < x < x_d$, due to parabola smoothness we have $y(x, t) \approx 0$, while in $x > x_d$ the point of tangency is $y(x, t) = x$. Therefore, the positive pulse at sufficiently large times has the form of a triangle,

$$v(x, t) = \begin{cases} \dfrac{x}{t}, & 0 < x < x_d, \\ 0, & x < 0,\ x > x_d, \end{cases} \tag{3.2}$$

where for the discontinuity coordinate we obtain

$$x_d = \sqrt{2At}. \tag{3.3}$$

The amplitude of such a pulse decreases with time as $x_d/t = \sqrt{2A/t}$. The shape of its right-hand front is described by (2.19), where $v_+ = v_- = \sqrt{2A/t}$, and has width $\delta = \sqrt{2\mu^2 t/A}$. The relative width of the shock front varies inversely with Reynolds number: $\delta/x_d = 1/R_0$, and remains constant with

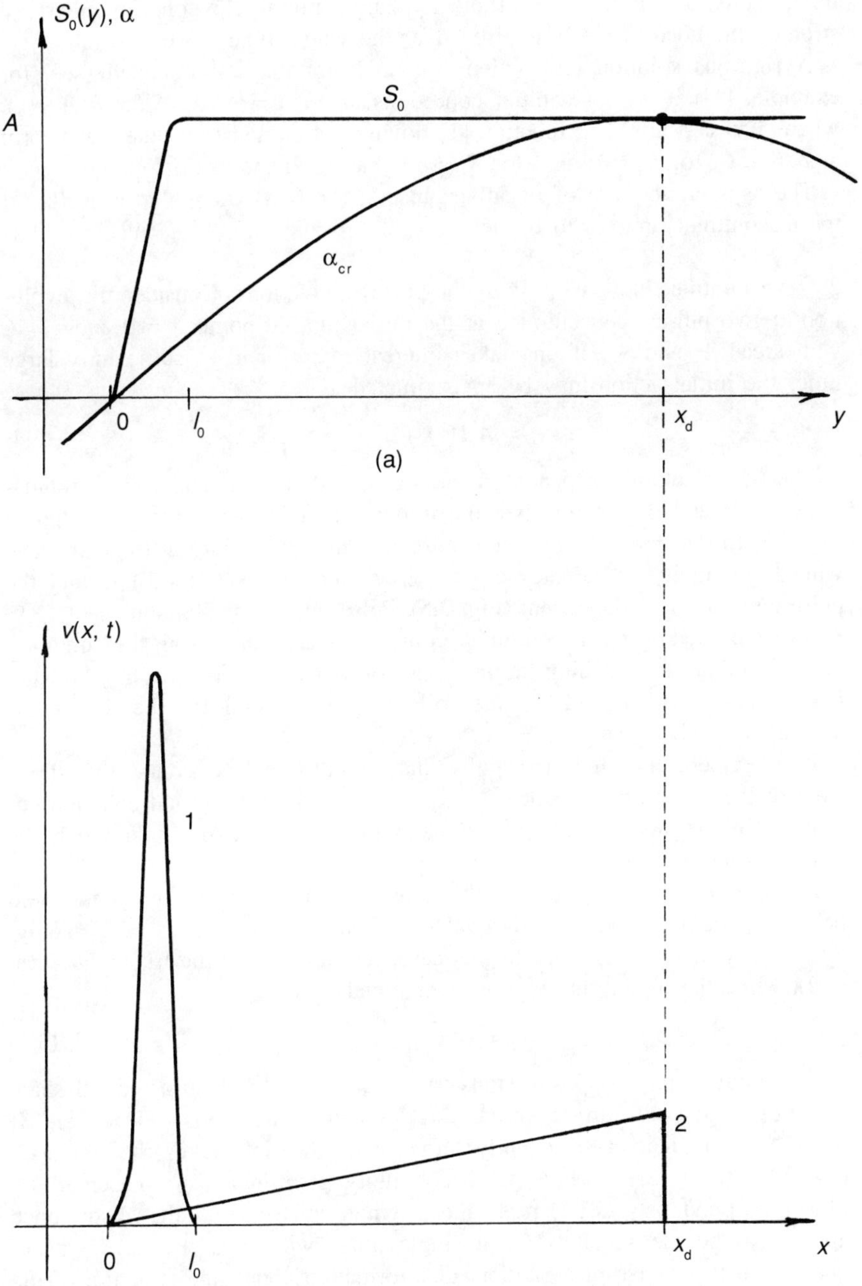

Fig. 2.4 Monopolar-pulse evolution: $t = 0$, 1; $t > t_n$, 2

time. Thus, at finite but sufficiently large initial Reynolds numbers, a strongly nonlinear stage is realised over the whole time $t > 0$.

A rigorous solution (convenient for analysis) may be obtained (see, for example, [1]) if we take initial conditions as $v_0(x) = A\delta(x)$, $S_0(x) = A\,\mathbf{1}(x)$, where $\mathbf{1}(x)$ is a unit step function [1]. Figure 2.4 demonstrates the wave form at $t \gg l_0^2/A$ for $R_0 = \infty$ and for $R_0 \gg 1$, based on the exact solution.

The asymptotic form of negative pulses ($A < 0$) is triangular as well, but the discontinuity moves to the left, $v(x, t) \leq 0$ and $x_d = -\sqrt{2|A|t}$.

2. We continue now with study of pulse interactions. Consider the evolution of two pulses, concentrated at the initial time at points $x = 0$ and $x = l_2$ with areas A_1 and A_2. If one takes interest in the field at sufficiently large times the initial action may be approximated as

$$S_0(x) = A_1\mathbf{1}(x) + A_2\mathbf{1}(x - l_2). \tag{3.4}$$

The field evolution depends on the signs of A_1 and A_2 and their combinations. If A_1 and A_2 are positive, the initial action has the form presented in Fig. 2.5. In the initial stage, each pulse is transformed into a triagular wave with discontinuity coordinates $x_{d1} = \sqrt{2A_1t}$ and $x_{d2} = l_2 + \sqrt{2A_2t}$, and the pulse motions are independent (Fig. 2.5). Pulse interaction starts at $t_* = l_2^2/2A_1$. At times exceeding t_*, discontinuity coordinates are found from the condition of the parabola α_{cr} touching the initial action twice, at points $x = 0$ and $x = l_2$ (Fig. 2.6), which leads to the following expression for the first pulse discontinuity coordinate: $x_{d1} = A_1t/l_2 + l_2/2$.

It is evident that the first pulse discontinuity will overtake the shock front of the second pulse, since $x_{d1} = \sim t$, $x_{d2} \sim \sqrt{t}$. Time t_c and collision coordinate x_c are determined from the condition that the parabola touch the initial action at three points: $x = 0$, $x = l_2$, $x = x_c$.

At times $t > t_c$, the nature of the tangency point and, hence, of the field behaviour, is described by the value of initial action at $x = \pm\infty$ only: $S_0(-\infty) = 0$, $S_0(\infty) = A_1 + A_2$. In this case, the pulse takes the triangular form (3.2), while the discontinuity coordinate equals

$$x_d = \sqrt{2(A_1 + A_2)t} \tag{3.5}$$

If the pulses have opposite signs: $A_1 > 0$, $A_2 < 0$, then in an initial stage the pulse discontinuities move in opposite directions. However, if $A_1 + A_2 > 0$, then at sufficiently large times the field takes the universal triangular form (3.2), while its discontinuity coordinate is represented by (3.5), as previously. Therefore, the asymptotic form of the disturbance appears to be the same as for a single pulse with area $A_1 + A_2$. Consequently, in the course of evolution all information about fine structure of the initial disturance is lost.

However, if $A_1 + A_2 = 0$, but $A_1 = A > 0$, $A_2 < 0$, the pulses initially move in opposite directions. At $t_c = l_2^2/2A$ the discontinuities coalesce into a

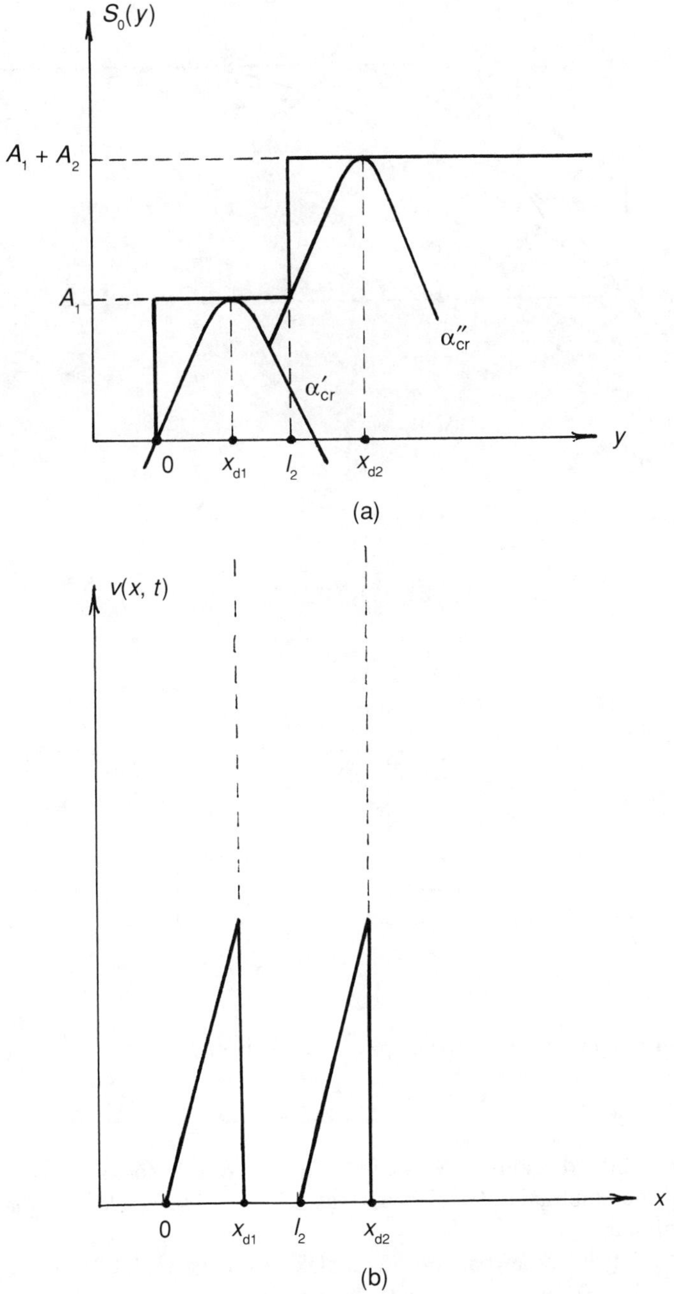

Fig. 2.5 Initial stage of the two-monopolar-pulse evolution for $t < t_*$

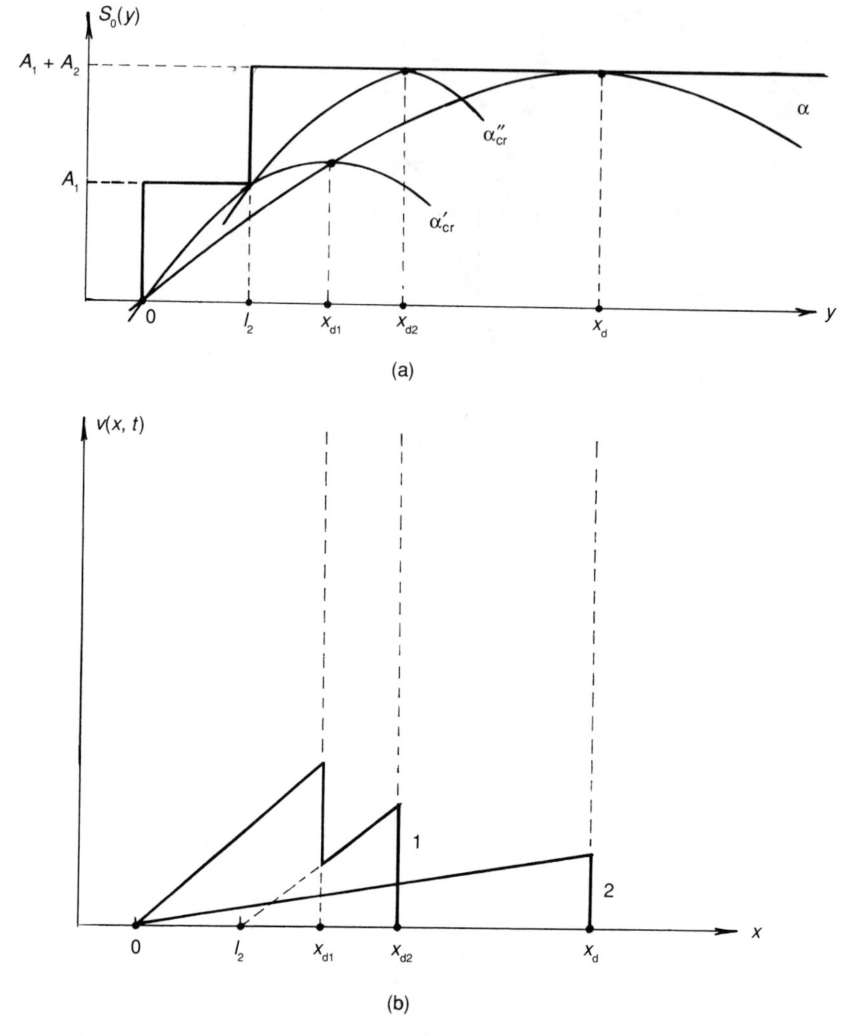

Fig. 2.6 Interaction of two pulses: $t_* < t < t_c$, 1; $t > t_c$, 2

single immobile discontinuity, and when $t > t_c$ the so-called S-wave is generated, the amplitude of which decreases, and is independent of the initial pulse amplitude.

The situation is somewhat more complex when the initial pulses separate, e.g. when $A_1 < 0$, but $A_2 = A > 0$. In the limiting case $\mu \to 0$ such pulses will not interact. In particular, if $A_1 + A_2 = 0$ and $l_2 \to 0$, the asymptotic form will be represented by an N-wave, which consists of two adjacent triangular pulses with diverging discontinuities at $x_d = \pm\sqrt{2At}$. However, this solution at finite viscosity will not be structurally stable, since the effective Reynolds number decreases with time as ~ $1/\ln t$ [1]. It should be noted that for a noisy finite pulse, regularisation of its fine structure takes place, and under sufficiently weak restrictions on the noise statistics such a pulse is asymptotically transformed into a practically deterministic N-wave [119].

3. Let us discuss the transformation of periodic signal with zero mean

$$\begin{gathered} v_0(x + l_0) = v_0(x) \\ S_0(x + l_0) = S_0(x), \quad v_0(x) = S_0'(x). \end{gathered} \tag{3.6}$$

An example of such a signal is given by the harmonic disturbance

$$v_0(x) = a \sin k_0 x, \quad S_0(x) = -\frac{a}{k_0} \cos k_0 x, \quad k_0 = \frac{2\pi}{l_0}, \tag{3.7}$$

the evolution of which is discussed in many articles (see, for example, [51–53]) in detail.

In the general case one may single out several scales in the initial profile (3.6). Let σ_0 be the amplitude of the initial disturbance $v_0(x)$ and l_* being the smallest spatial scale, described as $l_* = \sigma_0 / |\min v_0'(x)|$. For $t < t_* = l_*/\sigma_0$, parabola α and initial action $S_0(y)$ have a single contact point, while field $v(x, t)$ is continuous. At this stage, nonlinearity leads to profile distortion and to generation of new spatial harmonics, but discontinuities are still absent, and as $\mu \to 0$ the wave energy is conserved.

At $t = t_*$ and at a point where $v_0'(x)$ is minimal, the parabola curvature becomes equal to the curvature of the initial action, and the first discontinuity occurs. When $t > t_*$ each period may have several discontinuities, which may move and interact with each other. If the times and discontinuity occurrence coordinates are defined by an initial velocity gradient, their asymptotic behaviour (for $t \gg t_*$) will depend on the form of the initial action, namely, on the distribution and value of the local minima of $S_0(y)$ (see Fig. 2.7). Indeed, at sufficiently large times, the smooth parabola $\alpha(x, y, t)$ touches the initial action $S_0(y)$ in the neighbourhood of its local minima y_k, $k = 1, \ldots, n$, while when $t \gg t_*$ the absolute minimum coordinates $y(x, t)$ coincide with a local minimum y_k. Field $v(x, t)$ is described in this case by a system of critical parabolas each having two points of tangency with $S_0(y)$. The centres (axes) of these parabolas define the discontinuity location x_k; in the intervals between discontinuities the field has the universal structure

$$v(x, t) = \frac{x - y_k}{t}, \tag{3.8}$$

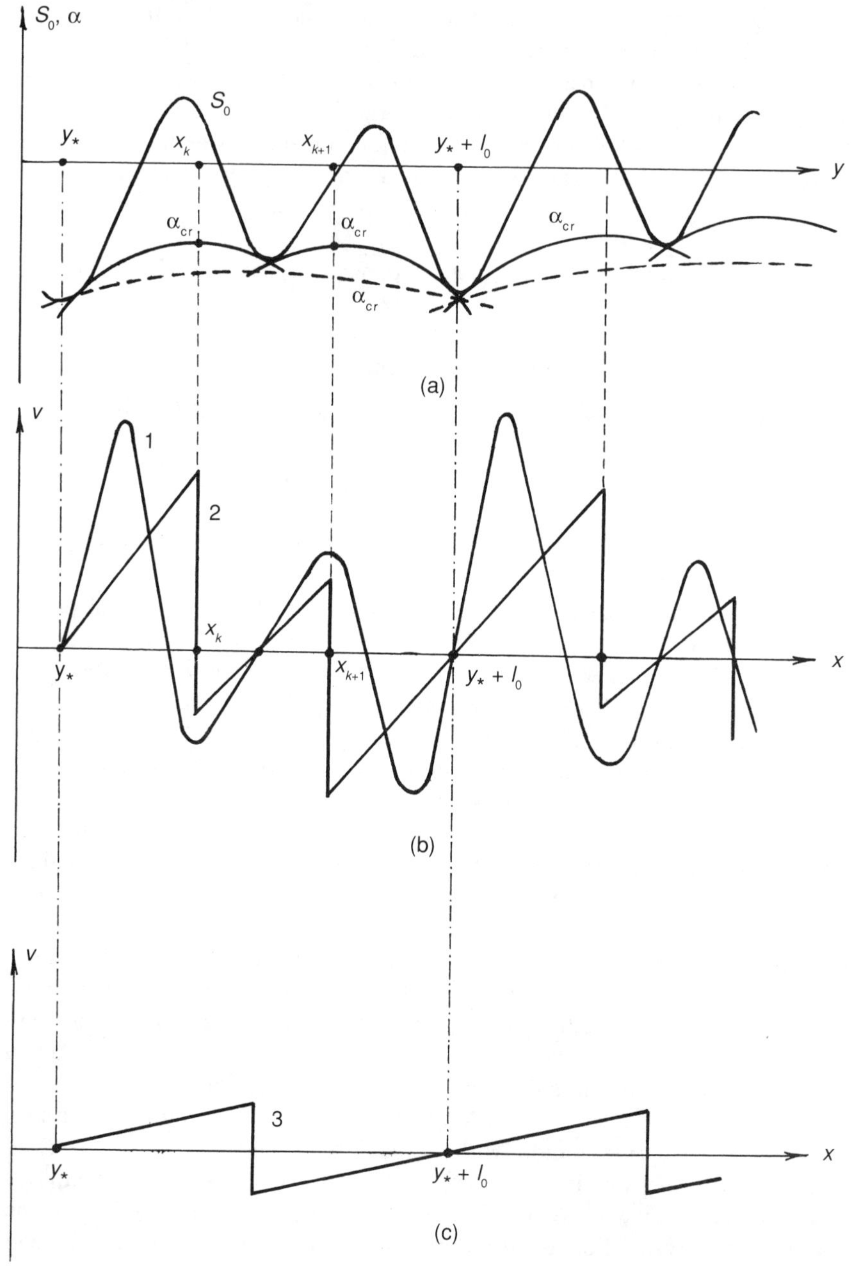

Fig. 2.7 Periodic signal evolution: $t = 0$, 1; $t > t_*$, 2; $t \gg t_*$, 3

while the discontinuity coordinates are given in (2.11), where $y_k^+ \approx y_k^- \approx y_k$ and do not depend on time:

$$x_k = \frac{y_{k+1} + y_k}{2} + V_k t, \quad V_k = \frac{S_0(y_{k+1}) - S_0(y_k)}{y_{k+1} - y_k}. \tag{3.9}$$

If $S_0(y_{k+1}) \neq S_0(y_k)$ the discontinuities are moving with constant velocity $V_k \neq 0$. The manner of discontinuity motion and coalescence may be quite complex, but the asymptotic behaviour of the field appears to be surprisingly simple.

We confine ourselves to the case when one of the minima of $S_0(y)$ in the period is deepen than the others, $S_0(y_*) \ll S_0(y_k)$. Denote this minimum coordinate in the jth period as $y^j = y_* + jl_0$. Then, resorting to the graphical representation (Fig. 2.7), one may easily see that the asymptotically critical parabola connects absolute minima of the neighbouring periods, y^j and y^{j+1}, while the field has the form of a periodic pulse sequence with the same slope $1/t$ and zeros at points y^j:

$$v(x, t) = \frac{x - y^j}{t}, \quad x_j < x < x_{j+1}.$$

Here the discontinuities are stationary and located in the middle between zeros of the field $x_j = (y^j + y^{j-1})/2$. The discontinuity amplitude decreases as l_0/t and does not depend on the initial disturbance amplitude or shape.

In a medium with finite viscosity the form of the shock front is described by (2.19), where $v_+ = 0$, $v_- = l_0/2t$ and the shock front width is linearly increasing with time: $\delta = \mu t/l_0$. If we consider the field harmonic amplitudes

$$v(x, t) = \sum_{n=1}^{\infty} A_n \sin nk_0(x - y_*), \tag{3.10}$$

then, accounting for the shock front shape (2.19), we have for them:

$$A_n(t) = \frac{\mu k_0(-1)^n}{\sinh(nk_0^2 \mu t)}. \tag{3.11}$$

Here it is obvious that there are two regions in the spectrum. For $n < (k_0^2 \mu t)^{-1}$ the harmonic amplitudes decrease in accordance with the power law $A_n \sim n^{-1}$, which is related to the presence of discontinuities, while for $n > (k_0^2 \mu t)^{-1}$ they are decreasing according to an exponential law.

With the course of time the shock front width increases, and when $t > t_d = l_0^2/\mu$, when the effective Reynolds number $R(t) \approx l_0/\delta$ becomes of the order of unity, the linear stage of dissipation begins.

At times $t > t_d$ the field becomes a harmonic (sinusoidal) one again and its amplitude does not depend on the initial profile shape or amplitude. That is, because of the joint nonlinear and dissipative effects all information on the initial periodic field fine structure is lost and only the memory of the field period is retained.

4. We now consider nonlinear interaction of the waves. Consider two other special, but very important cases of wave interaction. Let $\tilde{v}(x, t)$ be a solution of BE, subject to initial condition $\tilde{v}(x, t = 0) = \tilde{v}(x)$. In accordance with (1.8), (1.2), (1.14), it is written as

$$\tilde{v}(x, t) = -2\mu \frac{\partial}{\partial x} \ln \tilde{\varphi}(x, t),$$

$$\tilde{\varphi}(x, t) = \frac{1}{\sqrt{4\pi\mu t}} \int_{-\infty}^{\infty} \exp[-\tilde{\Phi}(x, y, t)/2\mu]\, dy, \tag{3.12}$$

$$\tilde{\Phi}(x, y, t) = \tilde{S}_0(y) + \frac{(x - y)^2}{2t}, \quad \tilde{S}_0(y) = \int^{y} \tilde{v}_0(x)\, dx.$$

Now we discuss the interaction of this wave with a constant component, specifying an initial condition to BE (1.1) as

$$v_0(x) = \tilde{v}_o(x) + V. \tag{3.13}$$

For function $\Phi(x, y, t)$ of (1.14) we therefore obtain

$$\Phi(x, y, t) = \tilde{\Phi}(x - Vt, y, t) + Vx,$$

and hence the solution of BE takes the form

$$v(x, t) = \tilde{v}(x - Vt, t) + V. \tag{3.14}$$

Thus, the constant component only shifts the profile $\tilde{v}(x, t)$ by the interval Vt, without changing its shape.

As a second example, let us consider the interaction of an initial field $v_0(x)$ with a linear profile:

$$v_0(x) = \beta x + \tilde{v}_0(x). \tag{3.15}$$

For the function $\Phi(x, y, t)$ corresponding to (3.15), we obtain

$$\Phi(x, y, t) = \frac{\beta y^2}{2} + \tilde{S}_0(y) + \frac{(x - y)^2}{2t} = \frac{\beta x^2}{(1 + \beta t)2} + \tilde{\Phi}\left(\frac{x}{1 + \beta t}, y, \frac{t}{1 + \beta t}\right).$$

Introducing new variables

$$x' = \frac{x}{1 + \beta t}, \quad t' = \frac{t}{1 + \beta t}, \tag{3.16}$$

it follows from the exact solution of BE, that

$$v(x, t) = \frac{\beta x}{1 + \beta t} + \tilde{v}\left(\frac{x}{1 + \beta t}, \frac{t}{1 + \beta t}\right) \frac{1}{1 + \beta t}. \tag{3.17}$$

Here the first term describes the evolution of the linear profile, while the second one depicts the distortion of the field $\tilde{v}(x, t)$ due to interaction with the profile. From (3.17) one can see that nonlinear interaction leads to three effects: field spatial scale alteration, amplitude variation, and change in time of the wave evolution tempo.

The form of evolution will be qualitatively different for $\beta > 0$ and $\beta < 0$. If $\beta > 0$ the slope of the linear profile is monotonically decreasing, and for $\beta t \gg 1$ it would not depend on the initial slope β. The interaction of the field $\tilde{v}(x, t)$ with such a profile leads to an additional monotone decrease of the field amplitude and to the increase of its spatial scale. In this case, within infinite time $t \in (0, \infty)$ only a part of the field evolution $\tilde{v}(x, t)$ occurs, which corresponds to the time interval $(0, t^*)$ where $t^* = 1/\beta$. For $t \gg t^*$ the field evolution is stabilised, and at any time $t \gg t^*$ the wave profile is similar to field $\tilde{v}$ at time t^*.

If $\beta < 0$, then within finite time $t \in (0, t^*)$, $t^* = 1/|\beta|$, the linear profile gradient becomes infinite. With that, within the finite time interval $t \in (0, t^*)$ all the three evolution stages of the field $\tilde{v}$ take place which correspond to the infinite time interval $(0, \infty)$ for a non-perturbed wave ($\beta = 0$). Nonlinear interaction with a linear profile leads, therefore, to an increase of the field amplitude $\tilde{v}$ and to a decrease of its scale. Peculiarities of the noise and regular signal interaction with such a profile are analysed in [75].

2.4 Burgers equation, hydrodynamics of noninteracting particles, and the parabolic equations of quasioptics

1. Let us first consider the relation of the BE asymptotic solution to the motion of a noninteracting particle hydrodynamic flow. If, referring to (2.1), we introduce new functions

$$S_m(x, t) = \Phi(x, \xi_m, t) = S_0(\xi_m) + \frac{(x - \xi_m)^2}{2t} = S_0(\xi_m) + \frac{1}{2} v_0^2(\xi_m) t,$$

$$v_m(x, t) = \frac{\partial S_m(x, t)}{\partial x} = \frac{x - \xi_m}{t}, \tag{4.1}$$

$$j_m(x, t) = \Phi_{yy}(x, \xi_m, t) t = 1 + v_0'(\xi_m) t,$$

the solution (2.2) may be written as

$$v(x, t) = \frac{\sum_m v_m(x, t) \left| j_m(x, t)^{-1/2} \right| \exp\left[-s_m(x, t)/2\mu\right]}{\sum_m \left| j_m(x, t) \right|^{-1/2} \exp\left[-s_m(x, t)/2\mu\right]}. \tag{4.2}$$

Here $s_m(x, t)$, $v_m(x, t)$, $j_m(x, t)$ may be interpreted as the branches of some multi-valued function $S(x, t)$, $v(x, t)$, $j(x, t)$. Taking into consideration that ξ_m is the extremum point of $\Phi(x, \xi_m, t)$, where $\Phi_y = 0$, one finds that $S(x, t)$ satisfies the Hamilton–Jacobi equation

$$\frac{\partial S}{\partial t} + \frac{1}{2}\left(\frac{\partial S}{\partial x}\right)^2 = 0, \quad S(x, t = 0) = S_0(x), \tag{4.3}$$

and hence the function $S(x, t)$ may naturally be called the *action* of the particle flow. The velocity of the particle flow $v(x, t) = \partial S(x, t)/\partial x$ satisfies, as is seen from (4.3), the RE

$$\frac{\partial v}{\partial t} + v\frac{\partial v}{\partial x} = 0, \qquad v(x, t = 0) = v_0(x), \tag{4.4}$$

while the function $j(x, t)$, which we call the flow divergence, satisfies the equation

$$\frac{\partial j}{\partial t} + v\frac{\partial j}{\partial x} = j\frac{\partial v}{\partial x}, \qquad j(x, t = 0) = 1. \tag{4.5}$$

Equations (4.3)–(4.5) describe the particle hydrodynamic flow at point x and time t in the so-called Eulerian representation. These equations are equivalent to the set of characteristic equations for the flow parameters in Lagrangian representation, i.e. in the vicinity of a moving particle with an initial (Lagrangian) coordinate y:

$$\frac{\mathrm{d}X}{\mathrm{d}t} = V, \qquad \frac{\mathrm{d}V}{\mathrm{d}t} = 0, \qquad \frac{\mathrm{d}S}{\mathrm{d}t} = \frac{1}{2}V^2, \qquad \frac{\mathrm{d}J}{\mathrm{d}t} = Q, \qquad \frac{\mathrm{d}Q}{\mathrm{d}t} = 0. \tag{4.6}$$

From (4.3)–(4.5) it follows that initial conditions for (4.6) have the form

$$\begin{gathered} X(y, 0) = y, \quad V(y, 0) = v_0(y), \quad S(y, 0) = S_0(y), \\ J(y, 0) = 1, \quad Q(y, 0) = v_0'(y) \end{gathered} \tag{4.7}$$

The solutions of these equation describe the uniform and rectilinear motion of particles with different initial coordinates and velocities,

$$\begin{gathered} X(y, t) = y + v_0(y)t, \quad V(y, t) = v_0(y), \\ S(y, t) = S_0(y) + \tfrac{1}{2}v_0^2(y)t, \\ J(y, t) = 1 + v_0'(y)t, \; Q(y, t) = v_0'(y). \end{gathered} \tag{4.8}$$

Function $J(y, t)$ equals the Jacobian of the Lagrangian-to-Eulerian variable transformation, $J = \partial X/\partial y$, and describes the divergence of the particle flow. Provided $J(y, t) > 0$ the particle trajectories do not intersect, and the solution of (4.2) has only one term, which corresponds to the only particle arrival at point x, with Lagrangian coordinate being determined from the equation

$$x = y + v_0(y)t. \tag{4.9}$$

Here the asymptotic solution of BE as $\nu \to 0$ coincides with the solution of RE (4.4) and $v(x, t)$ is a single-valued function. With increase of t, Jacobian $J(y, t)$ vanishes at time $t_* = -1/\min v_0'(y)$, which in order of magnitude is equal to the characteristic time of nonlinear processes, $t_n = l_0/\sigma_0$. The sign change of $J(y, t)$ corresponds to the particle sequence order change and to the multistream-effect appearance, i.e. to the multivaluedness of the functions $s(x, t)$, $v(x, t)$, $j(x, t)$. Then, the sum (4.2) has several summands, among which, as $\mu \to 0$, the largest and the most significant contribution belongs to the term corresponding to the particle with the least action $S(y, t)$.

The main distinction of the BE solution as $\mu \to 0$ from the motion of the noninteracting particle hydrodynamic flow manifests itself in the fact that the BE solution takes into account the flow competition, under which only the least action flow particles 'survive', while the flows of noninteracting hydrodynamic particles move independently, eventually resulting in a multi-stream flow.

2. Now we discuss the relation of BE with the nonstationary Schrödinger equation and the parabolic equation of quasioptics. If we substitute $\mu = i\hbar/2m$ into the equation for the auxiliary field (1.7), where $\hbar$ is the Planck constant, m the particle mass, then (1.7) will turn into the Schrödinger equation describing particle motion in free space. Here the case of large Reynolds numbers in BE ($\mu \to 0$) corresponds to the quasi-classical approximation in the Schrödinger equation ($\hbar \to 0$).

However, if in (1.7) we assume x to be a transverse coordinate, replace variable t by a longitudinal coordinate z, and set $\varphi(x, t) = E(x, z)$, $\mu = 1/2ik_0$, where k_0 is the wave number, then (1.7) will be turned into the parabolic equation of quasioptics, well-known in the theory of wave propagation,

$$\frac{\partial E}{\partial z} = \frac{1}{2ik_0}\frac{\partial^2 E}{\partial x^2}. \tag{4.10}$$

Passing from the equation above to the equation for the complex phase S, $E = \exp(ik_0 S)$ and $v = S_x$ we obtain BE (1.1). The procedure of the BE exact solution construction using the Hopf–Cole transformation (1.4)–(1.6) in this case is inverse to the transition from the parabolic equation (4.10) to equations for the complex phase (1.5), while the asymptotics $\mu \to 0$ corresponds to the geometrical optics approximation (4.3) for phase and the angle of the wave front (4.4), in the limit $k_0 \to \infty$. Here in all three cases (BE, Schrödinger equation, and parabolic equation of quasioptics) the behaviour of the beams and particle trajectories remains the same until their interaction begins. However, while in quantum mechanics and optics interference occurs between beams after their intersection, in BE, because of the imaginary nature of the wave number $k_0 = 1/2i\mu$, particles compete and only the least action particles survive.

3. The above-mentioned analogy appears to be useful for analysis of the inhomogeneous BE as well,

$$\frac{\partial v}{\partial t} + v\frac{\partial v}{\partial x} = \mu\frac{\partial^2 v}{\partial x^2} + f(x, t),$$
$$f(x, t) = -\frac{\partial}{\partial x}F(x, t), \quad v(x, 0) = v_0(x). \tag{4.11}$$

This equation describes the process of nonlinear acoustic wave excitation by external distributed sources [71, 72]. At $\mu = 0$, (4.11) is a system to which the problem of finite-amplitude ocean waves excitation by a convected pressure wave [73] is reduced. With a random force $f(x, t)$, equation (4.11) describes the excitation of one-dimensional acoustic turbulence by external forces [74, 76].

As for BE, (4.11) is reduced by means of the Hopf–Cole transformation

$$v = \frac{\partial s}{\partial x}, \quad s = -2\mu \ln \varphi(x, t), \tag{4.12}$$

to the linear equation of diffusion

$$\frac{\partial \varphi}{\partial t} = \mu\frac{\partial^2 \varphi}{\partial x^2} + \frac{1}{2\mu}F(x, t)\varphi,$$
$$\varphi(x, 0) = \exp[-S_0(x)/2\mu]. \tag{4.13}$$

Taking t as a longitudinal coordinate and x as a transverse one and setting $\varphi = E$, $\mu = 1/2ik_0$, where k_0 is the wave number, we rewrite (4.13) as

$$2ik_0\frac{\partial E}{\partial t} = \frac{\partial^2 E}{\partial x^2} - 2k_0^2 F(x, t)E. \tag{4.14}$$

As a result, we have arrived at a well-known parabolic equation of quasi-optics for the small-angle wave propagation in a medium with index of refraction $n(x, t) = 1 + F(x, t)$ [61]. As $\mu \to 0$, which corresponds to the geometric-optical approximation $k_0 \to \infty$ in (4.14), it is convenient to use the beam description to analyse the field. However, in the solution of (4.13), because of the imaginary character of $k_0 = 1/2i\mu$, it is necessary, in contrast to optics, to consider only the beam with absolutely minimal action. Such an approach was used in [74] to investigate stationary Burgers turbulence excited by random external forces.

In the general case, however, the field $v(x, t)$ is expressed through the solution of (4.13). It is known that the parabolic equation of quasioptics (4.14) has an exact solution for some particular types of refraction index. Making use of the analogy between (4.13) and (4.14), one can consequently develop a series of exact solutions to (4.14) [72, 75].

Therefore, Burgers turbulence is intimately connected with the optical problem of field diffraction beyond a phase screen, while the stationary Burgers turbulence is connected with the problem of wave propagation in a medium with refraction index fluctuations. In the limiting case of large Reynolds numbers to describe the field, it seems convenient to use the ray description corresponding to the geometrical optics approximation in problems of diffraction. Concerning this, a very important problem arises which treats the change from the field description along the beams (the Lagrangian representation) over to the search for the field statistical properties at fixed space–time points (the Eulerian representation). These questions will be tackled in the chapters to follow, both with regard to the theory of nonlinear waves in nondispersive media and to the equations of geometrical optics.

2.5 Field evolution at low Reynolds numbers

In this section we shall be concerned with the BE solution at low Reynolds numbers. First let us discuss the limiting case of a linear medium to clarify later a mechanism of low Reynolds number influence on the propagating field shape and, finally, using a probabilistic interpretation of the BE general solution, to obtain an exact solution of BE written as a Reynolds number power series. The two examples given below will illustrate our procedure throughout the section.

1. In the limiting case of a linear medium ($R_0 \to 0$) in (1.1) one can neglect the nonlinear term. In this case the BE will acquire the form of the linear equation of diffusion

$$\frac{\partial v}{\partial t} = \mu \frac{\partial^2 v}{\partial x^2} \tag{5.1}$$

This solution of this equation subject to the initial condition

$$v(x, 0) = v_0(x)$$

has the form

$$v(x, t) = \int_{-\infty}^{\infty} v_0(y)\, W_G(x - y, t)\, \mathrm{d}y, \tag{5.2}$$

where $W_G(z, t)$ is the Green's function of (5.1)

$$W_G(z, t) = \frac{1}{\sqrt{4\pi\mu t}} \exp\left(-\frac{z^2}{4\mu t}\right) \tag{5.3}$$

characterised by a spatial scale $l_d = l_d(t) = \sqrt{4\mu t}$.

Expression (5.2) allows one to follow the initial perturbation evolution in a linear medium. Let, for example, the initial perturbation be a pulse localised near $x = 0$ possessing a non-zero area

$$\int_{-\infty}^{\infty} v_0(x)\, \mathrm{d}x = A$$

and duration equal to l_0. It comes out of (5.2) that the pulse will have almost no deformation at such low times for which the spatial scale $l_d(t)$ of the Green's function is still much smaller than the initial pulse duration l_0. Diffusion spreading of a pulse begins at such time t_d for which $l_d(t)$ is close to l_0, i.e. at $t_d \approx l_0^2/\mu$. At large times, when $t \gg t_d$, (5.2) implies that any unipolar pulse (for $A > 0$) acquires a universal Gaussian form

$$v(x, t) = \frac{A}{\sqrt{4\pi\mu t}} \exp\left(-\frac{x^2}{4\mu t}\right) = AW_G(x, t) \tag{5.4}$$

with amplitude falling off as $1/\sqrt{t}$ and proportional to the initial pulse area A. The pulse duration is prolonged due to diffusion in proportion to $\sqrt{t}$, so that the pulse area, as was anticipated (see (3.1)), is time-invariant.

If as an initial field we take a harmonic signal

$$v_0(x) = a \sin kx,$$

then, as follows from (5.2),

$$v(x, t) = a\, e^{-k^2\mu t} \sin kx.$$

Thereforce an initial harmonic signal retains its form and its amplitude alone decays exponentially with time such that the damping factor is proportional to a spatial frequency squared. The latter means that, should the initial field appear as a certain periodic function of a coordinate, then the signal shape becomes smoother as time elapses, since the signal harmonic damping rate is proportional to the harmonic index squared, and the 'dying away' signal is therefore a sinusoidal one.

2. At low Reynolds numbers the impact of nonlinearity on the field evolution is weak, as dissipation prevails over the medium nonlinearity and the wave is damped out before significant effects are realised. Consider a correction to the linear solution (5.2) assuming that $R \ll 1$.

Let a characteristic value of the initial action variation $S_0(x)$ be equal to A. Introduce a dimensionless action $\tilde{s}_0(x)$ as $S_0(x) = A\tilde{s}_0(x)$. Thus the initial condition for function $\varphi(x, t)$ (see (1.6)–(1.10)) will take the form

$$\varphi(x, 0) = \exp[-R_0\tilde{s}_0(x)], \quad R_0 = A/2\mu,$$

where R_0 is the initial Reynolds number.

Assuming that $R_0 \ll 1$ we shall have, when we confine attention to three terms,

$$\varphi(x, 0) = 1 - R_0\tilde{s}_0(x) + \tfrac{1}{2}R_0^2\tilde{s}_0(x).$$

Then (1.11) and (5.2) yield

$$\varphi(x, t) = 1 - R_0\varphi_1(x, t) + \tfrac{1}{2}R_0^2\varphi_2(x, t),$$

where

$$\varphi_1(x,t)=\int_{-\infty}^{\infty}\tilde{s}_0(y)W_G(x-y,t)\,dy,$$
$$\varphi_2(x,t)=\int_{-\infty}^{\infty}\tilde{s}_0^2(y)W_G(x-y,t)\,dy. \tag{5.5}$$

Returning with the aid of (1.8) from the auxiliary field $\varphi(x, t)$ to the sought one $v(x, t)$, and being restricted to the first two terms of the expansion, we obtain

$$v(x,t)=A\frac{\partial\varphi_1(s,t)}{\partial x}+\frac{1}{2}R_0A\frac{\partial}{\partial x}\left[\varphi_1^2(x,t)-\varphi_2(x,t)\right]. \tag{5.6}$$

The first term in this expansion, as is easily seen (if one takes into consideration that $v_0(x)=A\tilde{s}_0'(x)$), coincides with the solution of the linear equation of diffusion (5.2), while the second one describes nonlinear effects proportional to the square of the initial perturbation amplitude and appears as a correction to the linear solution at $R_0 \ll 1$.

In order to better demonstrate the non-zero Reynolds number influence on the propagating perturbation type let us resort to the aforementioned examples.

Let us for the sake of definiteness assume that the area of a pulse localised near $x = 0$ is $A > 0$. The initial action of this pulse increases from the value $S_0(-\infty) = 0$ up to $S_0(+\infty) = A$ over a region whose width coincides with the pulse duration l_0. The Reynolds number is $R_0 = A/2\mu$ and does not change with time. At time $t \gg t_d$, when $l_d \gg l_0$, (5.5) yields

$$\varphi_1(x,t)=\varphi_2(x,t)=\int_{-\infty}^{x}W_G(z,t)\,dz$$

and, thus, based on (5.4),

$$v(x,t)=AW(x,t)\left\{1-\frac{1}{2}R_0^2\left[\frac{1}{2}-\int_{-\infty}^{x}W_G(z,t)\,dz\right]\right\}. \tag{5.7}$$

Comparing this one with the linear solution (5.4), we can readily reveal that, first, an arbitrary initial localised perturbation of non-zero area acquires a universal asymptotic form with time for $R_0 \neq 0$, and second, at $R_0 > 0$ the nonlinearity brings about a pulse shape asymmetry, namely, a pulse apex shift which is equal to $l_d(t)R_0\sqrt{\pi}$, and a leading edge steepening effect.

For the second example, $v_0(x) = a \sin kx$, the action and Reynolds number are equal to $\tilde{s}_0(x) = -(a/k)\cos kx$ and $R_0 = a/2k\mu$, respectively. Consequently, for $R_0 \ll 1$, (5.5) and (5.6) yield

$$v(x,t)=b_1(t)\sin kx+b_2(t)\sin 2kx,$$

where

$$b_1(t) = a\,\mathrm{e}^{-k^2\mu t},$$

$$b_2(t) = -\tfrac{1}{2}aR_0(\mathrm{e}^{-2k^2\mu t} - \mathrm{e}^{-4k^2\mu t}).$$

Therefore at low Reynolds number, nonlinearity results in generation of a second harmonic whose amplitude grows linearly at the initial stage when $t \ll t_d = 1/\mu k^2$: $b_2(t) = at/2t_n$ where $t_n = 1/ak$. At $t \gg t_d$ the second harmonic amplitude falls off twice as fast as that of the first one. At the same time the second harmonic decays more slowly than a wave of frequency $2k$ in a linear medium, which is due to a continuous energy transfer from the fundamental harmonic to the second one. At $t \gg t_d$ both the relative amplitude of the second harmonic $b_2(t)/b_1(t) \approx R_0 \exp(-t/t_d)$ and the current Reynolds number $R(t) \approx b_1(t)/2k\mu = R_0 \exp(-t/t_d)$ are exponentially small, and nonlinear effects can be disregarded.

If the second example of the initial sinusoidal perturbation for which area $\int_{-\infty}^{\infty} v_0(x)\,\mathrm{d}x$ equals zero is compared with the first one, that of the localised pulse with a non-zero positive area for which the current Reynolds number $R(t) = R_0$ is conserved, a difference becomes obvious in the evolution of perturbations which have zero and non-zero area. Physical interpretation of these distinctions goes in parallel with the fact that in the energy spectrum of the unipolar pulse perturbation, as distinct from that with zero area, spectral components with a frequency as low as desired are always present which are proportional to the pulse area and decay as slowly as desired (a component with a spatial frequency k decays proportionally to $\exp(-k^2\mu t)$). It is exactly the latter circumstance that determines the field asymptotic behaviour type at $t \to \infty$.

It is worthwhile noting that in 'spectral term' the meaning of the field expansion with respect to the Reynolds number becomes physically plain. Introducing a Fourier transform of the field $v(x, t)$,

$$c(k, t) = \frac{1}{2\pi} \int_{-\infty}^{\infty} v(x, t)\,\mathrm{e}^{ikx}\mathrm{d}x,$$

one can easily obtain the spectral form of the BE

$$\frac{\partial}{\partial t}c(k, t) - \frac{ik}{2} \int_{-\infty}^{\infty} c(k - k_1, t)\, c(k_1, t)\,\mathrm{d}k_1 = -\mu k^2 c(k, t). \tag{5.8}$$

The field spectral component convolution operation entering (5.8) describes their nonlinear interaction, whereas the right-hand side of (5.8) embodies their linear damping. In a linear medium ($R_0 = 0$) the field spatial spectral components decay as

$$c^0(k, t) = c_0(k)\,\mathrm{e}^{-k^2\mu t}, \tag{5.9}$$

where

$$c_0(k) = \frac{1}{2\pi} \int_{-\infty}^{\infty} v_0(x)\, e^{ikx} dx$$

are the initial field spectral components.

In the case of low Reynolds number, the spectral components of the field $v(x, t)$ (being a solution to BE) can be written as

$$c(k, t) = c^0(k, t) + c^1(k, t), \tag{5.10}$$

where $c^1(k, t)$ is obtained from (5.8) by a perturbation techinque with respect to a small Reynolds number. The calculation yields.

$$c^1(k, t) = \frac{ik}{2} e^{-k^2\mu t} \int_{-\infty}^{\infty} \frac{c_0(k - k_1)\, c_0(k_1)\, [1 - e^{-2\mu t(k_1^2 - kk_1)}]\, dk_1}{2\mu(k_1^2 - k_1 k)}. \tag{5.11}$$

Thereby formulae (5.9), (5.10) and (5.11) offer an equivalent representation of (5.6) in spectral terms.

It can be clearly seen from (5.11) that due to nonlinear interaction of the initial field spectral components with frequencies k' and k'', those with frequencies $k' \pm k''$ emerge in the resulting field. This means that the second term of the expansions (5.6) and (5.10) describes a single-stage interaction of the initial field harmonics. At $t \ll t_d$

$$c^1(k, t) = \frac{ikt}{2} \int_{-\infty}^{\infty} c_0(k - k_1)\, c_0(k_1)\, dk_1,$$

and consequently at the initial stage of the new spatial harmonic generation the higher spectrum harmonics are formed most actively.

If one tries in the same way to allow for the next expansion term, proportional to R_0^2, new spatial spectral components of the field appear to be generated still more actively, thanks to a twofold nonlinear interaction of the initial field spectral components. In the general case, an n-tuple interaction of the spectral components will correspond to an expansion term proportional to R_0^n.

3. Now let us find an exact solution of BE appearing as a Reynolds number power series for an arbitrary initial perturbation $v_0(x)$. First we resort to the case of a linear medium ($R_0 = 0$) and write (5.2) as

$$v(x, t) = \int_{-\infty}^{\infty} v_0(x + z)\, W_G(z, t)\, dz.$$

This formula can be easily interpreted in probabilistic terms as a statistical average value of the function $v_0(x + z)$ averaged over a random value z, which has Gaussian probability distribution (5.3) with zero average and variance equal to $D_z = 2\mu t$. If we introduce angle brackets $\langle \ldots \rangle_z$ meaning statistical averaging over a random quantity z, the proceding formula will take the form

$$v(x, t) = \langle v_0(x + z)\rangle_z. \tag{5.12}$$

Using statistical averaging over a Gaussian variable allows one to similarly represent the general solution of BE as well, at an arbitrary value of the Reynolds number.

Replacing variable in (1.13) and writing action $S_0(x + z)$ as a sum:

$$S_0(x + z) = \int_x^{x+z} v_0(u)\, du + \int_{x_0}^{x} v_0(u)\, du$$

one can represent the general solution (1.13) in the form

$$v(x, t) = -\frac{\langle z f(z)\rangle_z}{t\,\langle f(z)\rangle_z}, \tag{5.13}$$

where

$$f(z) = \exp\left[-\frac{1}{2\mu}\int_0^z v_0(x + u)\, du\right].$$

For further advance let us turn to the following formula [19]:

$$\langle ug(v)\rangle_{u,\,v} = \sum_{k=0}^{\infty} \frac{1}{k!} \langle g^{(k)}(v)\rangle_v\, \langle u, v, v, \ldots, v\rangle, \tag{5.14}$$

which offers the expansion of the product $ug(v)$ averaged over the totality of random values $\{u, v\}$ having arbitrary probability distribution over the joint cumulants. Angle brackets with commas in (5.14) are called cumulant brackets. In the general form it is defined as a joint cumulant (semi-invariant) of the random variable totality $\{\mathbf{x}\} = \{x_1, x_2, x_3, \ldots, x_N\}$ according to the formula

$$\langle x_1, x_2, \ldots, x_N\rangle = (-\mathrm{i})^N \left.\frac{\partial^N \ln \Theta(\mathbf{u})}{\partial x_1\, \partial x_2 \ldots \partial x_N}\right|_{\mathbf{u}=\mathbf{0}},$$

where $\Theta(\mathbf{u}) = \Theta(u_1, u_1, \ldots, u_N)$ is the characteristic function of a random totality $\{\mathbf{x}\}$.

Introduction of special symbols such as cumulant brackets for joint cumulants is due to the fact that cumulant brackets possess the following important properties adding significantly to the simplification of statistical calculations [19]:

(1) $\langle x_1, x_2, \ldots, x_N\rangle$ is a symmetric function of its arguments.
(2) $\langle A_1x_1, A_2x_2, \ldots, A_Nx_N\rangle = A_1A_2 \ldots A_N \langle x_1, x_2, \ldots, x_N\rangle$.
(3) $\langle x_1, x_2, \ldots, x_{k_1} + x_{k_2}, \ldots, x_N\rangle = \langle x_1, x_2, \ldots, x_{k_1}, \ldots, x_N\rangle$
$+ \langle x_1, x_2, \ldots, x_{k_2}, \ldots, x_N\rangle$.

(4) $\langle x_1, x_2, \ldots, y, \ldots, x_N\rangle = 0$ provided a random quantity y is statistically independent of $\{\mathbf{x}\}$.
(5) $\langle x_1, x_2, \ldots, A_k, \ldots, x_N\rangle = 0$
(6) $\langle x_1 + A_1, x_2 + A_2, \ldots, x_N + A_N\rangle = \langle x_1, x_2, \ldots, x_N\rangle$.

Here $A_1, A_2, \ldots, A_N$ are deterministic quantities.

It should be noted that for a Gaussian totality of random values only the cumulant brackets of the first two orders are other than zero:
$\langle x\rangle$ is an average value,
$\langle x, y\rangle = \langle xy\rangle - \langle x\rangle\langle y\rangle$ is covariance.

Therefore, in a particular case of Gaussian totality $\{\mathbf{u}, \mathbf{v}\}$ formula (5.14) yields

$$\langle ug(v)\rangle = \langle u\rangle \langle g(v)\rangle + \langle u, v\rangle \langle g'(v)\rangle. \tag{5.15}$$

Now let us turn to the average value entering the numerator in (5.13). As z is a random Gaussian quantity with zero average value and the variance equal to $\langle z, z\rangle = D_z = 2\mu t$, (5.15) implies

$$\langle z f(z)\rangle\, z = D_z \langle f'(z)\rangle = -t \langle v_0(x + z) f(z)\rangle_z.$$

Thus, instead of (5.13) one obtains

$$v(x, t) = \frac{\langle v_0(x + z) f(z)\rangle_z}{\langle f(z)\rangle_z} \equiv \frac{\langle u\mathrm{e}^{-w}\rangle}{\langle \mathrm{e}^{-w}\rangle},$$

where

$$u = v_0(x + z), W = \frac{1}{2\mu}\int_0^z v_0(x + y)\,\mathrm{d}y$$

due to arbitrary nature of function $v_0(x)$, are in this case non-Gaussian random quantities. Using formula (5.14) one can readily find

$$\langle u\mathrm{e}^{-w}\rangle = \langle \mathrm{e}^{-w}\rangle \sum_{k=0}^{\infty} \frac{(-1)^k}{k!} \langle u, \underbrace{w, w, \ldots, w}_{k}\rangle.$$

Thereby, the field $v(x, t)$ given in (1.13) can be written in the form of the following series

$$v(x, t) = \sum_{k=0}^{\infty} \frac{(-1)^k}{k!(2\mu)^k} \langle v_0(x + z), \underbrace{\int_0^z v_0(x + y)\,\mathrm{d}y, \ldots, \int_0^z v_0(x + y)\,\mathrm{d}y}_{k}\rangle. \tag{5.16}$$

Now let us introduce Reynolds number explicitly and assume that the initial perturbation $v_0(x)$ has only the scales σ_0 and l_0; then the Reynolds number can be defined as $R_0 = \sigma_0 l_0/\mu$. Taking into account that

$$\int_0^z v_0(x + y)\,\mathrm{d}y = S_0(x + z) - S_0(x)$$

and introducing the dimensionless action

$$\tilde{S}_0(x) = S_0(x)/\sigma_0 l_0$$

one can write

$$\frac{1}{2\mu}\int_0^z v_0(x+y)\,dy = \frac{R_0}{2}\,\tilde{s}_0(x+z) - \frac{R_0}{2}\,\tilde{s}_0(x). \tag{5.17}$$

The second term in the latter expression does not contain random variable z. Consequently, this term from a cumulant-brackets point of view is a deterministic quantity. For this reason we can use the sixth property of the cumulant brackets.

Thereby, substituting (5.17) into (5.16) one obtains the final form for the general solution of RE

$$\begin{aligned} v(x,t) &= \langle v_0(x+z)\rangle_z - \frac{R_0}{2}\langle v_0(x+z), \tilde{s}_0(x+z)\rangle \\ &+ \frac{1}{2}\left(\frac{R_0}{2}\right)\langle v_0(x+z), \tilde{s}_0(x+z), \tilde{s}_0(x+z)\rangle + \ldots \\ &= \sum_{k=0}^{\infty}\frac{(-1)^k}{k!}\left(\frac{R_0}{2}\right)^k \langle v_0(x+z), \underbrace{\tilde{s}_0(x+z), \ldots, \tilde{s}_0(x+z)}_{k}\rangle \end{aligned} \tag{5.18}$$

Hence we have found an exact solution of the BE in the form of an infinite series with respect to the Reynolds number powers. The first term in the series with $k = 0$ corresponds to the linear equation of diffusion. Therefore, k-tuple nonlinear interaction of the initial perturbation $v_0(x)$ and its action $\tilde{s}_0(x)$ corresponds to the kth order of Reynolds number.

Before we proceed to the example let us show the values of cumulant brackets of the second, third and fourth order, expressed through statistical averages over the totality of random quantities

$$\begin{aligned} \langle u, w\rangle &= \langle uw\rangle - \langle u\rangle\langle w\rangle, \\ \langle u, w, w\rangle &= \langle uw^2\rangle - \langle u\rangle\langle w^2\rangle - 2\langle w\rangle\langle uw\rangle + 2\langle u\rangle\langle w\rangle^2, \\ \langle u, w, w, w\rangle &= \langle uw^3\rangle - \langle u\rangle\langle w^3\rangle - 3\langle w\rangle\langle uw^2\rangle - 3\langle uw\rangle\langle w^2\rangle \\ &+ 6\langle u\rangle\langle w\rangle\langle w^2\rangle + 6\langle w\rangle^2\langle uw\rangle - 6\langle u\rangle\langle w\rangle^3. \end{aligned} \tag{5.19}$$

4. Let us consider the first example of a pulse with area $A = \sigma_0 l_0$ localised near $x = 0$. We restrict ourselves to the case of large enough t when $l_d(t) \gg l_0$. Making use of (5.19) one can easily reveal that

$$\langle v_0(x+z), \tilde{S}_0(x+z)\rangle = AW_G(x,t)\,[\tfrac{1}{2} - F(x,t)],$$

$$\langle v_0(x+z), \tilde{S}_0(x+z), \tilde{S}_0(x+z)\rangle = AW_G(x,t)\,[\tfrac{1}{3} - 2F(x,t) + 2F^2(x,t)],$$

$$\langle v_0(x+z), \tilde{S}_0(x+z), \tilde{S}_0(x+z), \tilde{S}_0(x+z)\rangle$$
$$= AW_G(x,t)\,[\tfrac{1}{4} - \tfrac{7}{2}F(x,t) + 9F^2(x,t) - 6F^3(x,t)],$$

where

$$F(x,t) = \int_{-\infty}^{x} W_G(y,t)\,dy.$$

Therefore, using (5.18) and being confined to the third degree of Reynolds number, we shall have

$$v(x,t) = AW_G(x,t)\left\{1 - \frac{R_0}{2}\left[\frac{1}{2} - F(x,t)\right] + \frac{R_0^2}{8}\left[\frac{1}{3} - 2F(x,t) + 2F^2(x,t)\right]\right.$$
$$\left. - \frac{R_0^3}{48}\left[\frac{1}{4} - \frac{7}{2}F(x,t) + 9F^2(x,t) - 6F^3(x,t)\right]\right\}. \tag{5.20}$$

This formula extends (5.7) up to larger values of R_0. Inasmuch as $F(x, t)$ does not exceed unity, formula (5.20) can be employed to a sufficient degree of accuracy up to $R_0 = 1$.

5. Consider the second example of the initial perturbation

$$v_0(x) = a_0 \sin k_0 x,$$

dimensionless action and Reynolds number of which are equal to

$$\tilde{s}_0(x) = 1 - \cos k_0 x, \quad R_0 = a_0/k_0\mu.$$

Bearing in mind the sixth property of cumulant brackets one can view quantity $\tilde{s}_0(x) = -\cos k_0 x$ as dimensionless action. Using statistical averages

$$\langle \sin k_0(x+z)\rangle_z = e^{-k_0^2\mu t}\sin k_0 x,$$
$$\langle \cos k_0(x+z)\rangle_z = e^{-k_0^2\mu t}\cos k_0 x,$$

we can easily find the following values for cumulant brackets

$$\langle \sin k_0(x+z), \cos k_0(x+z) = \tfrac{1}{2}\sin 2k_0 x\left(e^{-4\theta} - e^{-2\theta}\right),$$

$$\langle \sin k_0(x+z), \cos k_0(x+z), \cos k_0(x+z)\rangle$$
$$= \tfrac{1}{4}\sin 2k_0 x\,[e^{-\theta} - 2e^{-3\theta} + e^{-5\theta}] - \tfrac{1}{4}\sin 3k_0 x\,[-2e^{-3\theta} + 3e^{-5\theta} - e^{-9\theta}],$$

$$\langle \sin k_0(x+z), \cos k_0(x+z), \cos k_0(x+z), \cos k_0(x+z)\rangle$$
$$= -\tfrac{1}{4}\sin 2k_0 x\,[-3e^{-2\theta} + 8e^{-4\theta} - 6e^{-6\theta} + e^{-10\theta}]$$
$$-\tfrac{1}{8}\sin 4k_0 x\,[6e^{-4\theta} - 12e^{-6\theta} + 3e^{-8\theta} + 4e^{-10\theta} - e^{-16\theta}],$$

where notation $\theta = k_0^2\mu t$ is introduced.

Therefore, confining ourselves to the cubic term in the Reynolds number for initially harmonic propagating signal we have, proceeding from (5.18)

$$v(x, t) = b_1(t) \sin kx + b_2(t) \sin 2kx + b_3(t) \sin 3kx + b_4(t) \sin 4kx, \tag{5.21}$$

harmonic amplitudes are equal to

$$b_1(t) = a_0 e^{-\theta} - a_0 \frac{R_0^2}{32} \left[e^{-\theta} - 2e^{-3\theta} + e^{-5\theta}\right],$$

$$b_2(t) = -a_0 \frac{R_0}{4} \left[e^{-2\theta} - e^{-4\theta}\right] + a_0 \frac{R_0^3}{192} \left[3e^{-2\theta} - 8e^{-4\theta} + 6e^{-6\theta} - e^{-10\theta}\right],$$

$$b_3(t) = a_0 \frac{R_0^2}{32} \left[2e^{-3\theta} - 3e^{-5\theta} + e^{-9\theta}\right],$$

$$b_4(t) = -a_0 \frac{R_0^3}{384} \left[6e^{-4\theta} - 12e^{-6\theta} + 3e^{-8\theta} + 4e^{-10\theta} - e^{-16\theta}\right].$$

Single interaction and generation of the second harmonic correspond to the second term in the series (5.18) proportional to R_0 and coinciding with the result obtained previously. The term of the series $\sim R_0^2$ corresponds to double interaction: generation of the third harmonic and contribution into the first one; the term $\sim R_0^3$ corresponds to three-tuple interaction: generation of the fourth harmonic and contribution to the second one, and so on.

At the first stage for $\theta \ll 1$ the amplitude of the second harmonic grows in proportion to θ, $b_3(t) \sim \theta^2$, $b_4(t) \sim \theta^3$. On the other hand, at a stage of degeneracy for $\theta \gg 1$ the amplitude of N-harmonic decays N times faster than the first harmonic but N times slower than a wave with frequency NK in a linear medium. The latter, as stated already, is associated with continuous energy transfer, due to the nonlinearity of the medium, from lower harmonics to higher ones.

The derived expression (5.21) is also valid to a sufficient degree of accuracy up to the values of $R_0 \sim 1$.

3 Lagrangian and Eulerian statistics of random fields

This chapter is devoted to the description of the main mathematical principles of the theory of random fields which are indispensible in order to efficiently solve statistical problems of the theory of nonlinear waves in nondispersive media. Along with a brief review of basic concepts of the statistical theory, included in the chapter to remind the reader, we discuss in detail the problem of the relation between the behaviour and the statistical properties of random realisations, as well as between two alternative statistical descriptions of random fields – the Lagrangian and the Eulerian. It has to be noted that the latter is important both for the theory of turbulence and for the description of the wave field behaviour in randomly inhomogeneous media.

3.1 Main statistical properties of random fields

1. Consider a random field $v(x, t)$ as a two-parameter random variable v. The random field is given by a realisation space v and a probability measure, i.e. probability densities:

$$W_v(v; x, t),\ W_v(v_1, v_2; x_1, t_1, x_2, t_2),\ \ldots,$$
$$W_v(v_1, v_2, \ldots, v_n; x_1, t_1, x_2, t_2, \ldots, x_n, t_n),$$

the arguments of which are $v_1, v_2, \ldots, v_n$, which are random variable values in the realisation space, while the parameters are given as spatial coordinates $x_1, x_2, \ldots, x_n$ and time instants $t_1, t_2, \ldots, t_n$.

It is convenient for the further discussion to find the probability density functions (we shall also call them probability distributions or, simply, as random variable distributions) as the delta function averages

$$W_v(v; x, t) = \langle \delta(v(x, t) - v) \rangle_v,$$
$$W(v_1, v_2, \ldots, v_n; x_1, t_1, \ldots, x_n, t_n) \tag{1.1}$$

$$= \left\langle \prod_{k=1}^{n} \delta(v(x_k, t_k) - v_k) \right\rangle_v .$$

Similarly we obtain the joint probability densities for the random fields $v(x, t)$ and $u(x, t)$, for example

$$W_{v,u}(v, u; x_1, t_1, x_2, t_2) = \langle \delta(v(x_1, t_1) - v)\, \delta(u(x_2, t_2) - u) \rangle.$$

Here $\langle \dots \rangle$ symbolises statistical averaging over the realisation ensemble for appropriate random values. Their effect on an arbitrary deterministic function of the random field $v(x, t)$ is presented as

$$\langle g[v(x, t)] \rangle = \int_{-\infty}^{\infty} g(z)\, W_v(z; x, t)\, dz.$$

The brackets of the statistical averaging are linear operators with respect to their arguments.

In unambiguous situations the random variable subscript attached to the averaging brackets will be omitted.

In addition to the probability densities we consider also the integral distribution functions, such as

$$F_v(v; x, t) = \int_{-\infty}^{v} W_v(z; x, t)\, dz = \langle 1(v - v(x, t)) \rangle_v, \qquad (1.2)$$

where $1(x)$ is a unit (step) function.

Let us recall that the random field $v(x, t)$ is defined as *stationary* provided each of its probability density functions remains the same after the replacement of the time instants $t_1, t_2, \dots, t_n$ with $t_1, + r, t_2 + r, \dots, t_n + r$ for any r. In analogy with this, the random field $v(x, t)$ is statistically *homogeneous* if all its probability density functions do not change after the replacement of coordinates $x_1, x_2, \dots, x_n$ with $x_1 + s, x_2 + s, \dots, x_n + s$ for any s.

2. Among the simplest but none the less important problems in the statistical theory of random processes and fields, is the problem of establishing the law of their probability density transformation by a given nonlinear (non-inertial) transformation of the processes and the fields themselves [12, 19, 78]. The solution to this problem will be especially easy if one makes use of the probability density definition as an average of the delta function (1.1), as well as of different features of the delta function and unit function. Inasmuch as in the ensuing discussions in this book we are planning to frequently resort to these properties, the appropriate formulae are placed in the Appendix for the convenience of the reader.

Let the random field $v(x, t)$ possess the probability density function $W_v(v; x, t)$, and let us define the random field $u(x, t) = \Psi[v(x, t)]$ where $u = \Psi(v)$ is the nonlinear deterministic non-inertial conversion given by a determinatistic function $\Psi(z)$. It is required to find the probability density function $W_u(u; x, t)]$ for the random field $u(x, t)$.

Applying the definition (1.1) we have

$$W_u(u;\, x,\, t) = \langle \delta(u(x,\, t) - u)\rangle_u = \langle \delta(\Psi[v(x,\, t)] - u)\rangle$$
$$= \int_{-\infty}^{\infty} \delta(\Psi(z) - u)\, W_v(z;\, x,\, t)\, \mathrm{d}z. \tag{1.3}$$

Therefore, using the 'filtering' property of the delta function (A.13) we obtain

$$W_u(u;\, x,\, t) = \sum_i \frac{W_v(v_i(u);\, x,\, t)}{|\Psi'(v_i(u))|}, \tag{1.4}$$

where $v_i(u)$ is an ith branch of the function inverse to $u = \Psi(v)$. In other words, $v_i(u)$ is an ith real root of the equation $\Psi(v) = u$.

3. Now let us investigate the situation when the random variable itself is subject to random transformation. Let $u = \Psi(v)$ be a random function of a random variable v given by the probability density $W_v(v)$. The randomness of the function Ψ means that at a fixed value of v the value of the function $u = \Psi(v)$ will be a random variable. The probability density for u is apparently given as $W_u(u;\, v)$, where v is a paramenter while the probability density $W_u(u)$ for the random variable u is described by the 'total probability formula':

$$W_u(u) = \int_{-\infty}^{\infty} W_u(u;\, z)\, W_v(z)\, \mathrm{d}z, \tag{1.5}$$

which generalises (1.3) over to the random transformation $v \to u$.

If we take the inverse transformation $v = \Psi^{-1}(u)$ with the aid of the inverse and random function $\Psi^{-1}(u)$ then, owing to the symmetry, instead of (1.5) we obtain

$$W_v(v) = \int_{-\infty}^{\infty} W_v(v;\, z)\, W_u(z)\, \mathrm{d}z,$$

where $W_v(v;\, u)$ is the probability density of the random variable v specified by the random function $\Psi^{-1}(u)$ at a fixed value of u.

For the forthcoming discussion it is of importance to make clear the interrelations between $W_u(u;\, v)$ and $W_v(v;\, u)$.

Let us cast the problem in more general definitions. Assume $y = y(x)$ as a random function with probability density $W_y(y;\, x)$. It is required that $y(x)$ be a strictly monotonic continuously differentiable function so that its derivative is positive at any x: $y'(x) > 0$.

Then, based on (A.14),

$$\delta(X - x(y)) = -\frac{\partial}{\partial X}\, \delta(y - y(X)), \tag{1.6}$$

where $x = x(y)$ is the random function inverse to $y(x)$. A requirement is to find the probability density $W_x(X;\, y)$ of the former. Averaging the left-hand side of (1.6) over the random variable x and the right-hand side over y, making use of (A.22) one can write

$$W_x(X; y) = -\frac{\partial}{\partial X}\int_{-\infty}^{y} W_y(z; X)\,\mathrm{d}z. \tag{1.7a}$$

Correspondingly

$$W_y(Y; x) = -\frac{\partial}{\partial Y}\int_{-\infty}^{x} W_x(z; Y)\,\mathrm{d}z. \tag{1.7b}$$

Exactly these formulae, vital for the subsequent calculations, offer the probability density relations under mutually inverse random transformations. In order to render the symmetry of $W_x(x; y) \rightleftarrows W_y(y; x)$ obvious, it is sufficient to differentiate both parts of (1.7) with respect to y, x respectively, giving

$$\frac{\partial}{\partial y} W_x(x; y) = -\frac{\partial}{\partial x} W_y(y; x)$$

4. If we decline the monotonicity requirement for the random function $y = y(x)$, the number of its intersections with the prescribed lever y^* will be random, and the problem of finding the $N(y^*)$ statistics, i.e. the statistics of the number of roots x_n of the equation $y(x) = y^*$, arises. One can find the average number of these roots using expression (A.16), which takes the form

$$\sum_{n=1}^{N(y^*)} \delta(X - x_n) = |y'(X)|\,\delta(y^* - y(x)).$$

Averaging this equality over the ensemble of random realisations $y(x)$ and making use of the formula for the total probability, we can write

$$\sum_{N=1}^{\infty} P(N; y^*) \sum_{n=1}^{N} W_n(X; y^*|N) = \langle |y'(X)| \delta(y^* - y(X))\rangle. \tag{1.8}$$

Here $P(N; y^*)$ stands for the probability that the equation $y(x) = y^*$ has N roots, and $W_n(X; y^*|N)$ is the probability density for the value of the nth root x_n provided the total number of roots is N. The average appearing on the right-hand side is expressed through the joint probability density of function $y(x)$ and its derivative,

$$W_{y,y'}(Y, Y'; x) = \langle \delta(y(x) - Y)\, \delta(y'(x) - Y')\rangle,$$

in the following way:

$$\langle |y'(x)|\, \delta(y^* - y(x))\rangle = \int_{-\infty}^{\infty} |z| W_{y,y'}(y^*, z; x)\,\mathrm{d}z.$$

After the integration of (1.8) with respect to x we acquire the desired average number of level y^* intersections by the random function $y(x)$:

$$\langle N(y^*)\rangle = \sum_{N=1}^{\infty} NP(N; y^*) = \iint_{-\infty}^{\infty} |z| W_{y,y'}(y^*, z; x)\, dz\, dx. \tag{1.9}$$

5. Now let us get busy discussing the characteristic functions of a random field $v(x, t)$:

$$\theta(\lambda; x, t) = \langle \exp(i\lambda v(x, t))\rangle_v,$$

$$\theta_v(\lambda_1, \ldots, \lambda_n; x_1, t_1, \ldots, x_n, t_n) = \left\langle \exp\left(i\sum_{k=1}^{n} \lambda_k v(x_k, t_k)\right)\right\rangle. \tag{1.10}$$

As is evident, they appear to be the Fourier transforms of the corresponding probability densities (1.1), and they also give a comprehensive description of the random field viewed at a finite number of spatial points and at a finite number of time instarts.

A comprehensive description of the random field for the continum of parameter (x and t) values is present by a characteristic functional. For example, with respect to the random function $v(x)$ the characteristic functional can be formulated as

$$\Theta_v[\mu(x)] = \left\langle \exp\left(i\int_{-\infty}^{\infty} \mu(z) v(z)\, dz \right)\right\rangle_v, \tag{1.11}$$

where $\mu(z)$ is any deterministic function. Assuming in (1.11), for instance, that

$$\mu(z) = \lambda\delta(z - x) + \mu\delta'(z - x)$$

we obtain a joint single-point characteristic function for the random function $v(x)$ and its derivative $v'(x)$:

$$\theta_{v,v'}(\lambda, \mu; x) = \langle \exp(i\lambda v(x) + i\mu v'(x))\rangle. \tag{1.12}$$

Among all kinds of probability distributions, the Gaussian probability distribution plays an outstanding role. In view of the Central Limit Theorem of probability theory, any random event resulting from a superposition of many independent elementary actions is known to have a Gaussian distribution. By their nature, Gaussian random variables have an invariance property with respect to linear operations: namely, an arbitrary sum of independent Gaussian random variables will be Gaussian as well. Accordingly, $v(x)$ is a Gaussian random function if its values at any x form a Gaussian population.

The characteristic function and probability density of the Gaussian variable v have the forms

$$\theta_v(\lambda) = \exp(i\lambda\langle v\rangle - \sigma_v^2\lambda^2/2),$$

$$W_v(v) = \frac{1}{\sqrt{2\pi\sigma_v^2}} \exp\left(-\frac{(v - \langle v\rangle)^2}{2\sigma_v^2}\right), \tag{1.13}$$

where $\langle v\rangle$ and $\sigma_v^2 = \langle(v - \langle v\rangle)^2\rangle = \langle v^2\rangle - \langle v\rangle^2$ are the average value and variance of the random variable.

The characteristic function of the Gaussian population $v(x_1, t_1)$, $v(x_2, t_2)$ is expressed as follows:

$$\theta_v(\lambda_1, \lambda_2; x_1, t_1, x_2, t_2) = \exp[i\lambda_1\langle v(x_1, t_1)\rangle + i\lambda_2\langle v(x_2, t_2)\rangle \\ - \tfrac{1}{2}\lambda_1^2\sigma_v^2(x_1, t_1) - \tfrac{1}{2}\lambda_2^2\sigma_v^2(x_2, t_2) - \lambda_1\lambda_2 B_v(x_1, t_1, x_2, t_2)]. \tag{1.14}$$

Here

$$B_v(x_1, t_1, x_2, t_2) = K_v(x_1, t_1, x_2, t_2) - \langle v(x_1, t_1)\rangle\langle v(x_2, t_2)\rangle \tag{1.15}$$

is the so-called covariance function, while

$$K_v(x_1, t_1, x_2, t_2) = \langle v(x_1, t_1)\, v(x_2, t_2)\rangle \tag{1.16}$$

is the correlation function of the random field $v(x, t)$.

The characteristic functional of the Gaussian random function $v(x)$ may be written as

$$\Theta_v[\mu(x)] = \exp\left(i\int_{-\infty}^{\infty}\mu(z)\langle v(z)\rangle\,dz - \frac{1}{2}\iint_{-\infty}^{\infty}\mu(z_1)\mu(z_2)\,B_v(z_1, z_2)\,dz_1\,dz_2\right). \tag{1.17}$$

Let us find, for example, the probability density of the function $v(x)$ and its derivative $v'(x)$. We confine ourselves to a particular case where $v(x) = \varphi(x) + v_0(x)$, with $\varphi(x)$ deterministic and $v_0(x)$ a statistically homogeneous random function with zero mean and a covariance $B_v(x_2 - x_1) = K_v(x_2 - x_1)$ which depends only on the distance between points x_2 and x_1. In this instance (1.12) results in

$$\theta(\lambda, \mu; x) = \exp(i\lambda\varphi(x) + i\mu\varphi'(x) - \sigma_v^2\lambda^2/2 - \sigma_u^2\mu^2/2). \tag{1.18}$$

Here $\sigma_u^2 = -B_v''(0)$ is the variance of the derivative $v'(x)$. Correspondingly, the joint single-point probability density of the random function $v(x)$ and its derivative $v'(x)$ is equal to

$$W_{v,v'}(v, u; x) = \frac{1}{2\pi\sigma_v\sigma_u}\exp\left(-\frac{(v - \varphi(x))^2}{2\sigma_v^2} - \frac{(u - \varphi'(x))^2}{2\sigma_u^2}\right). \tag{1.19}$$

It has broken down into the product of the one-dimensional probability densities. It means that a statistically homogeneous Gaussian function $v(x)$ and its derivative $v'(x)$ in one and the same point x are statistically independent.

6. Applications are generally restricted to measurements of the random field spectral–correlation properties. Therefore, we are ready to discuss their covariance functions and spectral densities in more detail. For simplicity we analyse only statistically homogeneous random field $v(x, t)$ with their covariance function having the form

$$\langle v(x, t)\, v(x + s, t)\rangle = K_v(s; t) = B_v(s; t), \quad (\langle v\rangle = 0). \tag{1.20}$$

The distribution of the energy $\sigma_v^2 = B_v(0; t)$ of the field $v(x, t)$ over the spatial harmonics is described by the spatial spectral density of the energy

$$G_v(k; t) = \frac{1}{2\pi} \int_{-\infty}^{\infty} B_v(s; t)\, e^{-iks}\, ds = \frac{1}{\pi} \int_{0}^{\infty} B_v(s; t) \cos ks\, ds. \qquad (1.21)$$

The covariance function is expressed through the spectral density by means of the inverse Fourier transform

$$B_v(s; t) = \int_{-\infty}^{\infty} G_v(k; t)\, e^{iks}\, dk. \qquad (1.22)$$

In practical applications, statistically inhomogeneous random function $v(x)$ with infinite variance are found, but their increments $v(x + s) - v(x)$ are still statistically homogeneous in x. When analysing such random functions it is convenient to use a structure function

$$d_v(s) = \langle [v(x + s) - v(x)]^2 \rangle,$$

which, for statistically homogeneous functions, takes the form

$$d_v(s) = 2\sigma_v^2 - 2B_v(s).$$

As well as the covariance function, the structure function is connected with the spectral density through the Fourier transform [12, 18, 79]:

$$d_v(s) = 4 \int_0^{\infty} (1 - \cos ks)\, G_v(k)\, dk,$$

$$G_v(k) = -\frac{1}{2\pi} \int_0^{\infty} d_v(s) \cos ks\, ds \quad (k \neq 0). \qquad (1.23)$$

3.2 Connection of the statistical properties of random functions with the behaviour of their realisations

1. Much information on the random function realisation behaviour is contained in the covariance functions and spectral densities. It is particularly obvious if we take, for example, the impulse functions

$$v(x) = \sum_n \Psi(x - x_n), \qquad (2.1)$$

which are considered to be a sequence of pulses of a given shape $\Psi(x)$ appearing at random points x_n. So, if the intervals $s_n = x_{n+1} - x_n$ between adjacent pulses are mutually independent, have the same probability distribution of lengths $W_s(s)$ and a characteristic function $\theta_s(\kappa)$, the spectral power density of $v(x)$ may be given by the equality [78]

$$G_v(k) = \mu \mathcal{E}_\Psi(k)\, \mathrm{Re}\left[\frac{1 + \theta_s(k)}{1 - \theta_s(k)}\right], \qquad (2.2)$$

where $\mu = 1/\langle s_n \rangle$ is the average frequency of pulse appearance, $\mathcal{E}_\Psi(k)$ is the spectral energy density of the pulse,

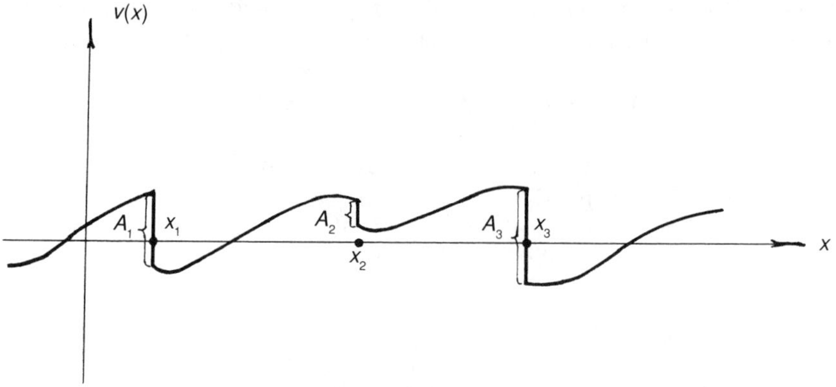

Fig. 3.1 Realisation of a random function with discontinuities

$$\mathcal{E}_\psi(k) = \frac{1}{2\pi} |c_\psi(k)|^2, \quad c_\psi(k) = \int_{-\infty}^{\infty} \Psi(x)\, e^{ikx}\, dx. \tag{2.3}$$

The spectral density (2.2) has a clear physical meaning. It is proportional to the spectral density of the pulse itself which is quite natural because the spectral density of the random function $v(x)$ can involve only those spectral components which are present in the single-pulse spectrum. The last factor in (2.2) depicts the effect of interference different pulse harmonics related to a certain ordering of their appearance and leading, at different k, to the amplification or suppression of the spectral components with respect to those of a single pulse. In the case of completely chaotic pulse occurrence (Poisson sequence) when there is no interference ($W_s(s) = \exp(-\mu s)$) (2.2) is change to

$$G_v(k) = \mu\, \mathcal{E}_\psi(k), \tag{2.4}$$

meaning that the pulse sequence spectrum is proportional to the single-pulse spectrum. The interference effect is lost also for $k > k^*$, where k^* is determined by $\theta(k^*) \ll 1$. At such $k > k^*$ the equality (2.2) is changed into (2.4). As an estimate of k^* one can resort to $k^* = 1/\sigma_s$, where $\sigma_s^2 = \langle s_n^2 \rangle - \langle s_n \rangle^2$ is the length variance of the random intervals betwen the pulses.

2. Expressions of the types (2.2) and (2.4), demonstrating explicitly the relation between the spectral density and the behaviour of the pulse function realisation, may prove helpful when estimating the spectral density asymptotic behaviour and other random functions, not necessarily of pulse type. It can be clarified by taking, as an example, the random function $v(x)$ with realisation discontinuities at certain points x_n (see Fig. 3.1). The immediate neighbourhood of the discountinuity may be thought of as a fragment of the discontinuity pulse. The discontinuity of the pulse leads to a universal form of the pulse spectral density as $k \to \infty$ [79]:

$$\mathcal{E}_\psi(k) = A^2/2\pi k^2. \tag{2.5}$$

Here A is a discontinuity amplitude (function jump). Knowing the average frequency of pulse appearance μ, and because of the absence of interference it is quite natural to evaluate the asymptotic behaviour of the discontinuous random function $v(x)$ spectral density at large k through a formula similar to (2.4),

$$G_v(k) = \mu\langle A^2\rangle/2\pi k^2. \tag{2.6}$$

The asymptotic behaviour of a random function spectral density at large k determines the behaviour of its corresponding covariance function at small s. Thus, from (2.6) and form Fourier transform properties it follows that the covariance and structure functions of the foregoing statistically homogeneous function $v(x)$ behave, as $s \to 0$, as

$$\begin{aligned} B_v(s) &= B_v(0) - \mu\,\langle A^2\rangle\,|s|/2, \\ d_v(s) &= \mu\,\langle A^2\rangle\,|s|. \end{aligned} \tag{2.7}$$

The same way of reasoning allows us to estimate the spectral density asymptotic behaviour of smooth random function $v(x)$ having realisation singularities of the type $\psi(x - x_n) = A_n(x - x_n)^\alpha$ in the vicinity points x_n. If we identify them with the singularities of the appropriate pulses we acquire the asymptotic behaviour of the pulse spectral density as

$$\mathcal{E}_\psi(k) = A_n^2\Gamma^2(\alpha + 1)/k^{2(\alpha+1)} \quad (k \to \infty, |\alpha| < 1/2).$$

Then, in analogy with (2.6), we write down the evaluation of the spectral density of the field $v(x)$ as

$$G(k) = \mu\langle A^2\rangle\,\Gamma^2(\alpha + 1)/k^{2(\alpha+1)} \quad (k \to \infty), \tag{2.8}$$

and of the corresponding structure function as

$$d_v(s) = \frac{2\pi\mu\,\langle A^2\rangle\,\Gamma^2(\alpha + 1)}{\Gamma(2\alpha + 2)\cos\pi\alpha}\,|s|^{2\alpha+1} \tag{2.9}$$

Here $\Gamma(z)$ is a gamma function.

Reversing the path of reasoning used above, it appears to be natural, from the form of the structure function $d_v(s)$ or spectral density $G_v(k)$ to advance reasonable hypotheses about peculiarities of realisations of the random function. Consider, for example, a random field of a passive impurity (e.g. temperature) in a turbulent atmosphere. It is well known that its structure function obeys the Kolmogororv–Obukhov law of two-thirds (2/3), and has the form $d_v(s) = c^2 s^{\frac{2}{3}}$. Comparing this structure function with (2.9), one can suggest that such an asymptotic behaviour of the turbulence structure function is due to turbulence realisation peculiarities such as $\Psi(x) \sim |x|^{-\frac{1}{6}}$.

An alternative way for the appearance of power spectral densities with power-law exponents is by the accumulation of singularities. Above we have assumed the asymptotic behaviour of the spectral density to be determined by the energy spectrum of an individual singularity. This is true when there is no interference at large wave numbers between the Fourier components of different singularities. But when the signal has a fractal structure, the behaviour of the spectrum at large wave number will be determined by the fractal dimension of the signal. This problem is discussed in references [161-166], and for example, in [166] it is shown that for step singularities the asymptotic behaviour of the energy spectrum is

$$G_v(K) \sim K^{-2+D_K},$$

where D_K is the capacity of the fractal set of singularities.

3. Probability theory attributes a purely probabilistic meaning to the integral distribution function (1.2): $F_v(v;\, x)$ is the probability that $v(x)$ has, at point x, a value less than the given v. The theory of random functions discovers a new aspect of these distribution functions, namely, their close relation to the realisation behaviour. Actually, if we integrate the equality which defines the integral distribution function

$$F(v;\, x) = \langle \mathbf{1}(v - v(x)) \rangle$$

over x we obtain

$$\int_a^b F(v;\, x)\,\mathrm{d}x = \langle L(v;\, a,\, b) \rangle, \tag{2.10}$$

where $\langle L(v;\, a,\, b) \rangle$ is the average length of the (line) segments within the interval $(a,\, b)$ where $v(x) < v$ (see Fig. 3.2). In the same way, the integral distribution function of a statistically homogeneous function $v(x)$ coincides with the relative (per unit length) segment length inside which $v(x) < v$:

$$F_v(v) = n_v(v) = \langle L(v;\, 0,\, 1) \rangle. \tag{2.11}$$

In analogy with the case of the spectral-correlation characteristics, the relation of the distribution functions to the realisation behaviour is particularly explicit with respect to the pulse random functions of type (2.1). Provided the pulses appear with rather low frequency μ and with no significant overlap, the relative length of the life of function $v(x)$ under leverl v and, therefore, its integral distribution function, are obviously connected with the pulse shape $\Psi(x)$ as follows:

$$F_v(v) = n_v(v) = \begin{cases} 1 - \mu \int_{-\infty}^{\infty} 1(\Psi(x) - v)\mathrm{d}x, \; v > 0, \\ \\ \mu \int_{-\infty}^{\infty} 1(v - \Psi(x))\mathrm{d}x, \; v > 0, \end{cases} \tag{2.12}$$

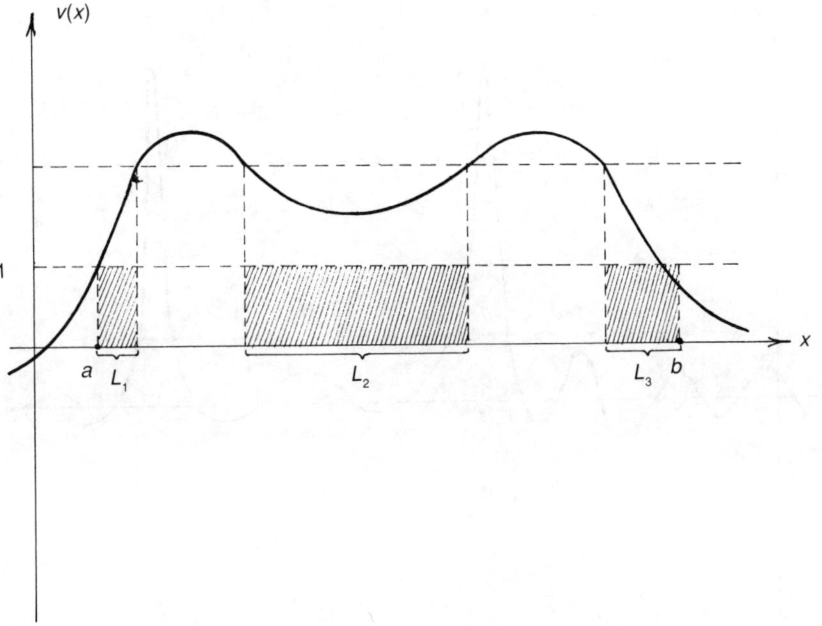

Fig. 3.2 The length of stay of the random function $v(x)$ realisation below the level v within the interval $x \in (a, b)$

Hence, from (A. 13) it follows that the continuous part of the impulse process probability density with non-overlapping impulses is equal to

$$W_v(v) = \mu \sum_{n=1}^{N} 1/|\Psi'(x_n(v))|. \tag{2.13}$$

Here we take the summation over all roots $x(v)$ of equation $\Psi(x) = v$.

In the ensuing analysis of random fields of hydrodynamic type we shall come across random functions which undergo, in the neighbourhood of some points x_n, large (much larger than σ_v which is a standard deviation with respect to a random function) excursions of short (shorter than a correlation length s_v) duration (as in Fig. 3.3). The shape of excursions $\Psi(x - x_n)$ is, as a rule, of the same type and is known, only their locations being random. Such random functions, for $v \gg \sigma_v$, can be identified with the impulse function (2.1), the pulses of which do not overlap. Accordingly, the probability density asymptotic behaviour of such random functions $v(x)$ for $v \gg \sigma_v$ being completely determined by the shape of separate excursions may be defined by such relations as (2.13).

As an example let us again consider a turbulent field with the structure function $d_v(s) = c^2|s|^{\frac{2}{3}}$. If the hypothesis stating that the law of two-thirds is based on the presence of singularities of type $\Psi(x) \sim |x|^{-\frac{1}{6}}$ in the turbulence

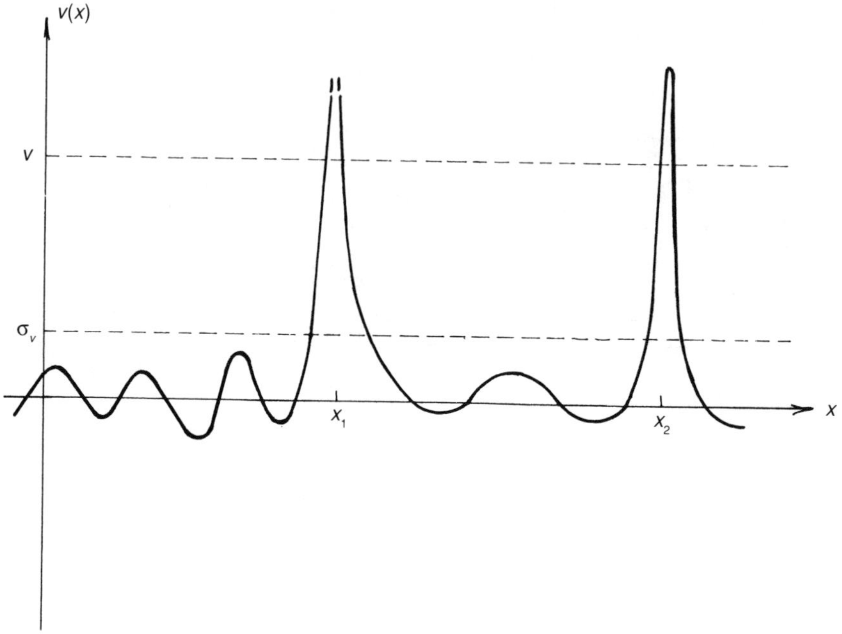

Fig. 3.3 Random function realisation overshoots beyond level v

realisations is valid, then (2.13) leads to the fact that the probability density of the turbulent field may have at large v a power 'tail': $W_v(v) \sim v^{-7}$.

4. It is also of interest to discover the general relation between the realisation behaviour of the statistically homogeneous function $v(x)$ and its probability density $W_v(v)$. We can show that

$$W_v(v) = \left\langle \int_0^1 \delta(v(x) - v)\, dx \right\rangle = \left\langle \sum_{n=1}^{N(0,1)} 1/|v'(x_n)| \right\rangle,$$

i.e. an average sum of the inverse moduli of the rates $v'(x_n)$ at which the level v is crossed by function $v(x)$, the sum taken over the interval of unit length, e.g. $x \in (0, 1)$. Hence, the probability density will increase with the number of terms in a sum, i.e. with the frequency of the level v intersection by the random function $v(x)$ realisation, though it will decrease with the intersection velocity.

Given the joint probability density of $v(x)$ and its derivative $v'(x)$: $W_{v,v'}(v, u; x)$, it seems to be easy, in analogy with (1.9), to calculate an average number of the level v crossing from below by a random function $v(x)$ within the interval (a, b) (i.e. intersections of lever v with positive derivative $v'(x) > 0$)

$$\langle M(v; (a, b))\rangle = \sum_{M=1}^{\infty} MP(M; v, a, b) = \int_a^b dx \int_0^{\infty} uW_{v,v'}(v, u; x)\, du. \qquad (2.14)$$

Here $P(M; v, a, b)$ means the probability that the number of lever v crossings from below within $x \in (a, b)$ equals M.

It seems to be important to underline that, at a fixed interval length $L = |a - B|$, the probability of level v being crossed from below is decreasing with v, so that at large v one can neglect the possibility of more than one intersection and rewrite the latter expression as follows:

$$P(1; v, a, b) = \langle M(v;(a, b))\rangle = \int_a^b dx \int_0^{\infty} uW_{v, v'}(v, u; x)\, du. \qquad (2.15)$$

5. Having counted the average number of level intersections from below v: $\langle M(v; (a, b))\rangle$, one can manage to formulate an asymptotic expression for another important characteristic of the random function realisation $v(x)$, namely, the absolute maximum probability density in a reasonably large interval $x \in (a, b)$, $|a - b| \gg s_v$, where s_v is the length of the random function statistical dependence. Let $v \gg \sigma_v$. Thus, there are few points of intersection of lever v from below by $v(x)$, and they satisfy the inequality $|x_{m+1} - x_m| \gg s_v$. Divide interval (a, b) into a set of adjoining, physically infinitesimal intervals dx_k:

$$s_v \ll dx_k \ll |x_{m+1} - x_m|, \quad |a - b|.$$

The probability that $v(x)$ in the given interval dx_k would not exceed level v can, in accordance with the asymptotic formula (2.15), be represented as

$$dP_k = 1 - P(1; v, dx_k) = 1 - \langle M(v; dx_k)\rangle,$$

where $\langle M(v; dx_k)\rangle \ll 1$. Moreover, when $v \gg \sigma_v$ the level v intersections from below in each of the intervals dx_k are, as a matter of fact, statistically independent. Therefore, the probability that $v(x)$ would not exceed level v in all intervals dx_k is equal to the product of the level v non-intersection probabilities in each of them,

$$P[v(x) < v; x \in (a, b)] = \prod_k dP_k = \prod_k (1 - \langle M(v; dx_k)\rangle) = \exp(-\langle M(v;(a; b))\rangle).$$

$P[v(x) < v; x \in (a, b)]$, however, is by definition equal to the integral distribution function of the aboslute maximum $v(x)$ within interval (a, b), and can be denoted as $F_{max}[v; (a, B)]$. Consequently, the foregoing equality yields

$$F_{max}[v;(a, b)] = \exp(-\langle M(v;(a, b))\rangle). \qquad (2.16)$$

Therefore, the asymptotic behaviour of the random function absolute maximum probability density $v(x)$ within the interval (a, b) takes the form of

$$W_{max}(v; (a, b)) = -\frac{\partial \langle M(v; (a, b))\rangle}{\partial v} \exp(-\langle M(v; (a, b))\rangle). \qquad (2.17)$$

A good example is to find the integral function of the absolute maximum distribution of the statistically homogeneous Gaussian function $v(x)$ (within the interval length L) with a zero average and covariance function $B_v(s)$. Substituting (1.19) at $\varphi \equiv 0$ into (2.14) gives the average number of the level v intersections from below within the interval length as

$$\langle M(v; L)\rangle = M_0 \exp(-v^2/2\sigma_v^2), \quad M_0 = L\sigma_u/2\pi\sigma_v. \tag{2.18}$$

Here M_0 is the average number of the intersections (from below) above zero level. Thus, in accordance with (2.16), (2.18),

$$F_{max}(v; L) = \exp[-M_0 \exp(-v^2/2\sigma_v^2)]. \tag{2.19}$$

If $L \gg s_v$, when $M_0 \sim L/s_v \gg 1$ and many local maxima of function $v(x)$ are 'contending' to become the absolute maximum, the distribution function F_{max} tends asymptotically to the universal, double exponential distribution. Let us demonstrate this. Consider a characteristic value v_*, when $F_{max} = e^{-1}$. As is obvious from (2.19), v_* satisfies the transcendental equation

$$M_0 \exp(-v_*^2/2\sigma_v^2) = 1.$$

Its solution

$$v_* = \sigma_v\sqrt{2\ln M_0}$$

determines the typical value of an absolute maximum. Making use of v_*, we rewrite (2.19) as

$$F_{max}(v; L) = \exp\left\{-\exp\left[-\sqrt{2\ln M_0}\,(v - v_*)/\sigma_v - (v - v_*)^2/2\sigma_v^2\right]\right\}. \tag{2.20}$$

from this it becomes clear that, provided $\sqrt{2\ln M_0} \gg 1$, $F_{max}(v; L)$ has within the interval

$$-1 < (v - v_*)\sqrt{2\ln M_0}/\sigma_v \ll 1$$

enough time to be changed essentially from zero to unity while $(v - v_*)^2/2\sigma_v^2$ in this interval appears to be negligibly small. Therefore, neglecting this term, we pass from (2.20) to the double exponential distribution, asymptotically true at $M_0 \gg 1$,

$$F_{max}(z) = \exp(-e^{-z}). \tag{2.21}$$

Here a dimensionless variable is introduced

$$z = (v - v_*)\sqrt{2\ln M_0}/\sigma_v.$$

The probability density of the absolute maximum dimensionless value corresponding to (2.21) is equal to

$$W_{max}(z) = e^{-z} \exp(-e^{-z}). \tag{2.23}$$

The latter expressions show that while the absolute maximum mean value, equal to $\langle v \rangle \approx \sigma_v \sqrt{2 \ln M_0}$, increases with M_0 as $\sqrt{\ln M_0}$, the absolute maximum variance is decreasing to the same extent (as $1/\sqrt{\ln M_0}$). This phenomenon, of the small variance of the absolute maxima, is applied, for example, in astronomy to measure the distance to a galaxy from the intensity of the brightest star, assuming that the radiation intensity of the brightest stars in different galaxies is actually the same [120].

3.3 Lagrangian and Eulerian statistics of random fields

1. In theoretical and experimental analysis of chaotic motion of continuous media and optical and acoustic wave propagation in randomly inhomogeneous media, two mutually complementary approaches are widely used. One of them is concerned with Lagrangian statistics of random fields and waves. With respect to the continuous medium it is the statistics of the fluid particle velocity and density fluctuation, while in geometrical optics it is understood as the fluctuation of the phase, arrival angle and intensity of a chosen ray tube. Another method is to determine the field and wave Eulerian statistics at a prescribed space point. Thus, for example, the Lagrangian statistics of a continuous medium are mearsured by probes and zero buoyancy buoys, while the Eulerian ones are estimated with the help of immobile sensors.

It is typical in many physical problems that the possibility exists to calculate or experimentally measure a certain type of random field statistical property, e.g. a Lagrangian statistic, while the need is for the Eulerian field property. Therefore, in the preceding chapters we have demonstrated that in the Lagrangian representation, the evolution of a Riemann wave (which can be represented as a hydrodynamic flow of noninteracting particles) appears to be trivial, i.e. each particle moves uniformly and it is quite simple to find the Lagrangian statistics for a random initial field. At the same time, the most attractive features prove to lie in the evolution of the Eulerian average field, spatial spectra and correlation functions, i.e. the Eulerian statistics. Later on it will be shown that there exist rather simple relations between the Lagrangian and Eulerian statistical properties of random fields which, in certain cases, make it possible to obtain the Eulerian statistics from the known Lagrangian ones, and vice versa.

Let us elucidate the common features of and distinctions the Lagrangian and Eulerian descriptions, using as an example the continuous medium hydrodynamic motion. With respect to incompressible fluid, the connections between the Lagrangian and Eulerian statistical descriptions were examined in [121]. For the majority of the problems under consideration in this book, however, the variation of the elementary volume of the medium plays a vital role, hence we shall adhere to the general case [118, 122, 123].

The Eulerian coordinate system is suited to the description of fields at fixed space points, where different particles of a continuous medium arrive in the course of time. A typical example of the Eulerian frame is the Cartesian system $\mathbf{x} = (x_1, x_2, x_3)$. As distinct from the Eulerian one, the Lagrangian frame is 'built' into the continuous medium, and moving with it appears to be rather handy to describe the behaviour of the medium in the vicinity of particles fixed in the medium. As Lagrangian coordinates one often takes the Eulerian coordinates of the particles at an initial time. It can be assumed, for example, that at time $t = 0$ a certain identified particle of the medium has coordinates $\mathbf{y} = (y_1, y_2, y_3)$, and the motion of this particle in the Eulerian coordinate system is depicted by a vector $\mathbf{x} = \mathbf{X}(\mathbf{y}, t)$. Within the Lagrangian frame the space point at which the given particle arrives is ascribed its initial coordinates $\mathbf{y}$. Equalities $\mathbf{x} = \mathbf{X}(\mathbf{y}, t)$ express the Eulerian coordinates in terms of the Lagrangian ones. Solving these equalities for $\mathbf{y}$ we represent the Lagrangian coordinates by way of the Eulerian ones: $\mathbf{y} = \mathbf{y}(\mathbf{x}, t)$. If, somehow, we have knowledge of the continuous medium velocity field in the Eulerian frame $\mathbf{v}(\mathbf{x}, t)$, the Eulerian coordinates of a fixed particle can be found from the solution of the following set of equations:

$$\frac{\mathrm{d}\mathbf{X}}{\mathrm{d}t} = \mathbf{v}[\mathbf{X}(\mathbf{y}, t), t] = \mathbf{V}(\mathbf{y}, t); \quad \mathbf{X}(\mathbf{y}, 0) = \mathbf{y}. \tag{3.1}$$

Here $\mathbf{V}(\mathbf{y}, t)$ is the velocity field in the Lagrangian coordinate system. In the following discussion, we shall, for the sake of brevity, call the fields in the Lagrangian frame 'Lagrangian fields' while those in the Eulerian system will be named 'Eulerian fields'. Thus $\mathbf{v}(\mathbf{x}, t)$is an Eulerian field and $\mathbf{V}(\mathbf{y}, t)$ is a Lagrangian velocity field, $\mathbf{X}(\mathbf{y}, t)$ is the Lagrangian field of the fixed particle Eulerian coordinates, $\mathbf{y}(\mathbf{x}, t)$ is the field of the Lagrangian coordinates of the particles arriving at point $\mathbf{x}$.

Consider an infinitesimal fluid particle of a continuous medium, with Lagrangian coordinates $\mathbf{y}$. Its volume at time t referred to an initial volume is equal to the Jacobian of the Eulerian-to-Lagrangian coordinate transformation

$$J(\mathbf{y}, t) = \left| \frac{\partial \mathbf{X}(\mathbf{y}, t)}{\partial \mathbf{y}} \right|, \quad J(\mathbf{y}, 0) = 1. \tag{3.2}$$

Provided $J < 1$ the fluid particle is compressed, while if $J > 1$ it expands. Therefore, we shall define $J(\mathbf{y}, t)$ as the Lagrangian divergence.

Introduce an Eulerian field of divergence as well,

$$j(\mathbf{x}, t) = J(\mathbf{y}(\mathbf{x}, t), t). \tag{3.3}$$

If, at a given time instant, the divergence field is positive everywhere, there proves to be a one-to-one correspondence between the Lagrangian and the Eulerian coordinate systems, and the continuous medium will be the single-stream one. If, however, $J(\mathbf{y}, t)$ is an alternating function of $\mathbf{y}$, then in some regions one can witness a multi-stream motion of a continuous medium, when several fluid particles simultaneously arrive at each point of those regions. In the ensuing analysis we shall examine the situation of the the multi-stream motion of a continuous medium in detail, while before that we shall treat the medium as a single-stream one.

The spatial and temporal variation of a continuous medium divergence field is intimately related to the density change of the medium or the change of concentration of a passive impurity moving together with the medium. Let $\rho^L(\mathbf{y}, 0) = \rho_0(\mathbf{y})$ be the concentration of a passive impurity at initial time $t = 0$; then the corresponding Lagrangian field at arbitrary time, while the medium remains a single-stream one, equals

$$\rho^{\mathrm{L}}(\mathbf{y}, t) = \rho_0(\mathbf{y})/J(\mathbf{y}, t).$$

The appropriate Eulerian field of a passive impurity concentration is as follows:

$$\rho^{\mathrm{E}}(\mathbf{x}, t) = \rho^{\mathrm{L}}(\mathbf{y}(\mathbf{x}, t), t) = \rho_0(\mathbf{y}(\mathbf{x}, t))/j(\mathbf{x}, t). \tag{3.4}$$

2. In the above we analysed some dynamic relations between the Lagrangian and Eurlerian fields. In the statistical investigation of random fields (for example, of turbulent fluid), it is important to know the relations between the statistical properties of the fields in the Lagrangian and Eulerian frames. Initially, we reveal the relationship between the one-point Lagrangian and Eulerian probability densities assuming, for the present, the continuous medium to be a single-stream one. As was stated, let $\mathbf{X}(\mathbf{y}, t)$, $\mathbf{V}(\mathbf{y}, t)$ $J(\mathbf{y}, t)$ be the random coordinates, velocity vector and divergence of a fixed fluid particle with Lagrangian coordinates $\mathbf{y}$. In accordance with (1.1) the joint probability density of the Lagrangian fields mentioned is equal to

$$W^{\mathrm{L}}_{\mathbf{x},\mathbf{v},J}(\mathbf{x}, \mathbf{v}, j; \mathbf{y}, t) = \langle \delta(\mathbf{X}(\mathbf{y}, t) - \mathbf{x})\, \delta(\mathbf{V}(\mathbf{y}, t) - \mathbf{v})\, \delta(J(\mathbf{y}, t) - j)\rangle_{\mathrm{L}}. \tag{3.5}$$

Here $\langle \dots \rangle_{\mathrm{L}}$ indicates statistical averaging over the ensemble of random Lagrangian field values. Making use of (A.14) for the case of one-to-one correspondence between the Lagrangian and Eulerian coordinate system ($J > 0$), we write the following equality:

$$\delta(\mathbf{X}(\mathbf{y}, t) - \mathbf{x}) = \delta(\mathbf{y}(\mathbf{x}, t) - \mathbf{y})/j(\mathbf{x}, t). \tag{3.6}$$

Substituting it into (3.5) we have

$$W^{\mathrm{L}}_{\mathbf{x},\mathbf{v},J}(\mathbf{x}, \mathbf{v}, j; \mathbf{x}, t) = \langle \delta(\mathbf{y}(\mathbf{x}, t) - \mathbf{y})\, \delta(\mathbf{v}(\mathbf{x}, t) - \mathbf{v})\, \delta(j(\mathbf{x}, t) - j)\rangle_{\mathrm{E}}/j, \tag{3.7}$$

where $\langle \dots \rangle_{\mathrm{E}}$ stands for averaging over the ensemble of the Eulerian random fieldues, in this case the Eulerian field $\mathbf{y}(\mathbf{x}, t)$, $\mathbf{v}(\mathbf{x}, t)$, $j(\mathbf{x}, t)$. On the other hand, the one-point joint probability density of the Eulerian fields mentioned is equal to

$$W^{\mathrm{E}}_{\mathbf{y},\mathbf{v},j}(\mathbf{y}, \mathbf{v}, j; \mathbf{x}, t) = \langle \delta(\mathbf{y}(\mathbf{x}, t) - \mathbf{y})\, \delta(\mathbf{v}(\mathbf{x}, t) - \mathbf{v})\, \delta(j(\mathbf{x}, t) - j)\rangle_{\mathrm{E}}.$$

Therefore, it follows from (3.7) that

$$W^{\mathrm{E}}_{\mathbf{y},\mathbf{v},j}(\mathbf{y}, \mathbf{v}, j; \mathbf{x}, t) = jW^{\mathrm{L}}_{\mathbf{x},\mathbf{v},J}(\mathbf{x}, \mathbf{v}, j; \mathbf{y}, t). \tag{3.8}$$

Later on, for brevity, we shall call the random Lagrangian field probability densities, describing the statistical properties of the fixed fluid particles of a continuous medium, 'Lagrangian probability densities', while the Eulerian field probability densities will be named 'Eulerian probability densities'. In such terms the equality (3.8) offers the relation between the Lagrangian probability density W^{L} and the Eulerian one W^{E}.

Integrating the equality (3.8) with respect to $\mathbf{y}$ one acquires the probability density of the Eulerian velocity and divergence fields, expressed through the Lagrangian probability density, as

$$W^{\mathrm{E}}_{\mathrm{v},\,j}(\mathbf{v},\, j;\, \mathbf{x},\, t) = j \int_{-\infty}^{\infty} W^{\mathrm{L}}_{\mathbf{X},\mathbf{V},J}(\mathbf{x},\, \mathbf{v},\, j;\, \mathbf{y},\, t)\, \mathrm{d}^3\mathbf{y}. \tag{3.9}$$

The last two equations imply that an important role in the relationship between the Lagrangian and Eulerian statistics is played by the divergence field $j(\mathbf{x}, t)$. In various physcial problems it is more natural when establishing such relations to employ, instead of the divergence, its reciprocal value, i.e. the continuous medium density or the passive impurity concentration. Let, for example

$$W^{\mathrm{L}}_{\mathbf{X},\mathbf{V},\rho}(\mathbf{x},\, \mathbf{v},\, \rho;\, \mathbf{y},\, t) = \langle \delta(\mathbf{X}(\mathbf{y},\, t) - \mathbf{x})\; \delta(\mathbf{V}(\mathbf{y},\, t) - \mathbf{v})\; \delta(\rho^{\mathrm{L}}(\mathbf{y},\, t) - \rho)\rangle_{\mathrm{L}}$$

be the joint Lagrangian probability density of coordinates $\mathbf{X}$, velocity $\mathbf{V}$ and passive impurity concentration ρ^{L} of the fixed fluid particle. Multiplying this equality by the initial concentration $\rho_0(\mathbf{y})$ which, for simplicity, is considered to be non-random, and making use of the equality yielded by (3.6)

$$\rho_0(\mathbf{y})\; \delta(\mathbf{X}(\mathbf{y},\, t) - \mathbf{x}) = \rho^{\mathrm{E}}(\mathbf{x},\, t)\; \delta(\mathbf{y}(\mathbf{x},\, t) - \mathbf{y}),$$

we cast a new formula relating the Lagrangian and Eulerian probability densities,

$$\rho W^{\mathrm{E}}_{\mathbf{y},\mathbf{v},\rho}(\mathbf{y},\, \mathbf{v},\, \rho;\, \mathbf{x},\, t) = \rho_0(\mathbf{y})\; W^{\mathrm{L}}_{\mathbf{X},\mathbf{V},\rho}(\mathbf{x},\, \mathbf{v},\, \rho;\, \mathbf{y},\, t) \tag{3.10}$$

where, instead of the divergence as in (3.8), the density of the medium is observed. Integrating (3.10) with respect to $\mathbf{y}$ one can see that

$$\rho W^{\mathrm{E}}_{\mathbf{v},\rho}(\mathbf{v},\, \rho;\, \mathbf{x},\, t) = \int_{-\infty}^{\infty} \rho_0(\mathbf{y}) W^{\mathrm{L}}_{\mathbf{X},\mathbf{V},\rho}(\mathbf{x},\, \mathbf{v},\, \rho;\, \mathbf{y},\, t)\, \mathrm{d}^3\mathbf{y}. \tag{3.11}$$

3. Now consider a few useful specific results of these relations. Thus, integrating (3.10) with respect to $\mathbf{v}$, $\mathbf{y}$ and ρ we are led to a well-known formula,

$$\langle \rho^{\mathrm{E}}(\mathbf{x},\, t)\rangle = \int_{-\infty}^{\infty} \rho_0(\mathbf{y}) W^{\mathrm{L}}_{\mathbf{X}}(\mathbf{x};\, \mathbf{y},\, t)\, \mathrm{d}^3\mathbf{y}, \tag{3.12}$$

which establishes the remarkable fact that the average density of a medium, or the average concentration of a passive impurity, at a fixed point of the space $\mathbf{x}$ is uniquely defined by the initial concentration $\rho_0(\mathbf{y})$ and the Lagrangian probability density of the fixed particle coordinates

$$W_{\mathbf{X}}^{L}(\mathbf{x}; \mathbf{y}, t) = \langle \delta(\mathbf{X}(\mathbf{y}, t) - \mathbf{x}) \rangle_{L}.$$

Another equality, related to (3.12), resulting from (3.9), has the form

$$\langle j(\mathbf{x}, t) \rangle = \int_{-\infty}^{\infty} W_{\mathbf{X}}^{L}(\mathbf{x}; \mathbf{y}, t)\, d^3\mathbf{y}. \tag{3.13}$$

The simplest relations between the Lagrangian and Eulerian statistics are observed with respect to the incompressible continuous medium for which $J = j = 1$ and (3.8) is equivalent to

$$W_{\mathbf{y},\mathbf{v}}^{E}(\mathbf{y}, \mathbf{v}; \mathbf{x}, t) = W_{\mathbf{X},\mathbf{V}}^{L}(\mathbf{x}, \mathbf{v}, \mathbf{y}; t),$$

which represents a peculiar reciprocal theorem. Under this theorem, if we interchange the Lagrangian and Eulerian coordinates of a random incompressible medium the Eulerian probability density will turn into the Lagrangian one.

Let us now clarify the form that the above relations take for the statistically homogeneous and in general compressible) medium. Inasmuch as the statistical properties of such a medium are not changed during the parallel translation of either Eulerian or Lagrangian coordinate systems, equations of the following type are valid:

$$\begin{aligned} W^{E}(\mathbf{x}, \mathbf{v}, j; \mathbf{y}, t) &= W^{E}(\mathbf{x} - \mathbf{y}, \mathbf{v}, j; t), \\ W^{L}(\mathbf{y}, \mathbf{v}, j; \mathbf{x}, \mathbf{t}) &= W^{L}(\mathbf{y} - \mathbf{x}, \mathbf{v}, j; t), \\ W^{E}(\mathbf{v}, j; \mathbf{x}, t) &= W^{E}(\mathbf{v}, j; t). \end{aligned} \tag{3.14}$$

Taking them into account a simple algebraic relation can be formulated from (3.9);

$$W^{E}(\mathbf{v}, j; t) = jW^{L}(\mathbf{v}, j; t). \tag{3.15}$$

An analogous expression follows from (3.11) provided the initial density of the medium $\rho_0(\mathbf{y}) = \rho_0$ is the same at all points of the space,

$$\rho W^{E}(\mathbf{v}, \rho; t) = \rho_0 W^{L}(\mathbf{v}, \rho; t). \tag{3.16}$$

Integrating it with respect to $\mathbf{v}$, we write a relation between the Lagrangian and the Eulerian probability density distributions of a statistically homogeneous continuous medium as

$$W_{\rho}^{E}(\rho; t) = \frac{\rho_0}{\rho} W_{\rho}^{L}(\rho; t). \tag{3.17}$$

From (3.15) it is apparent, in addition, that the Lagrangian and the Eulerian probability densities of the velocity field for the statistically homogeneous incompressible fluid coincide [121]:

$$W_{\mathbf{v}}^{E}(\mathbf{v}; t) = W_{\mathbf{V}}^{L}(\mathbf{v}, t). \tag{3.18}$$

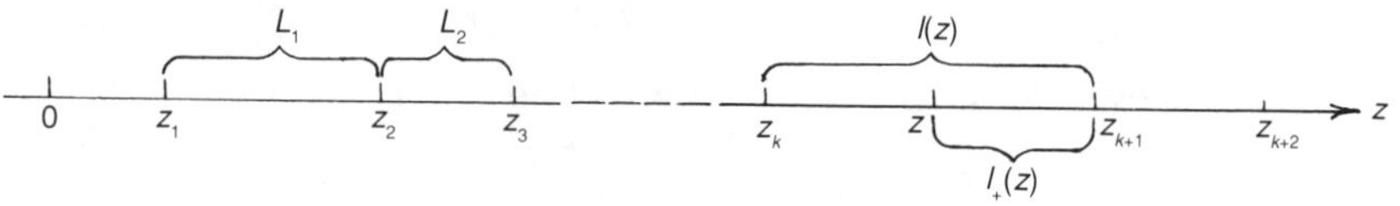

Fig. 3.4 Realisation of a random sequence of intervals

We draw attention to the fact that the Eulerian probability density in (3.15) differs from the Lagrangian one by a factor of j. In particular, the Eulerian and Lagrangian probability densities of the divergence field for the statistically homogeneous continuous medium are interrelated by way of the following equality:

$$W_j^{\mathrm{E}}(j;t) = jW_j^{\mathrm{L}}(j;t). \tag{3.19}$$

This is due to the fact that the expanded fluid particles for which $j > 1$ have within the Eulerian frame $\mathbf{x}$ a volume exceeding by a factor of j that within the Lagrangian frame $\mathbf{y}$. The statistical weight of such particles in the Eulerian ensemble is more than in the Lagrangian one by a factor of j, correspondingly.

4. We examime one more example, perhaps of special interest, displaying the geometrical essence of relationships like (3.19). Let a statistically homogeneous sequence of point $\{z_k\}$ be specified on the z-axis, the statistical properties of the intervals between these points $L_k = z_{k+1} - z_k$ being invariant to the shift of k. To be precise, we begin to enumerate point z_k starting from the first point to the right of the origin. (See Figure 3.4.) As a physical example leading to this problem one can consider a nonlinear wave satisfying the BE. As is evident from Chapter 2, $\mu \to 0$ and at rather large times it is understood as a sequence of sawtooth pulses, the inter-discontinuity distance statistics being one of its most important features.

In practice one can employ two seemingly equivalent approaches to measuring the statistical properties of such random sequences, e.g. interval lengths L_k. One of them is to determine the probability density on the statistical ensemble of the interval realisations with a given number $k \geq 1$. Let us denote the probability density of the interval lengths so obtained as $W(l)$. In another situation one can measure the interval lengths $l(z)$ including a designated point z. Denote the probability density determined on the ensemble of these interval lengths by $w(l)$. One might imagine that the statistical homogeneity of the sequence $\{z_k\}$ results in equalily of $W(l)$ and $w(l)$. This is not the case, however. To better understand the differences between $W(l)$ and $w(l)$ one needs to analyse the definition of $w(l)$ in more detail,

$$w(l) = \langle \delta(l(z) - l) \rangle \tag{3.20}$$

Since $w(l)$ does not depend on z, the following equality holds:

$$w(l) = \lim_{L\to\infty} \frac{1}{L} \int_0^L \langle \delta(l(z) - l) \rangle \, dz. \tag{3.21}$$

Ignoring end effects we write down the integral in (3.21) as follows:

$$\int_0^L \langle \delta(l(z) - l) \rangle \, dz = \left\langle \sum_{k=1}^{N(L)} \delta(L_k - l) L_k \right\rangle = l \left\langle \sum_{k=1}^{N(L)} \delta(l - L_k) \right\rangle.$$

Here $N(L)$ is the number of the last point z_k included in the interval $(0, L)$. As $L \to \infty$, it is possible, due to the law of large numbers to set $N(L) = \langle N(L) \rangle$, which gives us

$$\int_0^l \langle \delta(l(z) - l) \rangle \, dz = l \, \langle N(L) \rangle \, W(l), \tag{3.22}$$

where

$$W(l) = \langle \delta(l - L_k) \rangle$$

is the probability density of the length of the interval with a preset number. Substituting (3.22) into (3.20) yields

$$w(l) = l W(l) \lim_{L\to\infty} (\langle N(L) \rangle / L).$$

From the normalisation condition of $w(l)$ it is clear that

$$\lim_{L\to\infty} (\langle N(L) \rangle / L) = 1/\langle l \rangle_{\mathrm{L}},$$

where

$$\langle l \rangle_{\mathrm{L}} = \langle L_k \rangle = \int_0^\infty l W(l) \, dl$$

is the average length of an interval. Therefore, we finally obtain

$$w(l) = \frac{l}{\langle l \rangle_{\mathrm{L}}} W(l). \tag{3.23}$$

This expression is conceptually similar to (3.19). Moreover, it is quite natural to define $w(l)$ as the Eulerian and $W(l)$ as the Lagrangian probability density of the interval lengths. A factor $l/\langle l \rangle_{\mathrm{L}}$ in (3.23), corresponding to j in (3.19), takes into account the fact that the longer is an interval L_k the more probable is the inclusion of a given point z in it. Consequently, the average length of a random interval including the given point z always exceeds the average length of the interval with a given number

$$\langle l \rangle_{\mathrm{E}} = \langle l(z) \rangle = \frac{1}{\langle l \rangle_{\mathrm{L}}} \int_0^\infty l^2 W(l) \, dl = \frac{\langle l^2 \rangle_{\mathrm{L}}}{\langle l \rangle_{\mathrm{L}}} > \langle l \rangle_{\mathrm{L}}. \tag{3.24}$$

In analogy with the Eulerian and Lagrangian coordinates and fields, one can define z as Eulerian and k as Lagrangian coordinates with $l(z)$ as Eulerian and L_k as Lagrangian intervals, respectively. Thus $w(l)$ depicts the Eulerian statistics of the interval lengths and $W(l)$ describes the Lagrangian statistics, while equality (3.23) establishes the relationship between the Eulerian and Lagrangian probability densities of the interval lengths.

For practical use it is instructive to discuss another Eulerian statistical property of the interval lengths, namely the probability density $l_+(z)$ of a line segment length contained between point z and the right-hand end of the interval $l(z)$ (see Fig. 3.4). Keeping in mind that z may appear in any point of the interval $l(z)$ with the same probability, it is easy to find the probability density of length l as

$$w_+(l) = \int_l^\infty W(l')\,dl'/\langle l\rangle_L. \tag{3.25}$$

5. Come back now to the relationship between the Eulerian and the Lagrangian statistics of random fields characterising the continuous medium motion. It is not difficult to set up similar relations for the multi-point probability densities as well. Let the following two-point Lagrangian probability density of the fields **X**, **V** and J be given as an example:

$$W^L_{\mathbf{x},\mathbf{v},J}(\mathbf{x}_1, \mathbf{x}_2, \mathbf{v}_1, \mathbf{v}_2, j_1, j_2; \mathbf{y}_1, \mathbf{y}_2, t) = \langle\delta(\mathbf{X}(\mathbf{y}_1, t) - \mathbf{x}_1)\,\delta(\mathbf{X}(\mathbf{y}_2, t) - \mathbf{x}_2)$$
$$\times\,\delta(\mathbf{V}(\mathbf{y}_1, t) - \mathbf{v}_1)\,\delta(\mathbf{V}(\mathbf{y}_2, t) - \mathbf{v}_2)\,\delta(J(\mathbf{y}_1, t) - j_1)\,\delta(J(\mathbf{y}_2, t) - j_2)\rangle_L.$$

Intergrating this equality with respect to $\mathbf{y}_1$, $\mathbf{y}_2$ and making use of (3.6) we have

$$W^E(\mathbf{v}_1, \mathbf{v}_2, j_1, j_2; \mathbf{x}_1, \mathbf{x}_2, t) = j_1 j_2 \times$$
$$\int_{-\infty}^{\infty} W^L(\mathbf{x}_1, \mathbf{x}_2, \mathbf{v}_1, \mathbf{v}_1, \mathbf{v}_2; j_1, j_2; \mathbf{y}_1, \mathbf{y}_2, t)\,d^3\mathbf{y}_1\,d^3\mathbf{y}_2. \tag{3.26}$$

In the case of statistically homogeneous continuous media this formula becomes somewhat simpler. This can be illustrated by going over to the difference and centre-of-gravity coordinates

$$\mathbf{s} = \mathbf{x}_1 - \mathbf{x}_2,\quad \mathbf{S}_0 = \mathbf{y}_1 - \mathbf{y}_2,\quad \mathbf{Q} = (\mathbf{x}_1 + \mathbf{x}_2)/2,\quad \mathbf{Q}_0 = (\mathbf{y}_1 + \mathbf{y}_2)/2.$$

In a statistically homogeneous continuous medium the Eulerian probability density on the left-hand side of equality (3.26) depends only on **s** while the Lagrangian one on the right-hand side depends only on **s**, $\mathbf{s}_0$ and $\mathbf{Q} - \mathbf{Q}_0$. Passing in (3.26) to the integration with respect to $\mathbf{s}_0$, $\mathbf{Q}_0$, we obtain

$$W^E(\mathbf{v}_1, \mathbf{v}_2, j_1, j_2; \mathbf{s}, t) = j_1 j_2 \int_{-\infty}^{\infty} W^L(\mathbf{s}, \mathbf{v}_1, \mathbf{v}_2, j_1, j_2; \mathbf{s}_0, t)\,d^3 s_0. \tag{3.27}$$

Here the Lagrangian probability density is present,

$$W^{\mathrm{L}}(\mathbf{s}, \mathbf{v}_1, \mathbf{v}_2, j_1, j_2; \mathbf{s}_0, t) = \langle \delta(\mathbf{X}(\mathbf{s}_0, t) - \mathbf{X}(\mathbf{o}, t) - \mathbf{s})$$
$$\times\, \delta(\mathbf{V}(\mathbf{s}_0, t) - \mathbf{v}_1)\, \delta(\mathbf{V}(\mathbf{o}, t) - \mathbf{v}_2)\, \delta(J(\mathbf{s}_0, t) - j_1)\, \delta(J(\mathbf{o}, t) - j_2)\rangle_{\mathrm{L}}.$$

In the same way one can extend to a two-point situation the formulae of type (3.10) that connect the Lagrangian and Eulerian statistics through the passive impurity concentration,

$$\rho_1\rho_2 W^{\mathrm{E}}(\mathbf{v}_1, \mathbf{v}_2, \rho_1, \rho_2; \mathbf{x}_1, \mathbf{x}_2, t) = \int_{-\infty}^{\infty} \mathrm{d}^3\mathbf{y}_1\, \mathrm{d}^3\mathbf{y}_2 \,\times$$
$$\rho_0(\mathbf{y}_1)\, \rho_0(\mathbf{y}_2)\, W^{\mathrm{L}}(\mathbf{x}_1, \mathbf{x}_2, \mathbf{v}_1, \mathbf{v}_2, \rho_1, \rho_2; \mathbf{y}_1, \mathbf{y}_2, t). \tag{3.28}$$

If a continuous medium is statistically homogeneous while the initial concentration $\rho_0(\mathbf{y})$ is random, statistically homogeneous and statistically independent of the continuous medium motion, then (3.28) implies that the correlation function of the passive impurity concentration is equal to

$$K_\rho^{\mathrm{E}}(\mathbf{s}, t) = \langle \rho^{\mathrm{E}}(\mathbf{x} + \mathbf{s}, t)\, \rho^{\mathrm{E}}(\mathbf{x}, t)\rangle = \int_{-\infty}^{\infty} K_0(\mathbf{s}_0) W^{\mathrm{L}}(\mathbf{s}; \mathbf{s}_0, t)\, \mathrm{d}^3\mathbf{s}_0, \tag{3.29}$$

where

$$K_0(\mathbf{s}_0) = \langle \rho_0(\mathbf{y} + \mathbf{s}_0)\, \rho_0(\mathbf{y})\rangle,$$

and

$$W^{\mathrm{L}}(\mathbf{s}; \mathbf{s}_0, t) = \langle \delta(\mathbf{s} - \mathbf{X}(\mathbf{s}_0, t) + \mathbf{X}(\mathbf{o}, t))\rangle \tag{3.30}$$

is the Lagrangian probaility density of the relative displacement vector of the two fixed particles; their initial displacement vector was equal to $\mathbf{s}_0$.

6. To continue the investigation we shall require linear functionals of the Eulerian averages. Their expression through the Lagrangian averages can be illustrated by an example relating to the one-dimensional continuous medium. Let $v(x, t)$ be a random field related to the continuous medium motion, and suppose the following functional must be calculated:

$$q(t) = \int_{-\infty}^{\infty} \langle v(x, t)\rangle_{\mathrm{E}} f(x)\, \mathrm{d}x,$$

where $f(x)$ is a known deterministic function. Passing to the integration with respect to the Lagrangian coordinate one can write the following:

$$q(t) = \int_{-\infty}^{\infty} \langle V(y, t)\, f(X(y, t))\, J(y, t)\rangle_{\mathrm{L}}\, \mathrm{d}y.$$

Taking into account that in the one-dimensional case the Lagrangian divergence field is governed by the quality

$$J(y, t) = \frac{\partial X(y, t)}{\partial y}, \tag{3.31}$$

we obtain

$$q(t) = \int_{-\infty}^{\infty} \left\langle V(y, t) \frac{\partial F(X(y, t))}{\partial y} \right\rangle_{\mathrm{L}} \mathrm{d}y,$$

where

$$F(x) = \int^{x} f(z)\, \mathrm{d}z.$$

Integrating again by parts and assuming for simplicity that $V(-\infty, t) = V(\infty, t) = 0$ we finally find

$$q(t) = -\int_{-\infty}^{\infty} \left\langle F(X(y, t)) \frac{\partial V(y, t)}{\partial y} \right\rangle_{\mathrm{L}} \mathrm{d}y.$$

In accordance with this formula, for example, the Fourier transform of the average field $\langle v(x, t)\rangle$ is equal to

$$\int_{-\infty}^{\infty} \langle v(x, t)\rangle_{\mathrm{E}} \mathrm{e}^{ikx}\, \mathrm{d}x = \frac{i}{k} \int_{-\infty}^{\infty} \left\langle \frac{\partial V(y, t)}{\partial y} \exp\,(ikX(y, t)) \right\rangle_{\mathrm{L}} \mathrm{d}y. \qquad (3.32)$$

Analogous relationships are also valid for functionals of two-point averages. Thus for the spectral density of the Eulerian statistically homogeneous field $v(x, t)$ we have

$$G_v(k; t) = \frac{1}{2\pi} \int_{-\infty}^{\infty} B_v(s; t)\, \mathrm{e}^{-iks}\, \mathrm{d}s, \qquad (3.33)$$

where

$$B_v(s; t) = \langle v(x + s, t)\, v(x, t)\rangle - \langle v\rangle^2 \qquad (3.34)$$

is the Eulerian covariance function and is expressed through the Lagrangian average using a formula similar to (3.32),

$$G_v(k; t) = \frac{1}{2\pi k^2} \int_{-\infty}^{\infty} \left\langle \frac{\partial V(y, t)}{\partial y} \frac{\partial V(y + s, t)}{\partial y} \right. \times \exp\,\{ik[X(y, t) - X(y + s, t)]\}\Big\rangle_{\mathrm{L}}\, \mathrm{d}s. \qquad (3.35)$$

7. Up to now we have assumed a continuous medium to be single-stream. However, in many physical phenomerna (in the flows of noninteracting particles, in cold plasma, and in optics after the formation of caustics, to give several examples) multi-stream behaviour can be anticipated. Not one but several particles having different Lagrangian coordinates can simultaneously arrive at one and the same spatial point. Correspondingly, the Eulerian fields turn out to be multi-stream ones, i.e. non-single-valued functions of the Eulerian coordinates. The relations between the Lagrangian and Eulerian statistical properties of the continuous medium become complicated as well. Let us elucidate a set of such relations valid for multi-stream

situations. First it should be noted that at fixed $\mathbf{x}$ the euqality $\mathbf{x} = \mathbf{X}(\mathbf{y}, t)$ can be fulfilled by more than one set of Lagrangian coordinates $\mathbf{y}_1(\mathbf{x}, t), \ldots, \mathbf{y}_N(\mathbf{x}, t)$, where $N = N(\mathbf{x}, t)$ is the total number of particles arriving at time t at a point with Eulerian coordinates $\mathbf{x}$. In other words, when multi-stream behaviour is present the Eulerian field of the Lagrangian coordinates $\mathbf{y}(\mathbf{x}, t)$ is a multi-valued function having at point $\mathbf{x}$ not one but $N(\mathbf{x}, t)$ values. Other Eulerian fields become multi-valued as well. Therefore, the relationship (3.6) is changed to a more general one, identical to (A.12),

$$\delta(\mathbf{X}(\mathbf{y}, t) - \mathbf{x}) = \sum_{n=1}^{N(\mathbf{x}, t)} \delta(\mathbf{y}_n(\mathbf{x}, t) - \mathbf{y})/|j_n(\mathbf{x}, t)|, \tag{3.36}$$

where $j(\mathbf{x}, t) = J(\mathbf{y}_n(\mathbf{x}, t), t)$ is the divergence for the nth fluid particle.

Below, we are going to apply the following general topological properties of multi-stream field $\mathbf{y}(\mathbf{x}, \mathrm{t})$, $j(\mathbf{x}, t)$. If the Lagrangian field $\mathbf{X}(\mathbf{y}, t)$ is continuously differentiable with respect to $\mathbf{y}$ and the particle displacements $\mathbf{X}(\mathbf{y}, t) - \mathbf{y}$ are uniformly bounded at any $\mathbf{y}$, then everywhere, to the exclusion of the surfaces with zero volume measure (caustic surfaces where particle density tends to infinity while some values $j_n(\mathbf{x}, t)$ vanish) the number of particles arriving at a given point $\mathbf{x}$ will be odd. Further, with respect to $(N + 1)/2$ of them the divergences $j_n(\mathbf{x}, t)$ will be positive while the divergences of the other $(N - 1)/2$ particles will be negative.

After these preliminary remarks one can go over to establishing relations between the Lagrangian and Eulerian one-point probability densities. Examine first the Lagrangian probability density (3.5). Substituting relation (3.36) into the right-hand side of (3.5) we see that

$$W^{\mathrm{L}}(\mathbf{x}, \mathbf{v}, j; \mathbf{y}, t) = \frac{1}{|j|}\left\langle \sum_{n=1}^{N(x, t)} \delta(\mathbf{y}_n(\mathbf{x}, t) - \mathbf{y})\, \delta(\mathbf{v}_n(\mathbf{x}, t) - \mathbf{v})\, \delta(j_n(\mathbf{x}, t) - j)\right\rangle_{\mathrm{E}}.$$

Hence, according to the total probability formula,

$$W^{\mathrm{L}}(\mathbf{x}, \mathbf{v}, j; \mathbf{y}, t) = \frac{1}{|j|} \sum_{N=1}^{\infty} P(N; \mathbf{x}, t) \sum_{n=1}^{N} W_n^{\mathrm{E}}(\mathbf{y}, \mathbf{v}, j; \mathbf{x}, t \mid N), \tag{3.37}$$

where $P(N; \mathbf{x}, t)$ is the probability for the Eulerian fields at point $\mathbf{x}$ and at time t to have N streams, while $W_n^{\mathrm{E}}(\mathbf{y}, \mathbf{v}, j; \mathbf{x}, t|N)$is the joint probability density of the Eulerian fields in the nth stream *provided* that the total number of the streams is N. Having integrated both parts of the equality with respect to the Lagrangian coordinates we approach another useful relation,

$$\int_{-\infty}^{\infty} W^{\mathrm{L}}(\mathbf{x}, \mathbf{v}, j; \mathbf{y}, t)\, \mathrm{d}^3\mathbf{y} = \frac{1}{|j|} \sum_{N=1}^{\infty} P(N; \mathbf{x}, t) \sum_{n=1}^{N} W_n^{\mathrm{E}}(\mathbf{v}, j; \mathbf{x}, t \mid N). \tag{3.38}$$

An analogous extension of (3.11) over to a many-stream case has the form

$$\int_{-\infty}^{\infty} \rho_0(\mathbf{y}) W^{\mathrm{L}}(\mathbf{x}, \mathbf{v}, \rho; \mathbf{y}, t)\, \mathrm{d}^3\mathbf{y} = \rho \sum_{N=1}^{\infty} P(N; \mathbf{x}, t) \sum_{n=1}^{N} W_n^{\mathrm{E}}(\mathbf{v}, \rho; \mathbf{x}, t \mid N).$$

This, in particular, leads to the fact that

$$\int_{-\infty}^{\infty} \rho_0(\mathbf{y}) W^{\mathrm{L}}(x;\, \mathbf{y},\, t)\mathrm{d}^3\mathbf{y} = \sum_{N=1}^{\infty} P(N;\, \mathbf{x},\, t) \sum_{n=1}^{N} \langle \rho_n^{\mathrm{E}}(\mathbf{x},\, t)\rangle_N .$$

Here $\langle \dots \rangle_N$ denotes averaging by Eulerian statistics under the condition that the number of streams is N.

Inasmuch as the total density of a multi-stream medium at a point equals the sum of the densities of each stream,

$$\rho^{\mathrm{E}}(\mathbf{x},\, t) = \sum_{n=1}^{N(x,\, t)} \rho_n^{\mathrm{E}}(\mathbf{x},\, t),$$

it follows from the latter equality that the average density, in a many-stream case as well, is governed by caption (3.12) formulated for the one-stream case. The same can be said about the extension of (3.29) to a many-stream situation.

A significant physical feature of many-stream fields is the Eulerian average number of streams arriving at a given point,

$$\langle N(\mathbf{x},\, t)\rangle_{\mathrm{E}} = \sum_{N=1}^{\infty} N P(N;\, \mathbf{x},\, t).$$

Let us express it through the Lagrangian averages. Having multiplied the equality (3.38) by $|j|$ and integrated it with respect to $\mathbf{v}$ and j we are led to

$$\langle N(\mathbf{x},\, t)\rangle_{\mathrm{E}} = \int_{-\infty}^{\infty} |j|\, W^{\mathrm{L}}(\mathbf{x},\, j;\, \mathbf{y},\, t)\, \mathrm{d}^3\mathbf{y}\, \mathrm{d}j.$$

Consider separately the case of a statistically homogeneous many-stream continuous medium. The obvious extension of (3.15) over to this problem has the form

$$W^{\mathrm{L}}(\mathbf{v},\, j;\, t) = \frac{1}{|j|} \sum_{N=1}^{\infty} P(N;\, t) \sum_{n=1}^{N} W_n^{\mathrm{E}}(\mathbf{v},\, j;\, t \mid N). \tag{3.39}$$

The average number of the streams of the statistically homogeneous medium

$$\langle N(\mathbf{x},\, t)\rangle_{\mathrm{E}} = \langle |J(\mathbf{y},\, t)|\rangle_{\mathrm{L}} \tag{3.40}$$

is equal to the Lagrangian average of the divergence modulus.

8. In addition to the average 'many-streamness' the important data on the many-streamness behaviour is contained in the probabilities of the stream number. Thus $P(1;\, t)$ is equal to the relative space volume fraction where one-streamness is still conserved, $P(3;\, t)$ is the specific weight of the three-stream regions, and so on. Generally, these probabilities are presented through the many-point Lagrangian averages. At the initial stage of many-

stream formation, however, when $P(1;\ t)$ is close to unity one can neglect the possibility of 5 or more streams appearing and deduce simple approximate formulae for $P(1;\ t)$ and $P(3;\ t)$. In this approximation the equality

$$\langle N\rangle_{\mathrm{E}} = \langle |J|\rangle_{\mathrm{L}} = P(1;\ t) + 3P(3;\ t)$$

is valid and the approximate normalisation condition is $P(1;\ t) + P(3;\ t) = 1$. The solution of these equations yields

$$P(1; t) = \frac{3 - \langle N\rangle_{\mathrm{E}}}{2}, \quad P(3; t) = \frac{\langle N\rangle_{\mathrm{E}} - 1}{2}. \tag{3.41}$$

4 Random waves of hydrodynamic type

In this chapter we analyse the statistical properties of hydrodynamic-type random waves using three examples. These include Riemann waves describing the propagation of intense acoustical noise at stages before discontinuity formation, density perturbation waves in a gas of noninteracting particles, and finally, fluctuations of optical wave intensity beyond a phase screen, considered in the approximation of geometrical optics. Making use of the mathematical method connecting the Lagrangian and Eulerian descriptions of random fields developed in Chapter 3, we manage to obtain for all these waves quite a comprehensive statistical description for the single-flow mode. Further, some characteristics of multi-flow motions of noninteracting particle hydrodynamic beams will be revealed, and the role of the particle velocity thermal scatter in the formation of the density distribution will be discussed.

4.1 Probabilistic properties of random Riemann waves

1. As was stated in Chapter 1, the basic equation of the theory of nonlinear waves in nondispersive media is the Riemann equation (RE):

$$\frac{\partial v}{\partial t} + v\frac{\partial v}{\partial x} = 0, \quad v(x, 0) = v_0(x). \tag{1.1}$$

Here $v_0(x)$ is a random function, the statistical properties of which are assumed given. In particular, one can reduce to equation (1.1) the description of nonlinear acoustical waves at large Reynolds numbers at any stage before discontinuity formation. In the following discussion, when considering statistical properties of random Riemann waves we shall, as a rule, resort to a more demonstrative interpretation, assuming $v(x, t)$ to be the Eulerian velocity field of a one-dimensional hydrodynamic flow of noninteracting particles. In electronics this model of noninteracting particles is employed to describe electron clusters in klystron-type devices within the framework of the so-called kinematic theory [152, 153].

Nonlinearity of RE hampers direct analysis of the Eulerian velocity field statistics. At the same time, in the Lagrangian representation the description of the particle motion has the utmost simple form (2.4.6). Therefore, exploiting the mathematical tools of section 3.3 which establish the connections between Lagrangian and Eulerian statistics of random fields, we shall manage to comprehensively solve the problem of the statistical description of random Riemann waves.

Equation (1.1) can be represented as a system of characteristic equations for the coordinate of a fixed flow-particle $X(y, t)$ and the Lagrangian velocity field $V(y, t)$:

$$\frac{dX}{dt} = V, \quad \frac{dV}{dt} = 0, \quad X(y, 0) = y, \quad V(y, 0) = v_0(y). \tag{1.2}$$

Hence there is no problem in finding the Lagrangian statistics of noninteracting particle flow. To pass from Lagrangian statistics to Eulerian, we shall also need the statistical properties of the Lagrangian divergence field (3.3.31). As can be readily shown, this is governed by the following set of equations:

$$\frac{dJ}{dt} = U = \frac{\partial V}{\partial y}, \quad \frac{dU}{dt} = 0, \quad J(y, 0) = 1, \quad U(y, 0) = v_0'(y). \tag{1.3}$$

The solutions to (1.2) and (1.3) are obvious,

$$X = y + v_0(y)t, \quad V = v_0(y), \quad J = 1 + v_0'(y)t, \quad U = v_0'(y). \tag{1.4}$$

The joint probability density of the fields $X(y, t)$, $V(y, t)$, $J(y, t)$ is equal to

$$W_{x,v,J}(x, v, j; y, t) = \langle \delta(X(y, t) - x)\delta(V(y, t) - v)\, \delta(J(y, t) - j)\rangle.$$

Substituting here equalities (1.4) we obtain

$$W_{x,v,J}(x, v, j; y, t) = W_0(v, (j - 1)/t; y)\, \delta(x - y - vt)/t, \tag{1.5}$$

where $W_0(v, u; x)$ is the joint one-point probability density of initial fields $v_0(x)$ and $v_0'(x)$.

2. Let us find, first, the one-point Eulerian probability density of a velocity field $v(x, t)$ at such times t when $v(x, t)$ remains, in fact, a single-flow field. such Eulerian probability density is expressed through a Lagrangian one by an equality analogous to (3.3.9). Multiplying (1.5) by j and integrating it over y and j, we can write the required Eulerian probability density of a Riemann wave as

$$W_v^{\mathrm{E}}(v; x, t) = \int_{-\infty}^{\infty} W_0(v, (j - 1)/t; x - vt)\, j\, dj/t.$$

Going now over to a new integration variable $u = (j - 1)/t$ we shall have

$$W_v^{\mathrm{E}}(v; x, t) = \int_{-\infty}^{\infty} (1 + ut)\, W_0(v, u; x - vt)\, du. \tag{1.6}$$

Transform this equality to a more convenient form using (A.25),

$$\int_{-\infty}^{\infty} uW_0(v, u; x - vt)\,\mathrm{d}u = \left\langle \frac{\partial v_0(x - vt)}{\partial x}\,\delta(v_0(x - vt) - v) \right\rangle$$

$$= -\frac{\partial}{\partial x}\int_{-\infty}^{v} \langle \delta(v_0(x - vt) - z) \rangle\,\mathrm{d}z = -\frac{\partial}{\partial x}\int_{-\infty}^{v} W_0(z; x - vt)\,\mathrm{d}z.$$

Here $W_0(v; x)$ is the probability density of the initial velocity field $v_0(x)$. In view of the latter equality, (1.6) will be changed into

$$W_v^{\mathrm{E}}(v; x, t) = W_0(v; x - vt) - t\int_{-\infty}^{v} \frac{\partial}{\partial x} W_0(z; x - vt)\,\mathrm{d}z,$$

or finally,

$$W_v^{\mathrm{E}}(v; x, t) = \frac{\partial}{\partial v}\int_{-\infty}^{v} W_0(z; x - vt)\,\mathrm{d}z. \tag{1.7}$$

Now let us turn to the discussion of the evolution of the Eulerian probability density of the velocity field in various particular cases. First, we treat the statistically homogeneous random field $v_0(x)$, the probability density $W_0(v)$ of which does not depend on x. Then instead of (1.7) one can observe that

$$W_v^{\mathrm{E}}(v; t) = W_0(v), \tag{1.8}$$

i.e. the one-point probability density of the statistically homogeneous field of the noninteracting particle flow velocity remains invariant at the single-flow stage. Nonlinear self-action of the wave does not affect the shape of its one-point probability density. This result (quite unexpected at first sight) is easy to understand using the relationship between the probability density, the integral function of the random function distribution and its realisation behaviour described in section 3.2: the probability density of a statistically homogeneous (and ergodic in x) field $v(x, t)$ may be represented through the limit of the relative length of stay of the realisation $v(x, t)$ within the interval $(v, v + \Delta v)$:

$$W_v^{\mathrm{E}}(v; t) = \lim_{L\to\infty} \frac{1}{L\Delta v}\sum_k \Delta x_k, \tag{1.9}$$

where Δx_k are the lengths of the intervals inside which $v(x, t)$ is changed from v to $v + \Delta v$ (see Fig. 4.1). Due to the uniform motion (1.4) of noninteracting particles, the length of each interval Δx_k varies with time as

$$\Delta x_k(t) = \Delta x_k(0) \pm \Delta v t, \tag{1.10}$$

where the minus sign corresponds to the intervals on steepening sections of the velocity field profile (interval Δx_2 in Fig. 4.1) whereas the plus sign corresponds to stretching sections. From (1.10) it is clear that, for the field

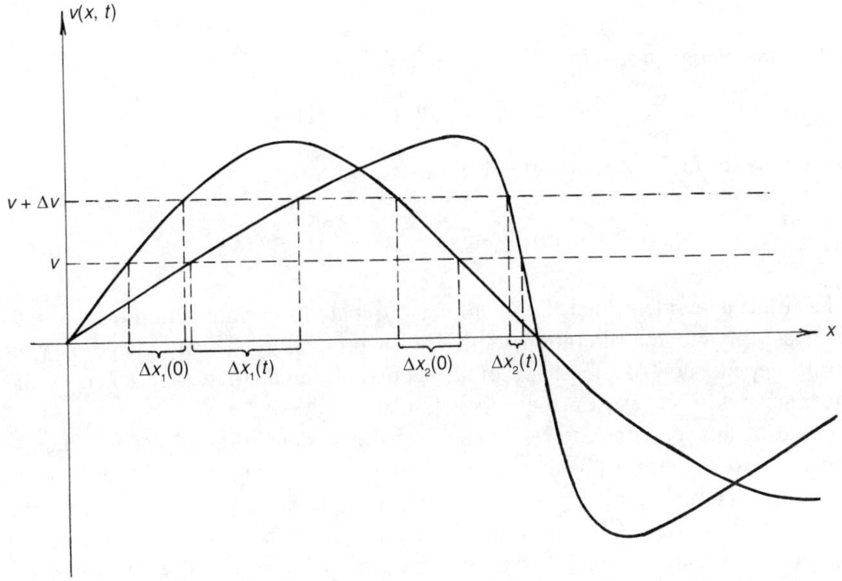

Fig. 4.1 Relative stay length of a realisation $v(x, t)$ within interval $[v, v + \Delta v]$

still remaining one-stream, the sum of any two neighbouring interval lengths is the same:

$$\Delta x_k + \Delta x_{k+1} = \text{const}, \tag{1.11}$$

the relative length of stay $v(x, t)$ in any given interval $(v, v + \Delta v)$ does not change and, therefore, the one-point probability density of the field $v(x, t)$ remains invariant as well. It should be noted, however, that the conclusion of the time-independence of the one-point probability density of the statistically homogeneous Riemann wave cannot be extended over to its other multi-point statistical properties, e.g. to the correlation function and the spectrum of the field $v(x, t)$ which do indeed change in time due to the Riemann wave nonlinearity. It is obvious that the one-point probability density conservation is valid also for statistically homogeneous acoustic noise at any stage prior to discontinuity formation over which their evolution is defined by Re. After emergence of discontinuities the lengths of the discontinuity-captured intervals Δx_k vanish, dynamic equilibrium (1.11) is destroyed, and the one-point probability density of discontinuous acoustic noise becomes time-dependent.

3. Now we make use of (1.7) to analyse signal-noise interaction. Let, at the initial time, $v_0(x)$ be equal to the sum of a deterministic signal $v_s(x)$ and statistically homogeneous noise $v_n(x)$ the probability density $W_n(v)$ of which is known:

$$v_0(x) = v_s(x) + v_n(x). \tag{1.12}$$

Then the initial probability density is

$$W_0(v;\ x) = W_n(v - v_s(x))$$

and equality (1.7) is transformed into

$$W_v^E(v;\ x,\ t) = W_n(v - v_s(x - vt))\ \frac{\partial}{\partial v}\,(v - v_s(x - vt)). \tag{1.13}$$

Hence it is evident that the nonlinear signal–noise interaction leads with increase of t to the alteration of the noise probability density shape as well as to dependence of the shape of the coherent signal (naturally taken as the average field $\langle v(x,\ t)\rangle$) on the noise statistics.

One of the specific features of the probability density (1.13) of the signal–noise mixture, with a physical interpretation akin to that of the invariance of the statistically homogeneous Riemann wave probability density, is that the probability density $W_v^E(v;\ x,\ t)$ obtained does not depend on the form of the initial noise realisation $v_n(x)$: for the same one-point probability density $W_n(v)$ of noise the probability density (1.13) of a signal–noise mixture will be the same for quasi-harmonic and wide-band cases, as well as for the case when all realisations $v_n(x)$ do not depend on x. The only thing that is changed is the range of applicability of (1.13), which is valid for one-stream hydrodynamic flows, or, in the case of acoustic waves, before discontinuity formation. This insensitivity of the probability density (1.13) to the noise realisation behaviour permits us to write the probability density of the signal–noise mixture in another form, equivalent to (1.13). The intention is to prescribe a simple initial condition instead of (1.12), namely

$$v_0(x) = v_s(x) + v_n, \tag{1.14}$$

where v_n is just a random value with probability density $W_n(V)$. The solution of RE (1.1) with initial condition (1.14) is connected through a simple relation (2.3.14) with the field in the absence of noise, that is, with $v_s(x,\ t)$ which is the solution of (1.1) with initial condition $v(x,\ o) = v_s(x)$. Indeed, in terms of the noninteracting particle hydrodynamic flow the term v_n in (1.14) means simply velocity increase of all particles by v_n. In this case, in the frame of reference $z = x - v_n t$ moving with velocity v_n the profile of the particle velocity field will be the same as if $v_n = 0$, i.e. equal to $v_s(z,\ t)$. Consequently, in an immobile coordinate system,

$$v(x,\ t) = v_n + v_s(x - v_n t,\ t).$$

Hence, the probability density of $v(x,\ t)$ is equal by definition to

$$W_v^E(v;\ x,\ t) = \langle \delta(v - v_n - v_s(x - v_n t,\ t))\rangle,$$

where the averaging is performed over an ensemble of variable random values v_n. Explicit presentation of the averaging operation yields

$$W_v^E(v; x, t) = \int_{-\infty}^{\infty} W_n(z)\, \delta(v - z + v_s(x - zt, t))\, dz. \tag{1.15}$$

Then it follows, in particular, that for $\langle v_n \rangle = 0$ the average field has the form

$$\langle v(x, t) \rangle = \int_{-\infty}^{\infty} W_n(v)\, v_s(x - vt, t)\, dv. \tag{1.16}$$

The formula (1.15), equivalent to (1.13), for the probability density of the signal–noise mixture is convenient when we know the solution, i.e. the signal shape without noise $v_s(x, t)$. It is worth mentioning that, as distinct from (1.13), representations (1.15) and (1.16) are valid for acoustic waves also at times when discontinuities are formed in the signal profile provided only that the noises changes slowly enough, as compared with the signal, in x. In this case the signal profile blurring described in (1.16) is caused by the effect of a random drift of the signal $v_s(x, t)$, averaged over all possible drift velocities v_n.

4. It is instructive to consider in more detail an example of an initially sinusoidal signal

$$v_s(x) = v_0 \sin k_0 x$$

and Gaussian noise with covariance σ_n^2. In this situation the probability density (1.13) will take the form

$$W_v^E(v; x, t) = [1 - v_0 k_0 t \cos (x - vt)] \\ \times \frac{1}{\sqrt{2\pi}\, \sigma_n} \exp\left\{-\frac{[v - v_0 \sin k_0(x - vt)]^2}{2\sigma_n^2}\right\}. \tag{1.17}$$

The right-hand side of this equality is positive and has the meaning of the probability density while the velocity field $v_s(x, t)$ of the particle hydrodynamic flow remains single-stream, i.e. until $\tau \leq 1$ where $\tau = v_0 k_0 t$ is a parameter characterising the degree of nonlinear signal distortion.

In (1.17) it is implied that due to signal–noise interaction the velocity field probability density becomes non-Gaussian with the increase of t. We are interested in the investigation of the probability density evolution at points $s = 0$ and $s = \pi$, where $s = k_0 x$ is a dimensionless coordinate, i.e. at the points of the maximum stretching and steepening of the signal, respectively. At these points, for the probability distribution of normalised velocity field $\alpha = v/\sigma_n$ we have

$$W_{\pm}(\alpha; \tau) = \frac{1}{\sqrt{2\pi}} (1 \pm \tau \cos \alpha\tau/\mu) \exp\left[-\tfrac{1}{2}(\alpha \pm \mu \sin \alpha\tau/\mu)^2\right], \tag{1.18}$$

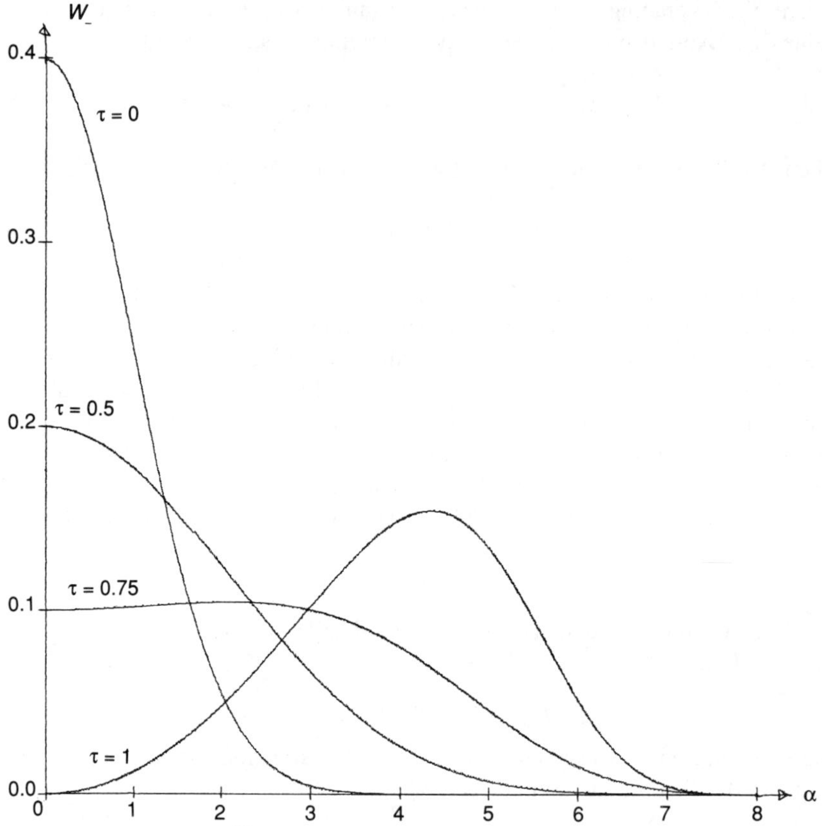

Fig. 4.2 Signal–noise mixture probability densities at the points of maximum steepening and stretching of a signal, at signal-to-noise ratio $\mu = 4$: $\tau = 0$, 1; $\tau = 0.5$, 2; $\tau = 0.75$, 3; $\tau = 1$, 4

where the minus sign corresponds to the point $s = \pi$ whereas the plus sign refers to $s = 0$. Parameter $\mu = v_0/\sigma_n$ included in (1.18) appears as a signal-to-noise ratio. Let us analyse the evolution of the signal–noise mixture probability density when $\mu \gg 1$. Here, not too close to the time of the wave overturning and many-stream formation ($\tau < 1$), one can expand the sine in a Taylor series and confine attention to the first term of the expansion. In such an approximation the probability density for α proves to be Gaussian, with covariance

$$\sigma_{\pm}^2(\tau) = (1 \pm \tau)^{-2}. \tag{1.18'}$$

from this, as well as from the analysis of the more general expression (1.18), it is obvious that on steepening parts of the signal noise variance growth is

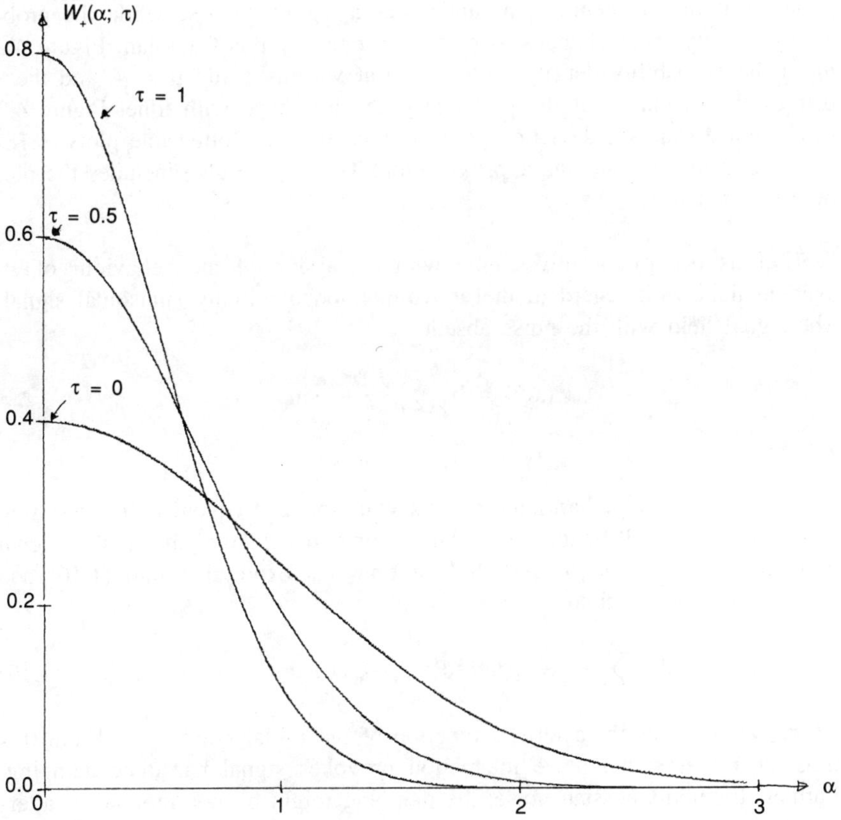

observed with increase of τ. This is due to the fact that on a steepening front even a small drift of a signal attributed to the noise velocity leads to a significant change of field $v(x, t)$. In contrary fashion, on the stretching parts of the signal the noise is suppressed by the signal, since on stretched parts a signal drifting with velocity v_n tends to compensate for the noise component of the wave.

This behaviour of the noise covariance can be readily comprehended from the example (2.3.15) of Chapter 2, where the interaction of some arbitrary field $v_0(x)$ with a linear profile βx was dealt with. Indeed, in the vicinity of zeros a sinusoidal initial perturbation may be expanded into the series $v_s(x) = \pm v_0 k_0 (x - x_*) + \dots$ and, hence, in (2.3.15) $|\beta| = v_0 k_0$. Then, allowing for the fact that for a statistically homogeneous Riemann wave the covariance is conserved, we immediately obtain expression (1.18) for $\sigma_\pm(\tau)$ from (2.3.17) and, moreover, taking into account that for such a field the one-point probability density is conserved as well one also formulates the probability distribution shape conservation.

Approaching the time instant of the signal overturning, $\tau = 1$, the noise variance at the steepening part undergoes a significant rise while the probability density itself changes to being substantially non-Gaussian. Figure 4.2 gives the probability density plots (1.18) at various τ and $\mu = 4$, and these display the evolution of the probability density shape with time. Figure 4.3 shows noise standard deviation plots $\sigma_+(\tau)$ at $\mu = 4$. Dotted line plots (1.18) correspond to the Gaussian approximation. This Figure also includes the plot of $\sigma_-(\tau)$ at $\mu = 1$.

5. Let us occupy ourselves now with discussion of the behaviour of an average field with regard to the above-mentioned initially sinusoidal signal. The signal field with the noise absent,

$$v_s(x, t) = 2v_0 \sum_{r=1}^{\infty} \frac{J_r(r\tau)}{r\tau} \sin rs$$

$$(s = k_0\, x, \quad \tau = v_0\, k_0\, t)$$

proves to be a set of harmonic signals with spatial frequencies rk_0 resulting from nonlinear self-action and, then, from the interaction of the signal harmonics themselves [51–53]. Substituting this expression into (1.16) one can find an average field

$$\langle v(x, t)\rangle_E = 2v_0 \sum_{r=1}^{\infty} \frac{J_r(r\tau)}{r\tau} [\sin rs \operatorname{Re} \theta_n(rk_0t) + \cos rs \operatorname{Im} \theta_n(rk_0t)], \qquad (1.19)$$

where $\theta_n(\kappa)$ is the characteristic function of the initial noise $v_n(x)$. From this it is clear that signal–noise interaction provokes signal harmonic damping. Further, for non-Gaussian noise, the damping might be non-monotone, alternating with a certain increase of the average field harmonics. Should the noise be distributed asymmetrically when $\operatorname{Im} \theta_n \neq 0$, it leads to the phase shifts of the average field harmonics as compared with the signal harmonics. In the case of a Gaussian initial noise we find

$$\langle v(x, t)\rangle_E = 2v_0 \sum_{r=1}^{\infty} \frac{J_r(r\tau)}{r\tau} \sin rs \exp[-(r\tau/\mu)^2/2].$$

As one should have expected, the higher harmonics of a signal are damped by noise more heavily than the low ones, since they are more sensitive to the noise variations of phase, and those variations are the same for all signal harmonics.

In the above we treated just the simplest statistical properties of a wave equal, at output, to a harmonic signal plus stationary noise superposition, and restricted consideration to the initial stage of interaction where neither noise nor harmonic signal possess discontinuities. With respect to physical applications, it is instructive to understand energy redistribution over the spectrum due to nonlinear interaction, and to visualise the interaction pattern change as a result of discontinuity formation. These problems are tackled in a series of

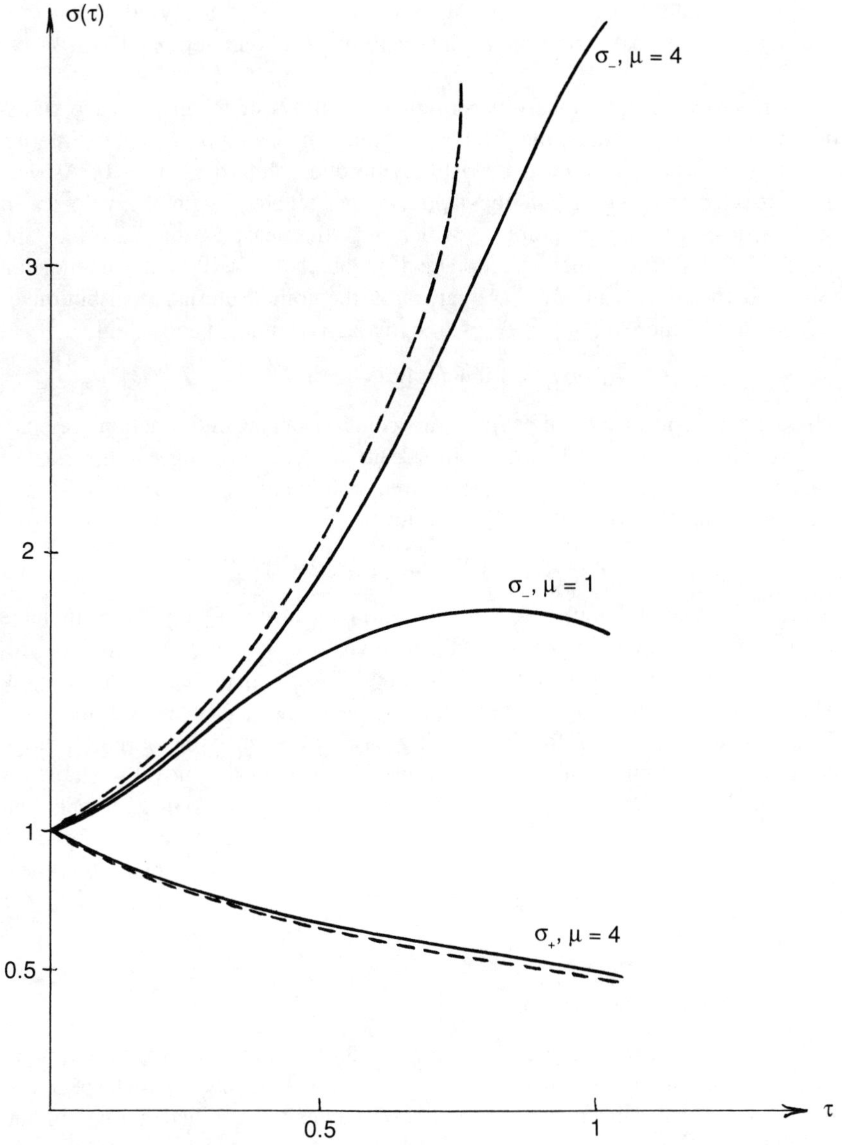

Fig. 4.3 Noise r.m.s. variation at the points of the signal corresponding to maximum steepening and stretching

works where the spectral composition of a wave was analysed [126, 127] and a general classification of the interaction types was compiled [128].

6. The procedure previously proposed for analysis of Riemann wave statistics, resting on the transition from a Lagrangian description to the Eulerian one may also be extended to the BE asymptotic solution as $\mu \to 0$. As was demonstrated in Section 2.4, the field $v(x, t)$ coincides with the velocity of that particle arriving at point x, which has the least action S. Using the solution (2.4.8) for velocity, action and Jacobian as well as the connection (3.3.8) between W^E and W^L, we have for the joint Eulerian distribution of velocity and action, $W^E(s, v; t)$, of a statistically homogeneous field

$$W^E(s, v; t) = \int |1 + ut|\, W_0(s - v^2t/2, v, u)\, du,$$

where $W_0(s, v, u)$ is the single–point distribution of the initial action, velocity and velocity derivative. In the multi-stream region, allowing for the rule of selecting a particle with the least action, one obtains the equation for the Eulerian probability density [74] as follows

$$W^E(v; t) = \int W^E(s, v; t)\, P(s, v; t)\, ds,$$

where P is the probability for all the particles arriving at point x to have actions $S_k > s$ in comparison with one particle reaching this point with parameters s and v. Generally, P is expressed through a functional integral, which can be reduced to a finite-dimensional one under assumption of a Gaussian field $v_0(y)$ and for the limiting cases of a small or extremely large number of competing particles. In Chapter 5 we plan to study the statistical properties of discontinuous waves using a somewhat different approach (but one close to the above-mentioned approach) on the basis of the field expression (2.2.4) for $v(x, t)$, directly by means of the Lagrangian coordinate $y(x, t)$ of a particle with the least action.

4.2 Riemann wave spectrum

1. Much more exhaustive information on the evolution of spatial-temporal properties of random Riemann waves is contained in their spatial spectrum, as compared with the one-point probability densities studied above. In the case of a statistically homogeneous Riemann wave its spectral density is expressed in terms of the Lagrangian average by the equality (3.3.35). Substituting $X(y, t)$ and $V(y, t)$ from (1.4) into this equation, one comes to

$$G_v(k; t) = -\frac{1}{2\pi\, k^4 t^2} \int_{-\infty}^{\infty} e^{-iks} \frac{\partial^2}{\partial s^2} \langle \exp\{ikt\,[v_0(y) - v_0(y+s)]\}\rangle\, ds. \tag{2.1}$$

Integrating by parts we obtain the final form of the spectral density of a random Riemann wave

$$G_v(k, t) = \frac{1}{2\pi(kt)^2} \int e^{-iks}[\theta_2(kt, -kt; s) - \theta_1(kt)\,\theta_1(-kt)]\, ds. \qquad (2.2)$$

Here

$$\theta_2(\kappa_1, \kappa_2; s) = \langle \exp\{i[\kappa_1 v_0(x) + \kappa_2 v_0(x+s)]\}\rangle,$$
$$\theta_1(\kappa) = \langle \exp[i\kappa v_0(x)]\rangle$$

are the two-point and one-point characteristic functions of the initial statistically homogeneous field $v_0(x)$.

Thus, while the one-point probability density of a statistically homogeneous Riemann wave at a stage prior to discontinuity and multi-stream formation remains unchanged, its spectral density, as follows from (2.2), changes substantially with time due to nonlinear self-action and interaction of harmonics. We analyse now some peculiarities of the Riemann wave spectral density evolution, using as an example the initial Gaussian field $v_0(x)$ with zero average and a given covariance function $B_0(s)$, $B_0(0) = \sigma_0^2$. In this case spectral density (2.2) takes the form [124, 125]:

$$Gv(k; t) = \frac{\exp[-(\sigma_0 kt)^2]}{2\pi(kt)^2} \int_{-\infty}^{\infty} [\exp(B_0(s)\, k^2 t^2) - 1]\, e^{-iks}\, ds. \qquad (2.3)$$

2. First, it is worth mentioning that from (2.3), and from (2.2) as well, the following invariant results:

$$G_v(0; t) = G^0 = \frac{1}{2\pi} \int_{-\infty}^{\infty} B_0(s)\, ds = \text{const}, \qquad (2.4)$$

i.e. the spectral density at zero wave number is invariant. Statistical invariant (2.4) appears as a consequence of the dynamical invariant

$$\int_{-\infty}^{\infty} v(x, t)\, dx = \text{const}. \qquad (2.5)$$

For acoustic waves described by BE, invariants (2.4) and (2.5) remain valid after discontinuity formation also. Invariant (2.4) plays a specific role in the theory of Burgers turbulence and in the statistical theory of nonlinear random acoustic waves since, as will be shown in the ensuing chapters, at the discontinuous stage the evolution of the spectral–correlation and probabilistic properties of random acoustic waves and of Burgers turbulence are qualitatively depending an whether $G^0 \neq 0$ or $G^0 = 0$.

It can be pointed out also that the rate of nonlinear self-action and harmonic generation depends on the spatial frequency (wave number) k. The smaller k the slower are these processes. From this point of view, invariant (2.4) proves to be the consequence of the infinity of the characteristic self-action time and of the harmonic generation at zero frequency. Therefore, when $\sigma_0 kt \ll 1$ we may expand the exponents in (2.3) in Taylor series and confine ourselves to the first few terms, e.g.

$$G_v(k; t) = G_0(k) + \tfrac{1}{2}(kt)^2 \, [G_0(k) * G_0(k) - 2G^0 G_0(k)] + \ldots, \qquad (2.6)$$

where $G_0(k)$ is the spectrum of the initial field $v_0(x)$, the sign] indicating the convolution operation. If we only consider the first term, it means that nonlinear effects are neglected whereas restriction to two terms allows for nonlinear interaction of the initial field harmonic pairs, which leads to the appearance of spectral components with difference and sum wave numbers (single-stage interaction), and so on.

Summation of the harmonic wave numbers leads, with time, to the appearance of a slowly decreasing tail of the Riemann wave spectral density due to the energy redistribution towards large k, which is physically associated with the appearance of still smaller scales in the steepening wave spectrum. If, however, the initial spectral density $G_0(k)$ has its maximum at $k_* \neq 0$ (e.g. when $v_0(x)$ is a quasi-harmonic function) then, apart from the formation of the short-wave tails of the spectrum, the flow of the spectral density towards low wave numbers $k < k_*$ due to the parametric generation of difference harmonics is observed, and the low-frequency part of the spectrum is described by (2.6) over the whole range of RE applicability.

Consider now the behaviour of the Riemann wave spectral density (2.3) at large k. In this case one can, when calculating integral (2.3), make use of the saddle-point approach. If in (2.3) we are confined to the first two terms of the expansion of the covariance function,

$$B_0(s) = \sigma_0^2(1 - k_1^2 s^2/2! + k_2^2 \quad s^4/4! - \ldots), \qquad (2.7)$$

$$k_n^2 = \int_{-\infty}^{\infty} k^{2n} G_0(k)\, \mathrm{d}k \Bigg/ \int_{-\infty}^{\infty} G_0(k)\, \mathrm{d}k, \qquad (2.8)$$

we arrive at

$$G_v(k; t) = \frac{\sigma_0^2 \exp(-1/2\tau^2)}{k_1 \sqrt{2\pi}\, (\kappa\tau)^3}, \quad (\tau = \sigma_0\, k_1\, t, \kappa = k/k_1). \qquad (2.9)$$

This implies that as $k \to \infty$ the Riemann wave spectral density decrease according to the universal power law $G_v \sim k^{-3}$. In the previous chapter we discussed the relation between power-law asymptotic roll-off of the spectral densities and the realisation behaviour of the appropriate random functions. It becomes clear, therefore, that this type of spectral asymptotic decay can be attributed to the presence of $\sqrt{x}$ type singularities in the wave realisation. The velocity profile of noninteracting particle hydrodynamic flow, at times when formation of several streams is possible, does in fact possess such singularities [83]. For this reason the spectral decay testifies to multi-stream appearance. For acoustic waves, the appearance of multi-stream behaviour in a Riemann wave indicates that the approximation of the Riemann waves has become inappropriate, because when analysing the acoustic wave in this case one has to consider shock front formation. As is evident from Chapter 3, shock fronts lead to another, slower asymptotic roll-off of the acoustic wave spectral density at large k: $G_v \sim k^{-2}$. It is this asymptotic behaviour of the spectrum that results from the asymptotic analysis, given in Chapter 5, of the acoustic wave spectral density at the stages of discontinuity formation and multiple coalescence. Nevertheless, an asymptotic behaviour k^{-3} still correctly reflects the behaviour of the acoustic wave spectral density over a certain range of wave numbers at the stage before shock front formation, and also at the initial stage of shock appearance since, directly before formation and immediately after shock front appearance, the acoustic wave profile has parts where $\sqrt{x}$-type singularities have been formed and not yet destroyed completely. The detailed relationship between the spectral density asymptotic behaviours k^{-2} and k^{-3} at the initial stage of discontinuity formation will be tackled in the ensuing chapters.

3. Let us illustrate the time evolution of the random Riemann wave spectral density for three characteristic initial spectra

$$G_0(k) = \gamma\alpha_n (k/k_*)^{2n} \exp(-k^2/2k_*^2),$$
$$\gamma = \sigma_0^2/k_*^2\sqrt{2\pi}, \quad \alpha_0 = \alpha_1 = 1, \alpha_2 = 1/3. \tag{2.10}$$

Here $n = 0, 1, 2$, whereas constants α_n are chosen so that all three spectra have the same variance σ_0^2. Introduce a dimensionless spatial frequency $\kappa = k/k_*$, time $\tau = \sigma_0 k_* t$ and new integration variable $z = k_* s$ into (2.3). Then, for the normalised spectrum $g(\kappa; \tau) = G_v/\gamma$ we obtain from (2.3)

$$g_n(\kappa; \tau) = \sqrt{\frac{2}{\pi}} \frac{1}{(\kappa\tau)^2} \exp[-(\kappa\tau)^2] \int_0^\infty \{\exp[(\kappa\tau)^2 R_n(z)] - 1\} \cos \kappa z \, dz, \tag{2.11}$$

where $R_n(z)$ are the correlation coefficients for the three types of initial spectrum (2.10), namely

$$R_0(z) = \exp(-z^2/2), \; R_1(z) = (1 - z^2) \exp(-z^2/2),$$
$$R_2(z) = (1 - 2z^2 + z^4/3) \exp(-z^2/2). \tag{2.12}$$

Common plots and log plots of spectra $g_n(\kappa; \tau)$ are each depicted in Figs 4.4–6. These plots indicate that nonlinear interaction of the spectral components without change of the spectrum value at zero frequency affects its behaviour in the vicinity of $k = 0$ in different ways depending on the initial spectrum type. Provided $G_0(0) = G^0 \neq 0$ and provided spectrum $G_0(k)$ decreases monotonically with k (Fig. 4.4), then the process of energy transfer from the low-frequency range to the high spatial frequencies will dominate. If, however, $G^0 = 0$, then parametric generation of low-frequency components occurs along with it (Figs 4.5, 4.6) and if in the initial spectrum $G_0(k) \sim k^{2n}$, $n > 1$, then, as follows from (2.6), in the low-frequency range the nonlinear interaction leads to a universal asymptotic behaviour,

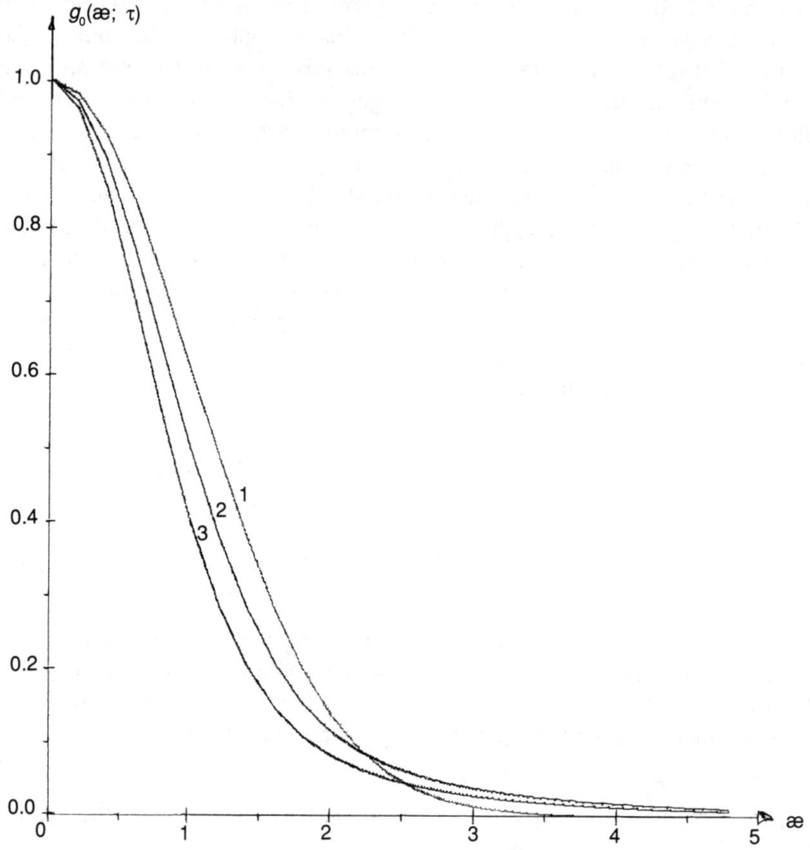

Fig. 4.4 Riemann wave spectrum evolution g_0(æ; τ); $\tau = 0$, 1; $\tau = 0.6$, 2; $\tau = 0.9$, 3

$$G_v(k; t) = \frac{1}{2} k^2 t^2 \int_{-\infty}^{\infty} G_0^2(k)\, dk. \tag{2.13}$$

Parametric generation of low-frequency components and appearance of the said universal law $g(\kappa; \tau) \sim \kappa^2$ are most distinctly manifested in Fig. 4.6.

Within the high-frequency range for all the three spectrum types (2.10), the appearance of the universal slow damping power tails $g(\kappa; \tau) \sim \kappa^{-3}$ (which are in agreement with (2.9)) is observed. Recall that the emergence of such asymptotic behaviour is accompanied by the appearance of non-single-valued parts in the Riemann wave profile. In this situation the right-hand side of (2.3) ceases, rigorously speaking, to be a spectral density and, in particular, does not satisfy the law of conservation $\langle v^2(x, t)\rangle = \sigma^2(t) = \sigma_0^2$ which is valid for single-valued Riemann waves. Deviation of $\sigma^2(t)$ from σ_0^2, calculated as an integral of $G(k; t)$ from (2.3) may, it seems, serve as an

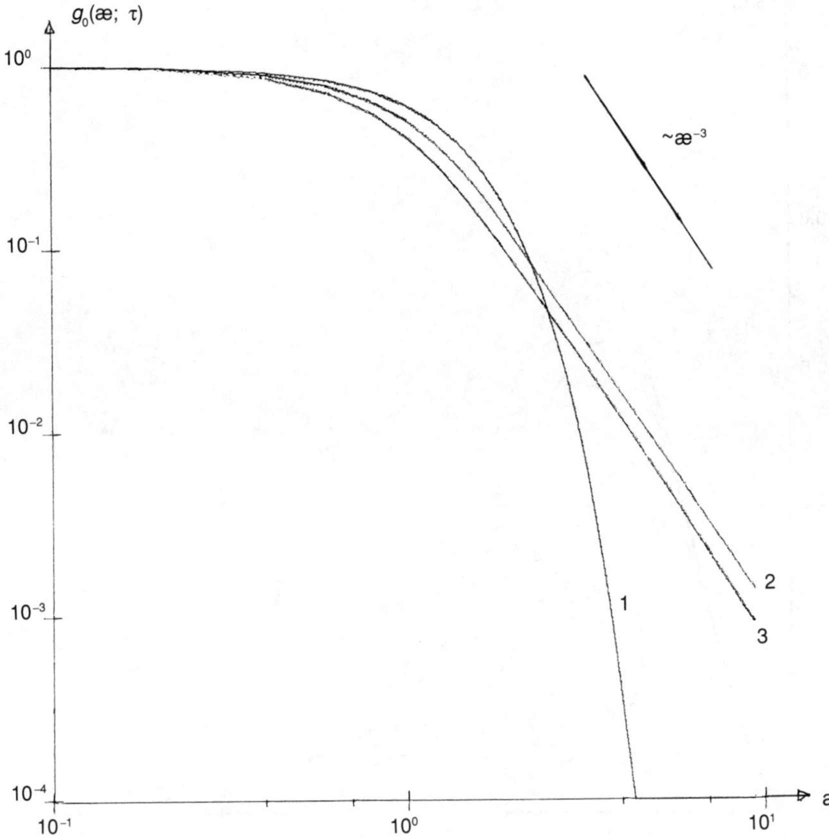

applicability criterion with respect to the formulae obtained [124]. Namely, from numerical calculation by means of (2.11) it follows that at $n = 0$ and $\tau = 0.9$, $\sigma^2 = 0.823\sigma_0^2$; at $n = 1$, $\tau = 0.6$, $\sigma^2 = 0.652\sigma_0^2$; and at $n = 2$, $\tau = 0.4$, $\sigma^2 = 0.682\sigma_0^2$.

4.3 Density fluctuations of noninteracting particle gas

1. When we study statistical properties of noninteracting particle hydrodynamic flow, it is quite natural to take interest not only in its velocity fluctuations but also in the particle density fluctuations appearing as a consequence of the velocity fluctuations. Therefore, we are going now to examine the spectral correlation properties of the density field $\rho(x, t)$ of noninteracting particle hydrodynamic flow assuming, for simplicity, the random initial velocity $v_0(x)$ to be statistically homogeneous and the initial flow density to be given and the same in all points: $\rho(x, 0) = \rho_0 = \text{const}$. Then, from the one-dimensional analysis of (3.3.29) it follows that the correlation function of the density field is equal to

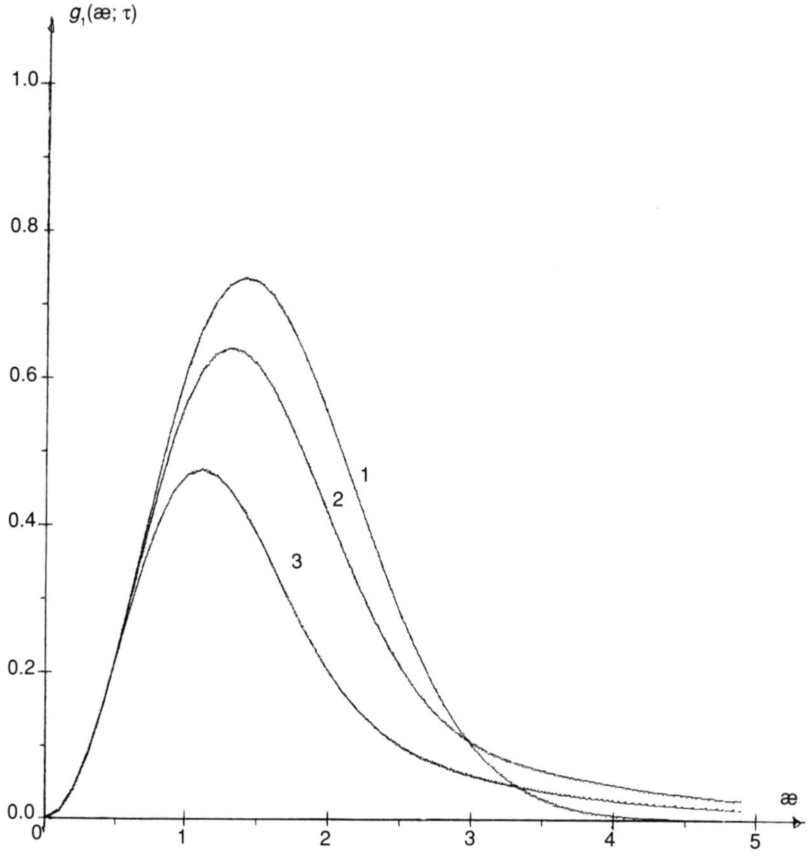

Fig. 4.5 Riemann wave spectrum evolution $g_1(æ;\ \tau)$: $\tau = 0$, 1; $\tau = 0.3$, 2; $\tau = 0.6$, 3

$$K_\rho(s; t) = \langle \rho(x+s, t)\, \rho(x, t)\rangle = \rho_0^2 \int_{-\infty}^{\infty} W^{\mathrm{L}}(s; s_0, t)\, \mathrm{d}s_0. \tag{3.1}$$

Here $W^{\mathrm{L}}(s;\ s_0,\ t)$ is the Lagrangian probability density of the distance between two particles of the flow with initial interval s_0 between them.

It is worth recalling that, as distinct from the statistical properties of the velocity field, discussed above, which make sense only before the multi-stream formation, correlation function (3.1) correctly describes the correlation properties of the density field both before and after multi-stream formation.

It is obvious from (1.4) that

$$W^{\mathrm{L}}(s;\ s_0,\ t) = \langle \delta(s - s_0 - (v_0(y + s_0) - v_0(y))t)\rangle, \tag{3.2}$$

where the averaging is performed over the statistical ensemble of a random field $u(y,\ s_0) = v_0(y + s_0) - v_0(y)$, the velocity diference of the said parti-

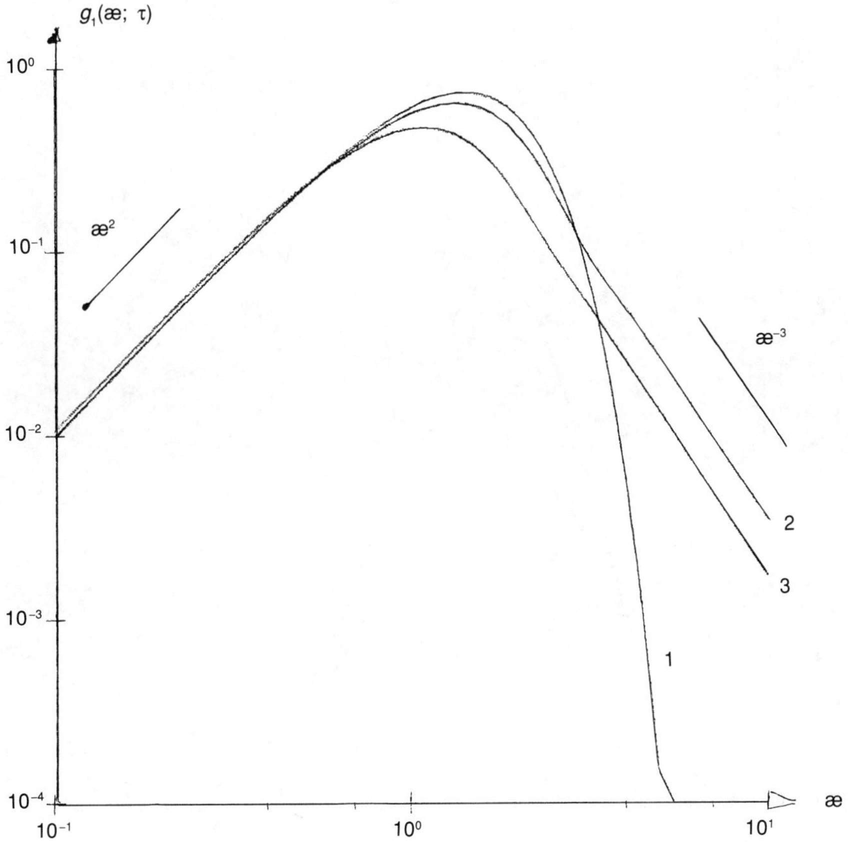

cles. Let $W_u(u;\ s_0)$ be the probability distribution of the (statistically homogeneous in y) field $u(y,\ s_0)$. Then

$$B_\rho(s;t) = K_\rho(s;t) - \rho_0^2 = \frac{\rho_0^2}{t}\int_{-\infty}^{\infty} W_u((s-s_0)/t;\ s_0)\,\mathrm{d}s_0 - \rho_0^2. \tag{3.3}$$

For the spectrum of the random density field we obtain

$$G_\rho(k;t) = \rho_0^2[\tilde{\theta}(kt, -k)/2\pi - \delta(k)], \tag{3.4}$$

where

$$\tilde{\theta}(\mu, \nu) = \int_{-\infty}^{\infty} w_u(u;\ s_0)\mathrm{e}^{\mathrm{i}\mu u + \mathrm{i}\nu s_0}\,\mathrm{d}u\,\mathrm{d}s_0$$

is a Fourier transform over s_0 of the characteristic function of the random field $u(y,\ s_0)$. Later on we shall restrict attention to the case of Gaussian homogeneous field $v_0(x)$ with a covariance function $B_0(s)$. Then

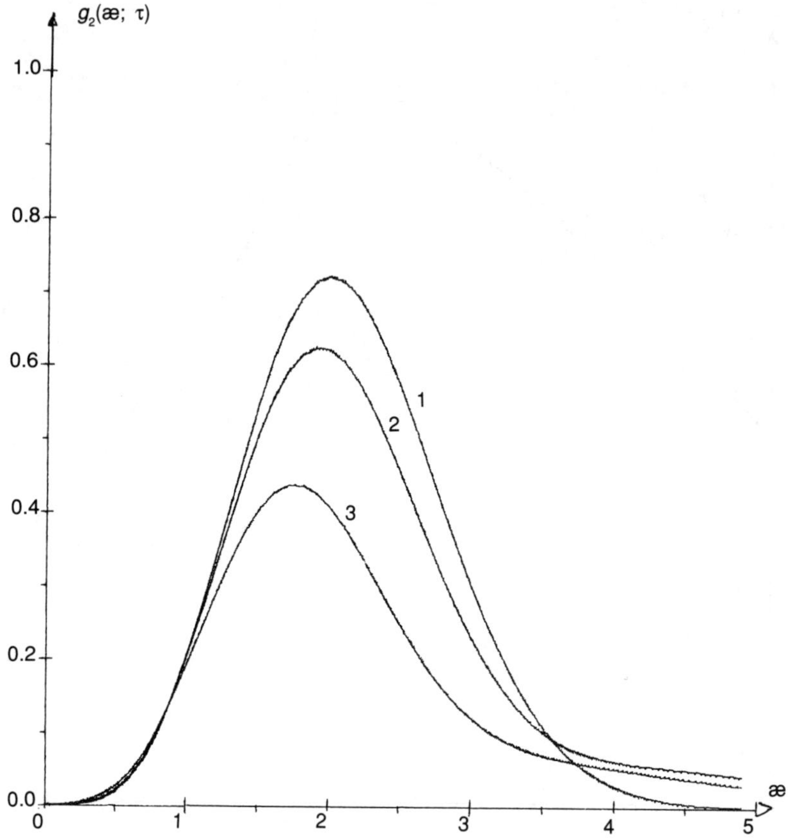

Fig. 4.6 Riemann wave spectrum evolution g_2(æ; τ): τ = 0, 1; τ = 0.2, 2; τ = 0.4, 3

$$W_u(u; s_0) = \frac{1}{\sqrt{4\pi[\sigma_0^2 - B_0(s_0)]}} \exp\left\{-\frac{u^2}{4[\sigma_0^2 - B_0(s_0)]}\right\}. \tag{3.5}$$

whereas the spectral density (3.4) is converted into

$$G_\rho(k; t) = \frac{\rho_0^2}{2\pi} \mathrm{e}^{-(\sigma_0 kt)^2} \int_{-\infty}^{\infty} (\mathrm{e}^{B_0(s_0)k^2t^2} - 1)\, \mathrm{e}^{-iks_0}\, \mathrm{d}s_0. \tag{3.6}$$

From (3.6) one can deduce a familiar invariant of Loitsianski [18]

$$G_\rho(0; t) = \frac{1}{2\pi} \int_{-\infty}^{\infty} B_\rho(s; t)\, \mathrm{d}s = 0,$$

which is valid for the density fluctuations of any statistically homogeneous continuous medium, and expresses a specific character of the density

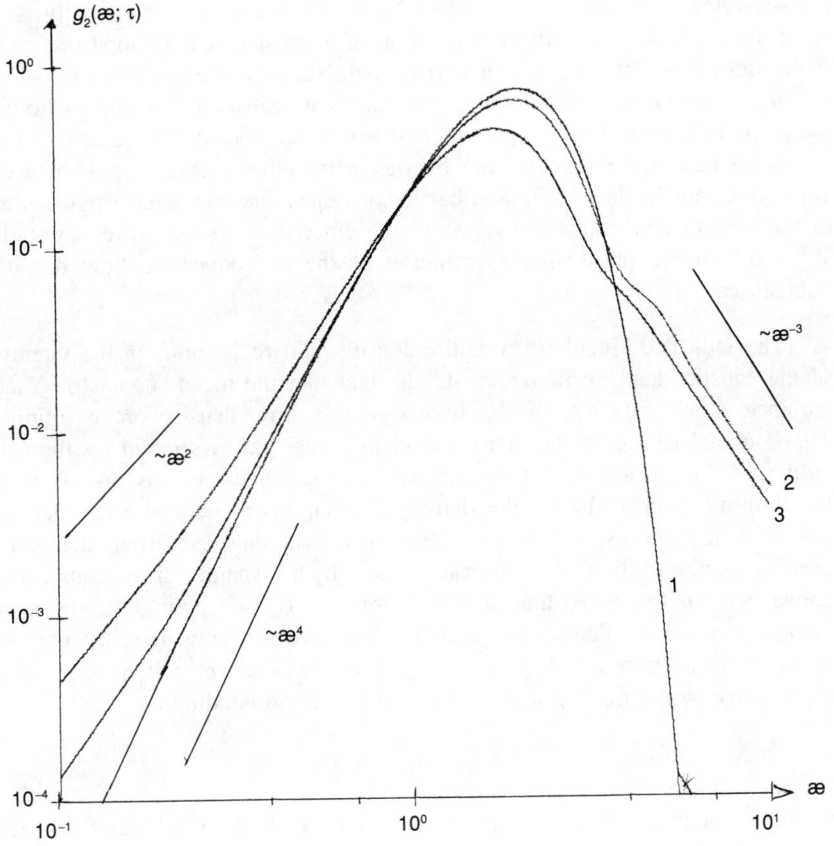

fluctuations: medium condensation ($\rho > \rho_0$) is accompanied by a rarefaction ($\rho < \rho_0$) in adjacent regions.

In analogy with the spectrum of the random velocity field, the spectrum of density fluctuations decreases at large k (or for sufficiently large time t) according to a universal power law. Let us find it. By calculating the integral in (3.6) using the saddle-point approach, we have

$$G_\rho(k; t) = \frac{\rho_0^2 \exp(-1/2\tau^2)}{\sqrt{2\pi}\, k_1 \, |\kappa \, t|}, \quad (\sigma_0 \quad |k| \, t \to \infty). \tag{3.7}$$

Here, as in (2.9), $\kappa = k/k_1$, $\tau = \sigma_0 k_1 t$. At sufficiently large times, i.e. in the range of fully developed multi-stream motion, this universal spectrum is simplified to

$$G_\rho(k; t) = \frac{\rho_0^2}{\sqrt{2\pi}\, k_1 \, |\kappa \, \tau|},$$

and describes not just the spectrum 'tails' as before but, essentially, the spectrum itself at all k with the exception of a small neighbourhood of $k = 0$ with width $k \sim 1/\sigma_0 t$, becoming narrow with time. This slow decrease $G_\rho \sim k^{-1}$ of the spectrum is attributed to the fact that realisations of the hydrodynamic particle flow density, at a multi-stream stage and in the vicinity of the caustics where some particles are overtaken by others, have singularities of $1/\sqrt{x}$ type. In Section 4.4 it will be demonstrated that the same singularities of the density field realisations lead to the emergence of the power-law tails $W_\rho^E \sim \rho^{-3}$ of the probability distribution of the hydrodynamic flow density fluctuations.

2. The indicated singularities of the density field realisations in the vicinity of the caustics lead, in particular, to the fact that the mean square $\langle \rho^2 \rangle_E$ and variance $\sigma_\rho^2 = \langle \rho^2 \rangle_E - \rho_0^2$ of the hydrodynamic flow density prove infinite. The bounded nature of the density field in a real gas is secured by thermal scatter of the particle velocities and by pressure forces. Let us discuss now the limiting mechanism for the density fluctuations associated with thermal scatter of the microscopic particle velocities, assuming, as earlier, the particles as noninteracting. In contrast to the hydrodynamic flow considered above, we shall suppose that at initial time $t = 0$ each point x of the flow incorporates many microscopic particles with different velocities, the density of which in a phase space (x, v) is equal to $\rho_0 f_0(x, v)$: here ρ_0 is the initial gas density, while $f_0(x, v)$ satisfies the normalisation condition

$$\int_{-\infty}^{\infty} f_0(x, v)\, dv = 1.$$

Evolution of this type of 'warm' gas of noninteracting particles is described by a Liouville equation

$$\frac{\partial f}{\partial t} + v\frac{\partial f}{\partial x} = 0, \quad f(x, v, t = 0) = \rho_0 f_0(x, v).$$

Its solution has the form

$$f(x, v, t) = \rho_0 f(x - vt, v).$$

If one knows the gas density f in the phase space, the gas density at a given point x can be readily found as

$$\rho_w(x, t) = \int_{-\infty}^{\infty} f(x, v, t)\, dv.$$

In the ensuing discussion, for the sake of simplicity we tackle only the case when the initial density in phase space (x, v) is equal to $f_w(v - v_0(x))$. Therefore, the gas density can be formulated as

$$\rho_w(x, t) = \rho_0 \int_{-\infty}^{\infty} f_w(v - v_0(x - vt))\, dv.$$

Let us transform this equation to a more obvious form. To attain this we rewrite it as

$$\rho_w(x, t) = \rho_0 \int_{-\infty}^{\infty} f_w(v - v_0(y))\, \delta(y - x + vt)\, dv\, dy.$$

Passing to a new integration variable $c = v - v_0(y)$, we obtain

$$\rho_w(x, t) = \rho_0 \int f_w(c)\, \delta(x - X(y, t, c))\, dc\, dy. \tag{3.8}$$

Here

$$X(y, t, c) = X(y, t) + ct = y + (c + v_0(y))t \tag{3.9}$$

can be interpreted as the current coordinate of a particle of a hydrodynamic flow with the initial velocity field equal to $v_0(x) + c$. Making use of the filtering property of the delta-function in y, one comes to

$$\rho_w(x, t) = \rho_0 \int_{-\infty}^{\infty} \frac{f_w(c)}{|j(x, t, c)|}\, dc.$$

Here $j(x, t, c)$ is the divergence field of the said hydrodynamic flow. It is evident that

$$j(x, t, c) = j(x - ct, t),$$

where $j(x, t)$ is the divergence field of the hydrodynamic flow with initial velocity field $v_0(x)$. Recalling that the hydrodynamic flow density is equal to $\rho(x, t) = \rho_0/|j(x, t)|$, we finally have

$$\rho_w(x, t) = \rho_0 \int_{-\infty}^{\infty} f_w(c)\, \rho(x - ct, t)\, dc. \tag{3.10}$$

From this it is quite clear that thermal scatter of the gas microscopic particle velocities leads to the smoothing of the caustic singularities of the density realisations.

Consider another consequence of (3.10). Assume that the corresponding hydrodynamic flow is single-stream, and that

$$f_w(c) = \begin{cases} 1/2c_0; & |c| \le c_0, \\ 0; & |c| > c_0. \end{cases}$$

In this case

$$\rho_w(x, t) = \frac{1}{2c_0} \int_{-c_0}^{c_0} \rho(x - ct, t)\, dc.$$

Now in an one-stream mode,

$$\rho(x, t) = \rho_0 \frac{\partial y(x, t)}{\partial x},$$

where $y(x, t)$ is the Lagrangian coordinate of the hydrodynamic flow particle in the Eulerian representation. For a flow of noninteracting particles it equals

$$y(x, t) = x - v(x, t)t. \tag{3.11}$$

Therefore, in the case under consideration,

$$\rho_w(x, t) = \rho_0 - \frac{\rho_0 t}{2c_0} \int_{-c_0}^{c_0} \frac{\partial v(x - ct, t)}{\partial x} \mathrm{d}c.$$

Going over to a new integration variable $z = x - ct$, we obtain

$$\rho_w(x, t) = \rho_0 - \frac{\rho_0}{2c_0} \int_{x-c_0t}^{x+c_0t} \frac{\partial v(z, t)}{\partial z} \mathrm{d}z,$$

or, finally,

$$\rho_w(x, t) = \rho_0\left[1 - \frac{v(x + c_0t, t) - v(x - c_0t, t)}{2c_0}\right]. \tag{3.12}$$

3. Let us return to the investigation of the noninteracting particle gas density fluctuations allowing for 'thermal' scatter of their velocities. The correlation function of the gas density for a statistically homogeneous initial velocity field $v_0(x)$, as per (3.9), (3.10), is equal to

$$K_\rho^w(s; t) = \langle \rho_w(x + s, t)\, \rho_w(x, t)\rangle_{\mathrm{E}} = \rho_0^2 \int_{-\infty}^{\infty} f_w(c_1)\, f_w(c_2)$$
$$\times \langle \delta(x + s - c_1t - X(y_1, t))\, \delta(x - c_2t - X(y_2, t))\rangle_{\mathrm{L}}\, \mathrm{d}y_1\, \mathrm{d}y_2\, \mathrm{d}c_1\, \mathrm{d}c_2 \tag{3.13}$$

Represent the average included here via the probability density of the mass centre coordinates and inter-particle distances,

$$W_{Q,s}(Q, s; y_1, y_2, t) = \langle \delta(Q - (X(y_1, t) + X(y_2, t))/2) \times$$
$$\delta(s - X(y_1, t) + X(y_2, t))\rangle_{\mathrm{L}}.$$

Notice that due to statistical homogeneity

$$W_{Q,s} = W_{Q,s}(Q - Q_0, s; s_0, t).$$

Here $Q_0 = (y_1 + y_2)/2$, $S_0 = y_1 - y_2$. Evidently, the average in (3.13) appears as

$$\langle \delta(x + s - c_1t - X(y_1, t))\, \delta(x - c_2t - X(y_2, t))\rangle_{\mathrm{L}} =$$
$$W_{Q,s}(x + s/2 - t(c_1 + c_2) - Q_0, s - t(c_1 - c_2); s_0, t).$$

Substituting this equality into (3.13) and passing from integration with respect to y_1, y_2 to integration with respect to Q_0 and s we find

$$K_\rho^w(s; t) = \rho_0^2 \int_{-\infty}^{\infty} f_w(c_1)\, f_w(c_2)\, W_s^{\mathrm{L}}(s - t(c_1 - c_2); s_0, t)\, \mathrm{d}c_1\, \mathrm{d}c_2\, \mathrm{d}s_0. \tag{3.14}$$

Here

$$W_s^{\mathrm{L}}(s; s_0, t) = \langle \delta(s - X(Q_0 + s_0/2, t) + X(Q_0 - s_0/2, t)) \rangle_{\mathrm{L}}$$
$$= \int_{-\infty}^{\infty} W_{Q,s}(Q - Q_0, s; s_0, t)\, \mathrm{d}Q$$

is the probability density of the distances between two prescribed particles with initial distance s_0. Introduce, in addition, the function

$$F(c) = \int_{-\infty}^{\infty} f_w\left(c_+ - \frac{c}{2}\right) f_w\left(c_+ + \frac{c}{2}\right) \mathrm{d}c_+.$$

Denoting in (3.14), $c_1 - c_2 = c$, one can finally formulate

$$K_\rho^w(s; t) = \rho_0^2 \int_{-\infty}^{\infty} F(c)\, W_s^{\mathrm{L}}(s - ct; s_0, t)\, \mathrm{d}s_0\, \mathrm{d}c. \qquad (3.15)$$

This relation extends (3.1) quite naturally to the case of a 'warm' gas of noninteracting particles. In analogy with (3.3) it can be cast as

$$K_\rho^w(s; t) = \frac{\rho_0^2}{t} \int_{-\infty}^{\infty} F(c)\, w_u[(s - s_0)/t - c; s_0]\, \mathrm{d}s\, \mathrm{d}c. \qquad (3.16)$$

The spectral density of the warm gas density fluctuations is, then, equal to

$$G_\rho^w(k; t) = \theta_w(kt)\, G_\rho(k; t). \qquad (3.17)$$

Here $G_\rho(k; t)$ is the spectral density in the hydrodynamic approximation given by (3.4), and

$$\theta_w(\kappa) = \int_{-\infty}^{\infty} F(c)\, e^{-i\kappa c}\, \mathrm{d}c.$$

The first factor on the right-hand side of (3.17) takes into account the 'thermal' scatter of the gas microparticle velocities, suppressing the small-scale spectrum components due to the 'blurring' of the caustic singularities in the density realisations.

Let us delve in somewhat more detail into the spectrum and the density variance behaviour, in the specific case of a Gaussian field $v_0(x)$ with correlation function (2.7). In addition we set

$$f_w(c) = \frac{1}{\sqrt{2\pi}\, c_w} \exp\left(-\frac{c^2}{2c_w^2}\right), \quad \theta_w(kt) = \exp(-k^2 t^2 c_w^2).$$

In this case the density fluctuation spectrum, as is evident from (3.6), (2.7), (3.17), is described as

$$G_\rho^w(k; t) = \frac{\rho_0^2}{2\pi k_1} \exp[-(\varepsilon + 1)(\kappa\tau)^2] \int_{-\infty}^{\infty} [\exp(\kappa^2 \tau^2 \mathrm{e}^{-z^2/2}) - 1]\, \mathrm{e}^{-i\kappa z}\, \mathrm{d}z. \qquad (3.18)$$

Here, as above, we define $\kappa = k/k_1$, $\tau = \sigma_0 k_1 t$. Apart from this a dimensionless parameter $\varepsilon = (c_w/\sigma_0)^2$ – the 'gas temperature' – is introduced. Let us display also the appropriate expression for the variance of the gas density fluctuations. Assuming $s = 0$ in (3.16) and substituting (3.5), we have after simple transformations,

$$\sigma_\rho^2(t) = \langle \rho_w^2(x, t) \rangle - \rho_0^2$$

$$= \frac{2\rho_0^2}{\sqrt{\pi}} \int_{-\infty}^{\infty} \frac{\exp\{-z^2/[\varepsilon + 1 - \exp(-2z^2\tau^2)]\}\, dz}{\sqrt{\varepsilon + 1 - \exp(-2z^2\tau^2)}} - \rho_0^2. \tag{3.19}$$

If the gas is hot, i.e. if $\varepsilon \gg 1$, then it is possible to pass to a simpler approximate expression for the spectral density (3.18), expanding the exponential function under the integral in Taylor series and confining ourselves to the first nonvanishing term of the expansion:

$$G_\rho^w(k; t) = \frac{(\rho_0 \kappa\tau)^2}{\sqrt{2\pi}k_1} \exp[-\tfrac{1}{2}\kappa^2(1 + 2\varepsilon\tau^2)].$$

The variance of the gas density fluctuations is approximately equal to

$$\sigma_\rho^2(t) = \int_{-\infty}^{\infty} G_\rho^w(k; t)\, dk = \frac{\rho_0^2\tau^2}{(1 + \varepsilon\tau^2)^{3/2}}.$$

Hence it is plain that due to a strong blurring effect of the hot gas microparticle 'thermal' motion, the density fluctuation variance appears to be small, $\sigma_\rho^2 \ll \rho_\rho^2$, for any τ.

As distinct from that of hot gas, the spectrum of cold gas ($\varepsilon \ll 1$) density fluctuations follows all the characteristic features of the hydrodynamic flow density fluctuation spectrum (3.6). Namely, at $\kappa > 1/\tau$ the spectrum is governed by the approximate formula

$$G_\rho^w(k; t) = \frac{\rho_0^2}{\sqrt{2\pi}\, k_1\tau\, |\kappa|} \exp\left(-\varepsilon\kappa^2\tau^2 - \frac{1}{2\tau^2}\right),$$

which evidently demonstrates that over $1 \ll \kappa\tau \ll 1/\sqrt{\varepsilon}$ the spectrum possesses the universal power asymptotic behaviour (3.7), typical of the density hydrodynamic fluctuations. The variance of the cold gas density fluctuations can be estimated from the following approximate formula:

$$\sigma_\rho^2(t) = 2 \int_{1/\sigma_0 t}^{\infty} G_\rho^w(k; t)\, dk = \rho_0^2 \exp(-1/2\tau^2)\, E_1(\varepsilon)/\sqrt{2\pi}\,\tau. \tag{3.20}$$

Here

$$E_1(\varepsilon) = \int_1^{\infty} \frac{e^{-\varepsilon y}}{y}\, dy \approx \ln\left(\frac{1}{\varepsilon}\right) - \gamma \quad (\varepsilon \ll 1),$$

and $\gamma = 0.5772$ is the Euler constant.

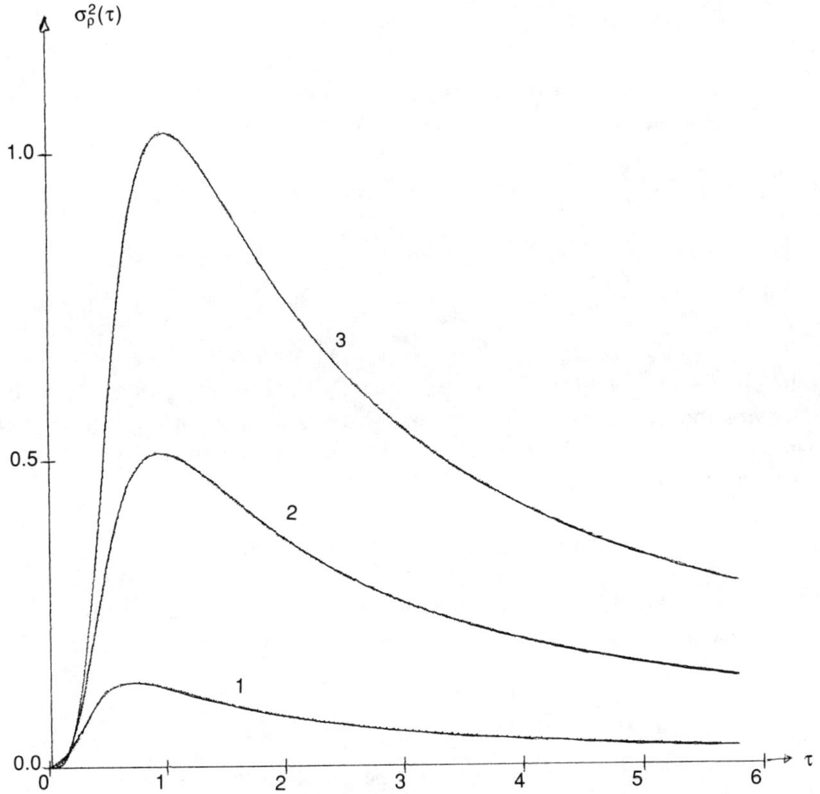

Fig. 4.7 Dependence of the gas density fluctuation variance on τ at various temperatures ε: $\varepsilon = 1$, 1; $\varepsilon = 0.1$, 2; $\varepsilon = 0.01$, 3

It is quite apparent from (3.19), (3.20), that even in the case of cold gas the variance of its density fluctuations vanishes for $\tau \gg 1$. This is attributed to the fact that the energy of the hydrodynamic motion of gas is, as time elapses, continuously converted into thermal form. Plots of the gas density fluctuation variance against τ at various temperatures ε are depicted in Fig. 4.7.

4. There exists an elegant analogy between the above evolution of the noninteracting particle heated gas density and the diffraction of monochromatic optical waves which have passed through a random phase screen. Let us illustrate it using the example of a plane wave incident on a one-dimensional phase screen. Denote a longitudinal coordinate along which the wave is essentially propagating as t, and a transverse coordinate as ρ. In the 'small angle' approximation of quasioptics, the complex wave amplitude $E(\rho, t)$ satisfies the equation

$$2ik\frac{\partial E}{\partial t} + \frac{\partial^2 E}{\partial \rho^2} = 0, \quad E(\rho, t = 0) = E_0(\rho),$$

where $k = \omega/c$ is the wave number, ω is frequency, and c is the wave propagation velocity. Let us introduce the Wigner function

$$f(x, v, t) = \frac{1}{2\pi}\int_{-\infty}^{\infty} E\left(x + \frac{z}{2k_0}, t\right) E^*\left(x - \frac{z}{2k_0}, t\right) e^{-ivz}\, dz.$$

As can be easily shown, this is governed by the Liouville equation

$$\frac{\partial f}{\partial t} + v\frac{\partial f}{\partial x} = 0.$$

Although, the Wigner function is, in general, an alternating one, it can be readily interpreted as the ray density of the wave arriving at point ($\rho = x$, t) at an angle v to the t-axis. The total wave intensity at point (x, t), as for the density of the noninteracting particle thermal gas, is equal to

$$I(x, t) = \left|E^2(x, t)\right| = \int_{-\infty}^{\infty} f(x, v, t)\, dv.$$

If a plant wave with intensity I_0 is incident on a phase screen located in plane $t = 0$ and bringing phase distortions $k_0\Psi(\rho)$ into the wave, then

$$f(x, v, 0) = \frac{I_0}{2\pi}\int_{-\infty}^{\infty} \exp\left\{ik_0\left[\psi\left(x + \frac{z}{2k_0}\right) - \psi\left(x - \frac{z}{2k_0}\right) - ivz\right]\right\} dz. \quad (3.21)$$

At sufficiently large k_0 one can expand the phase difference included in the above equation in a Taylor series, retaining only several terms of the expansion,

$$k_0\left[\psi\left(x + \frac{z}{2k_0}\right) - \psi\left(x - \frac{z}{2k_0}\right)\right] = v_0(x) + v_0''(x)\frac{z^3}{24k_0^2} + \cdots. \quad (3.22)$$

Here $v_0(x) = \Psi'(x)$. Taking only the first term on the right-hand side of (3.22) and substituting it into (3.21), we obtain

$$f_0(x, v) = I_0\delta(v(x) - v),$$

being the formulation of the ray intensity in the geometrical optics approximation or, as one might say concerning the gas of noninteracting particles, in the 'hydrodynamic flow approximation'. If both terms quoted in the right-hand side of (3.22) are retained, we have

$$f_0(x, v) = I_0 f_d(v - v_0(x), b(x)),$$

where

$$b(x) = \left(\frac{v_0''(x)}{8k^2}\right)^{1/3}, \quad \text{and} \quad f_d(v, b) = \frac{1}{b}\mathrm{Ai}\left(\frac{v}{b}\right)$$

is a function of width ~b taking into account the diffraction blurring when a wave is propagating beyond a phase screen, Ai(·) being the Airy function.

In analogy with the temperature of the noninteracting particle gas we may, in this case, introduce a parameter

$$\varepsilon = (\langle b^6\rangle)^{\frac{1}{3}}/\langle v_0^2\rangle = (\langle (v_0'')^2\rangle)^{\frac{1}{3}}/(8k^2)^{\frac{2}{3}}\langle v_0^2\rangle.$$

The smaller this parameter the more valid is the geometrical optics approximation (at not too large t).

4.4 Probability properties of density fluctuations

1. Up to now we have been concerned with the spectral–correlation properties of the density fluctuations. Let us examine their probability characteristics. In the case of statistically homogeneous hydrodynmaic flow with initially constant density ρ_0, he Eulerian probability distribution of a random field of the flow density is given by equality (3.3.17) as

$$W_\rho^{\mathrm{E}}(\rho; t) = \frac{\rho_0}{\rho} W_\rho^{\mathrm{L}}(\rho; t), \tag{4.1}$$

where $W^{\mathrm{L}}(\rho; t)$ is the Lagrangian probability distribution of the density field. Recalling that $\rho^{\mathrm{L}}(y, t) = \rho_0/J(y, t)$ and making use of the expression (1.4) for the divergence of the noninteracting particle hydrodynamic flow, we come to

$$W_\rho^{L}(\rho; t) = \langle \delta(\rho - \rho_0/|1 + v_0'(y)t|)\rangle.$$

Let $W_0(u)$ be the probability density of a random field $v_0'(y)$. Then

$$W_\rho^{\mathrm{L}}(\rho; t) = \int_{-\infty}^{\infty} W_0(u)\, \delta(\rho - \rho_0/|1 + ut|)\, du$$
$$= \frac{\rho_0^2}{t\rho^2}\left[W_0\left(\frac{1}{t}\left(\frac{\rho_0}{\rho} - 1\right)\right) + W_0\left(-\frac{1}{t}\left(\frac{\rho_0}{\rho} - 1\right)\right)\right].$$

Substitution of this expression into (4.1) yields the final probability distribution of the density,

$$W_\rho^{\mathrm{E}}(\rho; t) = \frac{1}{\rho t}\left(\frac{\rho_0}{\rho}\right)^2\left[W_0\left(\frac{1}{t}\left(\frac{\rho_0}{\rho} - 1\right)\right) + W_0\left(-\frac{1}{t}\left(\frac{\rho_0}{\rho} + 1\right)\right)\right]. \tag{4.2}$$

First, it is worth noting that as $\rho \to \infty$ the Eulerian probability density distribution vanishes following a power law,

$$W_\rho^{\mathrm{E}}(\rho; t) = 2\rho_0^2\, W_0(-1/t)/t\rho^3; \tag{4.3}$$

consequently, the mean square of the hydrohynamic flow density, for example, appears to be infinite: $\langle\rho^2\rangle = \infty$, as soon as $W_0(-1/t) \neq 0$. In the previous section we have already stated that the power-law asymptotic behaviour of the probability distribution (4.2) is caused by caustic singularities $\rho(x, t) \sim 1/\sqrt{x}$ in the vicinity of caustics where the flow divergence $j(x, t)$

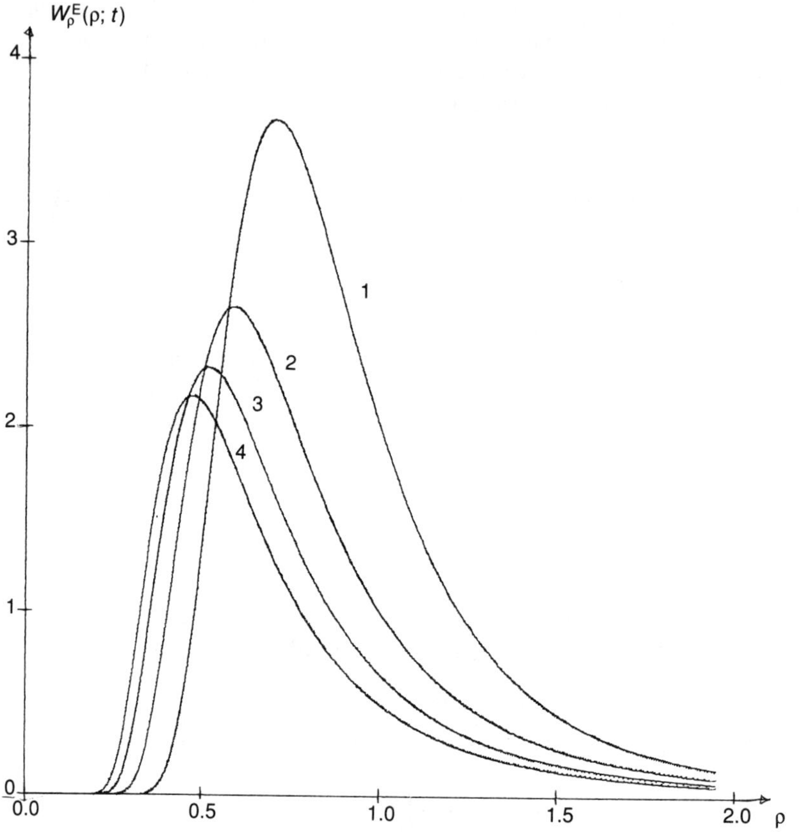

Fig. 4.8 Density probability distribution at various τ: τ = 0.2, 1; τ = 0.4, 2; τ = 0.6, 3; τ = 0.8, 4

vanishes. It is apparent from (1.4) that the Lagrangian coordinates of these points are the roots of the equation $tv_0'(y) = 1$. Provided, at a given time t, inequality $tv_0'(y) > -1$ holds at any y, the hydrodynamic flow is strictly a single-stream one, $W_0(-1/t) = 0$, and the density probability distribution is prescribed by an equality

$$W_\rho^E(\rho; t) = \rho_0^2\, W_0\left(\frac{1}{t}\left(\frac{\rho_0}{\rho} - 1\right)\right)\Big/ t\rho^3. \tag{4.4}$$

As usual, a more detailed analysis of the flow density statistics will be proposed for the case when the initial velocity field $v_0(x)$ is Gaussian and has a covariance function (2.7). Then, $v_0'(x)$ is also a Gaussian function with zero average, and variance $\sigma_0^2 k_1^2$, whereas the probability distribution of the density field (4.4) will take the form

$$W_\rho^E(\rho; t) = \frac{\rho_0^2}{\sqrt{2\pi\tau\rho^3}} \exp\left[-\frac{1}{2\tau}\left(\frac{\rho_0}{\rho} - 1\right)^2\right]. \tag{4.5}$$

The power-law asymptotic behaviour of the probability density as $\rho \to \infty$ in this case is defined by the equality

$$W_\rho^E(\rho; t) = \sqrt{\frac{2}{\pi}} \frac{\rho_0^2}{\tau\rho^3} \exp\left(-\frac{1}{2\tau^2}\right). \tag{4.6}$$

Plots of the probability distribution (4.6) at different τ values are depicted in Fig. 4.8. These plots suggest that the characteristic feature of the density probability distribution appears as a deviation of the distribution maximum towards the small values $\rho < \rho_0$ and the simultaneous emergence of slowly decreasing tails at large $\rho > \rho_0$. Both features are attributed to the increased concentration of matter in compressed zones and rareflaction of matter in the expanded ones.

2. The Eulerian probability densities of noninteracting particle hydro dynamic flows revealed and treated here do in fact have the meaning of probability densities while the flow remains essentially single-stream. Therefore, to establish the applicability range for the given analysis, one has to bear in mind the statistical properties of the stream number. In addition, the multi-stream aspects of a hydrodynamic flow or of an optical wave are considered to be vital physical properties in themselves. Because of this, we calculate here the average stream number, assuming the hydrodynamic flow to be statistically homogeneous. In this case, according to (3.3.40), the Eulerian average number of flows is equal to

$$\langle N(x, t)\rangle_E = \langle |J(y, t)|\rangle_L$$

or, allowing for (1.4),

$$\langle N(x, t)\rangle_E = \langle |1 + v_0'(y)t|\rangle.$$

Let, as above, $W_0(u)$ be the probability density of the function $v_0'(x)$; then

$$\langle N(x, t)\rangle_E = \int_{-\infty}^{\infty} W_0(u)\, |1 + ut|\, du. \tag{4.7}$$

Let us first pay attention to a sufficiently smooth initial field $v_0(x)$ such that at any x: $v_0'(x) \geq -u^*$. This indicates that $W_0(u) = 0$ for $u < -u_0^*$. Then at times $t < t^* = 1/u^*$ the expression inside the modulus in (4.7) remains positive everywhere when $W_0(u) \neq 0$, and one can omit the modulus and write

$$\langle N\rangle_E = \int_{-\infty}^{\infty} W_0(u)\, du + t \int_{-\infty}^{\infty} u\, W_0(u)\, du.$$

Here the first integral equals unity because of the normalisation condition and the second one is equal to zero because of the statistical homogeneity of $v_0'(x)$:

$$\int_{-\infty}^{\infty} u\, W_0(u)\, du = \langle v_0'(x)\rangle = \frac{d}{dx}\langle v_0(x)\rangle = 0.$$

Thus for $t < t^*$ the average number of flows is identically equal to unity, as was to have been expected.

In another limiting case, at sufficiently large t the average number of flows grows linearly with time,

$$\langle N\rangle_E = \gamma t, \tag{4.8}$$

where

$$\gamma = \int_{-\infty}^{\infty} |u|\, W_0(u)\, du \tag{4.8a}$$

describes the average scatter of the velocity gradient at the initial time. Apparently, the larger is γ the sooner a multi-stream flow develops, and the larger is the rate of growth of the flow number. To better understand the law $\langle N\rangle \sim t$, we mentally break down the flow at the initial time into segments with length $l_v \sim 1/k_1$ equal to a characteristic scale of variation of the velocity field $v_0(x)$. There is no doubt that after a sufficiently large time, each of these segments will be elongated and occupy the interval $L \sim \sigma_0 t$, where σ_0 is the standard deviation of the particle velocities. Here, evidently, the said segments will overlap. The number of segments covering a given point x is approximately equal to

$$N \sim L/l_v \sim (\sigma_0/l_v)t \sim \tau = \sigma_0 k_1 t.$$

Since $\sigma_0 k_1 \sim \gamma$, this formula gives a qualitative explanation of (4.8).

Calculate, as an example, the average number of flows when $v_0'(x)$ is a Gaussian field with varian $\sigma_0^2 k_1^2$. In this case we have from (4.7)

$$\langle N(x,t)\rangle_E = \Phi\left(\frac{1}{\tau}\right) + \sqrt{\frac{2}{\pi}}\,\tau \exp\left(-\frac{1}{2\tau^2}\right). \tag{4.9}$$

where $\Phi(z)$ is the probability integral (see A. 18)). The plot of it is given in Fig. 4.9. This figure also incorporates the probability plot $P(1; t)$ when the flow is a single-stream one (3.3.41). As is obvious from (4.9) and the plots, at $\tau =$ 1, when the nonlinear distortions of the hydrodynamic flow are already substantial and strongly affect the probability distribution of the density field (4.5), the spectral density (2.3) of the velocity field and other properties of the flow, the average flow number $\langle N\rangle_E = 1.167$ is quite close to unity, so that the hydrodynamic flow of noninteracting particles is virtually a single-stream one. Note in addition that at $\tau < 1$ the probability of multi-stream

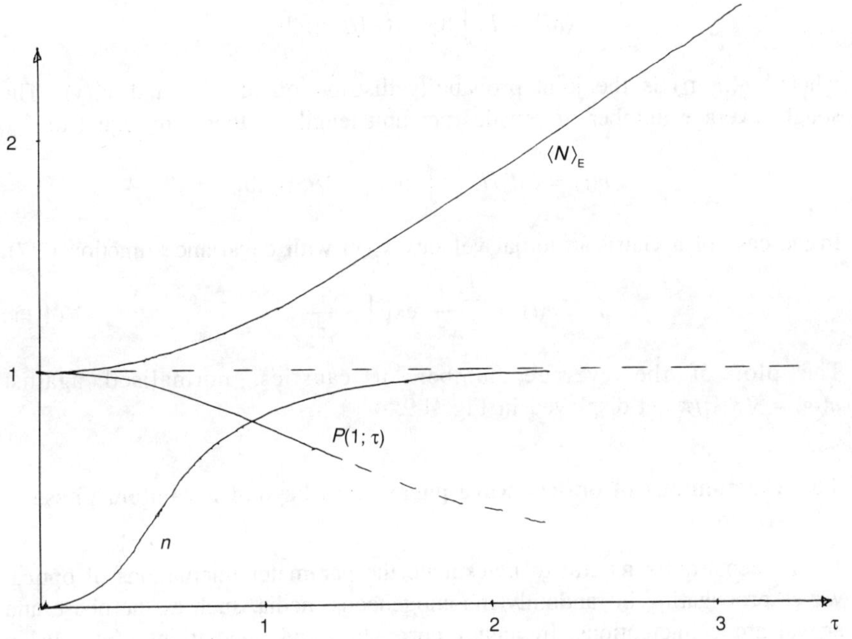

Fig. 4.9 Time-dependence plots of the average flow number $\langle N \rangle_E$, of the probability for the flow to be one-stream $P(1; \tau)$, and the normalised average caustic number per unit length n

behaviour, $P(3; t)$, (3.3.41), is low, being equal to the average relative length of the intervals where the flow is a multi-stream one. So, at $\tau = 1$ it follows from (3.3.41) that $P(3; t) = 0.083$.

Calculate another important characteristic of a multi-stream motion: $n(t)$, the average number of caustics per unit length. A caustic corresponds to a particle divergence $j(x, t)$ crossing zero. It is evident that in the case of statistically homogeneous flow, the average number of caustics per unit length is the same in both the Lagrangian and the Eulerian frame. Therefore, we calculate $n(t)$ in a simpler Lagrangian representation. According to (1.4), the average number of caustics coincides with the average number of intersections by the function $v_0'(y)$ of the level $(-1/t)$. To define it we introduce, with the aid of a delta function, the intersection number counter

$$\int_0^L |v_0''(y)|\, \delta(v_0'(y) + 1/t)\, dy = \sum_{m=1}^{M} |v_0''(y_m)|/|v_0''(y_m)| = M. \tag{4.10}$$

Here y_m are the coordinates of the level $(-1/t)$ intersection points, by the function $v_0'(y)$, belonging within the interval $y \in (0, L)$, M is the number of such points within the interval. Statistical averaging in (4.10) yields

$$\langle M \rangle = L \int_{-\infty}^{\infty} |\eta| \, w_0(-1/t, \eta) \, d\eta,$$

where $w_0(u, \eta)$ is the joint probability distribution of $v_0'(y)$ and $v_0''(y)$. The sought average number of caustics per unit length is, therefore, equal to

$$n(t) = \langle M \rangle / L = \int_{-\infty}^{\infty} |\eta| \, w_0(-1/t, \eta) \, d\eta.$$

In the case of a Gaussian initial velocity $v_0(y)$ with covariance function (2.7),

$$n(t) = \frac{\sqrt{3}\, k_1}{\pi} \exp\left(-\frac{1}{2\tau^2}\right). \tag{4.11}$$

The plot of the average number of caustics, normalised against $n(\infty) = \sqrt{3}\, k_1/\pi$, is displayed in Fig. 4.9.

4.5 Fluctuations of optical wave parameters beyond a random phase screen

1. It seems quite natural to investigate the parameter fluctuations of optical waves propagating in randomly inhomogeneous media, such as the phase and arrival angle fluctuations, frequency correlation and fluctuations of the wave intensity, using s sufficiently demonstrative but asymptotically rigorous geometrical approximation. However, the nonlinearity of the geometrical optics equations hampers their analysis to a large extent. Therefore, when studying the parameters of optical waves one frequently employs the linearised equations of geometrical optics (see, for example, [61]). Nevertheless, here as well, a Lagrangian approach to the Eulerian analysis of the field statistics, set forth in Section 3.3, proves effective, and permits us to examine the statistical properties of the foregoing parameters of optical waves, allowing for the appearance of caustic singularities in the wave field [154–156]. In this case we make use of the relations between the Lagrangian and Eulerian statistics to describe the Eulerian statistics (at a fixed spatial point) of the optical wave parameters beyond a random phase screen. It is worth noting that the probability distribution of the wave intensity beyond a random phase screen was found in [157].

Let an optical wave propagate in a homogeneous medium along the t-axis. Denote the transverse coordinates as $\mathbf{x} = (x_1, x_2)$. then in the small-angle approximation of geometrical optics, the eikonal $\Psi(x, t)$ and the intensity $I(x, t)$ of the wave satisfy the equations [62]:

$$\frac{\partial \psi}{\partial t} + \frac{1}{2}(\nabla \psi)^2 = 0, \quad \frac{\partial I}{\partial t} + \nabla . \, (I \nabla \psi) = 0.$$

Here ∇ is a gradient in the $\mathbf{x}$-plane. Let us introduce the vector $\mathbf{v} = \nabla \Psi$ which has a clear geometrical meaning: $\theta = |\mathbf{v}|$ being the angle with respect to the t-axis at which a wave arrives at the point ($\mathbf{x}$, t). Call this vector $\mathbf{v}$ a 'vector of wave arrival angles'. Acting with the operator ∇ on the equation of the eikonal we have

$$\frac{\partial \mathbf{v}}{\partial t} + (\mathbf{v}.\nabla)\,\mathbf{v} = 0 \tag{5.1}$$

coinciding with the equation of the vector velocity field of the two-dimensional hydrodynamic flow of noninteracting particles. In the new notation the equation of intensity

$$\frac{\partial I}{\partial t} + \operatorname{div}(I\mathbf{v}) = 0,$$

coincides with the continuity equation for density. Therefore, we can contemplate a complete analogy between the description of optical wave propagation under the geometrical optics approximation and the potential ($\mathbf{v} = \nabla\Psi$) motions of two-dimensional hydrodynamic flow of noninteracting particles. The role of the particle trajectories is played by the beams, whereas that of the Lagrangian coordinates is performed by the coordinates of the point at which the beam emanates from the initial plane $t = 0$.

2. In order to use the formulae relating the Lagrangian and Eulerian probability densities one should know the statistics of the Jacobian of the Eulerian-to-Lagrangian coordinate transformation ($j(\mathbf{x}, t)$), or, in other words, the divergence fields of the ray tubes. On this basis we introduce an additional set of equations governing $j(\mathbf{x}, t)$. To this end it is worth recalling that if, at $t = 0$, the wave intensity was the same at all points (x_1, x_2) and equal to I_0 then the divergence $j(\mathbf{x}, t) = I_0/I(\mathbf{x}, t)$ will be equal to the inverse of the wave intensity. Multiplying the continuity equation for I by I_0/I^2 and performing some simple calculations we arrive at a set of equations determining the divergence field:

$$\begin{aligned} &\frac{\partial j}{\partial t} + (\mathbf{v}.\nabla)j + 2l = 0, \\ &\frac{\partial l}{\partial t} + (\mathbf{v}.\nabla)l + m = 0, \\ &\frac{\partial m}{\partial t} + (\mathbf{v}.\nabla)m = 0. \end{aligned} \tag{5.2}$$

Here we introduce the auxiliary fields

$$l = -\frac{1}{2}\, j\Delta\psi, \quad m = j\left|\frac{\partial^2\psi}{\partial x_k\, \partial x_p}\right|, \quad k, p = 1, 2,$$

having a clear physical meaning: l and m multiplied by the divergence j constitute the average curvature and the Gaussian curvature of the wave front of a wave, respectively.

3. Let, at a point with coordinates $t = -R$, $\mathbf{x} = 0$, a spherical wave be emitted while the plane $t = 0$ contains a phase screen bringing random phase distortions $k\Psi_0(x)$ into the wave. Then one has to solve equations (5.1), (5.2) subject to initial conditions

$$\begin{aligned}
&\mathbf{v}(\mathbf{x}, 0) = \mathbf{v}_0(\mathbf{x}) = \frac{\mathbf{x}}{R} + \nabla\psi_0,\\
&j(\mathbf{x}, 0) = 1,\\
&l(\mathbf{x}, 0) = l_0(\mathbf{x}) = -\frac{1}{R} - \frac{1}{2}\Delta\psi_0,\\
&m(\mathbf{x}, 0) = m_0(\mathbf{x}) = \frac{1}{R^2} + \frac{1}{R}\Delta\psi_0 + \left|\frac{\partial^2\psi_0}{\partial x_k\,\partial x_p}\right|.
\end{aligned} \tag{5.3}$$

As in the previous case of hydrodynamic flow of noninteracting particles, it seems more convenient now to first find the Lagrangian probability density and then go over to the Eulerian statistics. In future we shall only be interested in the probability distribution of the arrival angles and the wave intensity. The governing equations in the Lagrangian representation are such that

$$\frac{\mathrm{d}\mathbf{X}}{\mathrm{d}t} = \mathbf{V}, \quad \frac{\mathrm{d}\mathbf{V}}{\mathrm{d}t} = 0, \quad \frac{\mathrm{d}J}{\mathrm{d}t} + 2L = 0, \quad \frac{\mathrm{d}L}{\mathrm{d}t} + M = 0, \quad \frac{\mathrm{d}M}{\mathrm{d}t} = 0. \tag{5.4}$$

The initial condition for the first one is $\mathbf{X}(\mathbf{y}, 0) = \mathbf{y}$, while those for the other equations are deduced from (5.3), performing a trivial replacement of $\mathbf{x}$ by $\mathbf{y}$:

$$\mathbf{V}(\mathbf{y}, 0) = \mathbf{v}_0(\mathbf{y}), \quad J(\mathbf{y}, 0) = 1, \quad L(\mathbf{y}, 0) = l_0(\mathbf{y}), \quad M(\mathbf{y}, 0) = m_0(\mathbf{y}).$$

The solutions to (5.4) are obvious. We only write out the solutions for $\mathbf{X}$, $\mathbf{V}$ and J, which are of direct concern,

$$\begin{aligned}
&\mathbf{X}(\mathbf{y}, t) = \mathbf{y}(1 + t/R) + t\nabla\psi_0(\mathbf{y}),\\
&\mathbf{V}(\mathbf{y}, t) = \mathbf{y}/R + \nabla\psi_0(\mathbf{y}),\\
&J(\mathbf{y}, t) = 1 - 2l_0 t + m_0 t^2 = (1 + t/R)^2 + t\,(1 + t/R)\Delta\psi_0 + t^2\left|\frac{\partial^2\psi_0}{\partial y_k\,\partial y_p}\right|.
\end{aligned} \tag{5.5}$$

Let $\Psi_0(\mathbf{y})$ be a Gaussian statistically isotropic field with a covariance function

$$B_\psi(s) = \frac{\sigma_0^2}{k_1^2}\exp\left(-\frac{1}{2}k_1^2 s^2\right).$$

In this case the random vector $\nabla\Psi_0(\mathbf{y})$ describing the probability properties of $\mathbf{X}$ and $\mathbf{V}$ does not depend statistically on the second partial derivatives $\partial^2\Psi_0/\partial y_k\,\partial y_p$ defining the statistics of J. In view of this the Lagrangian joint probability density of $\mathbf{X}$, $\mathbf{V}$, J can be factorised into the product of the two probability densities

$$W^{\mathrm{L}}_{\mathbf{X},\mathbf{V},J}(\mathbf{x}, \mathbf{v}, j; \mathbf{y}, t) = W^{\mathrm{L}}_{\mathbf{X},\mathbf{V}}(\mathbf{x}, \mathbf{v}; \mathbf{y}, t)\, W^{\mathrm{L}}_J(j; t). \tag{5.6}$$

4. Let us calculate first $W^{\mathrm{L}}_{\mathbf{X},\mathbf{V}}$, the joint probability density of the coordinates and the arrival angles of the fixed beam. Inasmuch as, in this instance, the probability distribution of the vector $\nabla\Psi_0$ equals

$$W_0(\mathbf{p}) = \frac{1}{2\pi\sigma_0^2}\exp\left(-\frac{\mathbf{p}^2}{2\sigma_0^2}\right),$$

then utilising (5.5) we obtain

$$W^{L}_{\mathbf{x},\mathbf{v}}(\mathbf{x}, \mathbf{v}; \mathbf{y}, t) = \frac{1}{2\pi\sigma_0^2}\int_{-\infty}^{\infty}\exp\left(-\frac{\mathbf{p}^2}{2\sigma_0^2}\right)$$

$$\times\,\delta\left(\mathbf{v} - \frac{\mathbf{y}}{R} - \mathbf{p}\right)\delta\left(\mathbf{x} - \mathbf{y}\left(1 + \frac{t}{R}\right) - t\mathbf{p}\right)d\mathbf{p} \tag{5.7}$$

$$= \frac{\delta(\mathbf{x} - \mathbf{y} - \mathbf{v}t)}{2\pi\sigma_0^2}\exp\left[-\frac{(\mathbf{v} - \mathbf{y}/R)^2}{2\sigma_0^2}\right].$$

Go over now to the analysis of the Lagrangian statistics of the divergence field $J(\mathbf{y}, t)$. As is clearly seen from (5.5), it is determined by the statistics of the Gaussian totality of three random values: $\partial^2\psi_0/\partial y_1^2$, $\partial^2\psi_0/\partial y_2^2$, $\partial^2\psi_0/(\partial y_1\,\partial y_2)$. In order to make the forthcoming calculations convenient, we shall cast them in terms of linear combinations of mutually independent Gaussian values α, β, γ possessing the same variance $\sigma_0^2 k_1^2$:

$$\frac{\partial^2\psi_0}{\partial y_1^2} = \sqrt{2}\alpha + \beta, \quad \frac{\partial^2\psi_0}{\partial y_2^2} = \sqrt{2}\alpha - \beta, \quad \frac{\partial^2\psi_0}{\partial y_1\,\partial y_2} = \gamma.$$

In this notation the expression for the divergence (5.5) is changed into

$$J = [(1 + t/R) + \sqrt{2}\alpha t]^2 - t^2\rho, \tag{5.8}$$

where $\rho = \beta^2 + \gamma^2$. Hence, the divergence probability density has the form

$$W^{\mathrm{L}}_J(j; t) = \langle\delta(j - [(1 + t/R) + \sqrt{2}\alpha t]^2 + t^2\rho)\rangle,$$

where the averaging is performed over an ensemble of statistically independent values α and ρ, the joint probability distribution of which reads

$$\mathrm{W}_0(\alpha, \rho) = \frac{1}{\sqrt{8\pi}\,(\sigma_0 k_1)^3}\exp\left(-\frac{\alpha^2 + \rho}{2\sigma_0^2 k_1^2}\right) \quad (\rho > 0). \tag{5.9}$$

Utilising the latter we finally obtain

$$W_J^{\mathrm{L}}(j;t) = \frac{1}{4\sqrt{3}\tau^2}\exp\left[\frac{j}{2\tau^2} - \frac{1}{6}\left(\frac{1}{\tau}+\frac{1}{r}\right)^2\right]$$

$$\times\begin{cases} 2; \; j<0, \\ 2-\Phi\left[\sqrt{\dfrac{3j}{2\tau^2}} - \dfrac{1}{\sqrt{6}}\left(\dfrac{1}{\tau}+\dfrac{1}{r}\right)\right] - \Phi\left[\sqrt{\dfrac{3j}{2\tau^2}} + \dfrac{1}{\sqrt{6}}\left(\dfrac{1}{\tau}+\dfrac{1}{r}\right)\right], & j>0 \end{cases} \tag{5.10}$$

Here we denote: $\tau = \sigma_0 k_1 t$, $r = \sigma_0 k_1 R$, $\Phi(z)$ being the probability integral (A.18).

5. Knowing the Lagrangian probability density (5.6) we can elucidate the Eulerian probability distributions of the fields $\mathbf{v}(\mathbf{x}, t)$ and $j(\mathbf{x}, t)$. Neglecting initially the wave many-stream composition we reveal the required Eulerian probability distribution making use of a formula similar to (3.3.9). In the case under consideration it can be formulated as follows:

$$W_{\mathbf{v},\mathbf{j}}^{\mathrm{E}}(\mathbf{v}, j; \mathbf{x}, t) = jW_J^{\mathrm{L}}(j;t)\int_{-\infty}^{\infty} W_{\mathbf{X},\mathbf{V}}^{\mathrm{L}}(\mathbf{x}, \mathbf{v}; \mathbf{y}, t)\, \mathrm{d}^2\mathbf{y}. \tag{5.11}$$

To begin with we treat the expression for the probability density of the arrival angles with respect to a given point $(\mathbf{x}, t)$. According to (5.11) it can be written as

$$W_{\mathbf{v}}^{\mathrm{E}}(\mathbf{v}; \mathbf{x}, t) = \langle J\rangle_{\mathrm{L}}\int_{-\infty}^{\infty} W_{\mathbf{X},\mathbf{V}}^{\mathrm{L}}(\mathbf{x}, \mathbf{v}; \mathbf{y}, t)\, \mathrm{d}^2\mathbf{y}.$$

The substitution of (5.7) into this, with allowance for $\langle J\rangle_{\mathrm{L}} = (1 + t/R)^2$, yields

$$W_{\mathbf{v}}^{\mathrm{E}}(\mathbf{v}; \mathbf{x}, t) = \frac{(1+t/R)^2}{2\pi\,\sigma_0^2}\exp\left\{-\frac{[\mathbf{v}(1+t/R) - \mathbf{x}/R]^2}{2\sigma_0^2}\right\}. \tag{5.12}$$

It should be emphasised that the evolution of the statistical properties of the arrival angle vector with t along a fixed beam as in (5.7) and at a fixed point $\mathbf{x}$ as in (5.12), is qualitatively different in the two cases. If the probability density along the beam does not depend on t, then $W_{\mathbf{v}}^{\mathrm{E}}$ is altered with distance from the screen. In particular, the variance of the arrival angle vectors

$$\sigma_{\mathbf{v}}^2(t) = \langle(\mathbf{v}(\mathbf{x}, t) - \langle\mathbf{v}(\mathbf{x}, t)\rangle)^2\rangle_{\mathrm{E}} = \sigma_0^2/(1+t/R)^2$$

diminishes with distance from the screen. this effect, of arrival angle fluctuation suppression, is akin to that described n Section 4.1 with respect to a random Riemann wave, namely an effect of noise variance drop along the stretching segment of a signal.

6. Let us now give our attention to the analysis of the expression for the average number of beams arriving at a given point $(\mathbf{x}, t)$. In accord with the

formula relating the Lagrangian and Eulerian statistics given in Section 3.3, the average number of spherical wave beams beyond the phase screen is equal to

$$\langle N(\mathbf{x}, t)\rangle_E = \langle |J|\rangle_L / \langle J\rangle_L = \langle |J|\rangle_L / (1 + t/R)^2.$$

The average modulus of the ray tube divergence, included here, can be found with the aid of (5.8),

$$\langle |J|\rangle_L = \langle |\, [(1 + t/R) + \sqrt{2}\alpha t]^2 - t^2\rho |\rangle.$$

Calculating this average and substituting it into the previous equality for the average beam number we obtain

$$N(x, t)\rangle_E = 1 + \frac{4}{\sqrt{3}}\left(\frac{\tau r}{\tau + r}\right)^2 \exp\left[-\frac{1}{6}\left(\frac{\tau + r}{\tau r}\right)^2\right]. \qquad (5.13)$$

This expression is in excess of unity at any point arbitrarily close to the phase screen. The latter means that multi-stream waves and caustics are formed immediately beyond a random phase screen, and implies a certain incompatibility between the Gaussian nature of the phase distortions introduced by the screen and the validity of the geometrical optics approximation. Meanwhile, as long as the average number of beams is close to unity, many statistical properties of a wave calculated in the geometrical optics approximation adequately reflect the behaviour of a wave beyond the phase screen.

Equation (5.13) implies that regular divergence of the spherical wave beams impedes the growth of the 'multi-stream index'. So, if for a plane wave, for which

$$\langle N(x, t)\rangle_E = 1 + \frac{4\tau^2}{\sqrt{3}} \exp\left(-\frac{1}{6\tau^2}\right),$$

the average multi-stream index shows an unlimited increase with τ, then the average multi-stream index of a spherical wave does not increase as $\tau \to \infty$, and indeed saturates at a level corresponding to the average number of the initially plane wave beams arriving at a point at distance R beyond the screen.

7. At last we come to the Eulerian probability density of the spherical wave intensity beyond a random phase screen. Prior to this, one should remark that the Eulerian probability density of the ray tube divergence, as follows from the formula relating the Lagrangian and Eulerian statistics, equals

$$W_j^E(j; t) = \frac{j}{\langle J\rangle_L} W_J^L(j; t) = \frac{j}{(1 + \tau/r)^2} W_J^L(j; t).$$

Substituting (5.10) here, and recalling that the intensity and divergence of a wave are connected by an equality $I(x, t) = I_0/|j(x, t)|$, we find the Eulerian probability density of a spherical wave beyond the phase screen as

$$W_I^{\mathrm{E}}(I;t) = \frac{I_0^2 r^2}{4\sqrt{3}\,(\tau r + \tau^2)^2\, I^3} \exp\left[-\frac{I_0}{2\tau^2 I} - \frac{1}{6}\left(\frac{1}{\tau} + \frac{1}{r}\right)^2 \right]$$
$$\times \left(1 + \exp\left(\frac{I_0}{I\tau^2}\right) \left\{ 2 - \Phi\left[\sqrt{\frac{3I_0}{2I\tau^2}} - \frac{1}{\sqrt{6}}\left(\frac{1}{\tau} + \frac{1}{r}\right) \right] \right. \right. \tag{5.14}$$
$$\left. \left. - \Phi\left[\sqrt{\frac{3I_0}{2I\tau^2}} + \frac{1}{\sqrt{6}}\left(\frac{1}{\tau} + \frac{1}{r}\right) \right] \right\} \right).$$

When $I \gg I_0$ the form of the probability density is determined by the intensity field singularities in the vicinity of caustics. In particular, as $I \to \infty$ an asymptotic formula results from (5.14), namely

$$W_I^{\mathrm{E}}(I;t) \sim \frac{I_0^2}{\sqrt{3}\, I^3 (1/\tau + 1/r)^2} \exp\left[-\frac{1}{6}\left(\frac{1}{\tau} + \frac{1}{r}\right)^2 \right],$$

according to which the probability density tail decreases as I^{-3}. This formula is similar to the corresponding asymptotic formula (4.3) for the one-dimensional hydrodynamic flow of particles and has a similar geometrical interpretation. It can be noted also that the asymptotic behaviour $W_I^{\mathrm{E}} \sim I^{-3}$ indicates that, beyond a random phase screen with statistically isotropic phase distortions, one-point focusing of the wave is not observed, but caustic surfaces are formed, in the vicinity of which the (actually one-dimensional) focusing of the wave across the caustic occurs. Such behaviour of the caustics one can see in fig I.1 and I.2 of the Introduction to the book, in which are represented the intensity of the optical wave and the density of gravitationally interacting particles. These two cases are not exactly the same as those considered above, but the main features of the intensity (density) fields are similar.

8. We examine this effect in more detail as regards the mainly one-dimensional character of the wave-focusing beyond a random phase screen. To this end we recall that in a homogeneous medium the cross-section of any ray tube of sufficiently small transverse dimensions is compressed or stretched with constant speed in two mutually perpendicular directions. We take these directions as the axes of a new Lagrangian frame z_1, z_2 while the origin will be placed in the ray tube centre. It can be assumed, to demonstrate the effect, that at $t = 0$ the cross-section of a ray tube forms a circle with radius r_0 while the initial area of its cross-section will be $S_0 = \pi r_0^2$. As a result of the compression or stretching of the ray tube along the axes $0z_1$ and $0z_2$, the circular cross-section transforms with t into an elliptical one having area

$$S(t) = \pi r_1(t)\, r_2(t) = \pi r_0^2\, (1 - \lambda_1 t)\,(1 - \lambda_2 t)$$

where λ_1 and λ_2 are the stretch rates of the ray tube along the axes $0z_1$ and $0z_2$. The divergence of this ray tube is equal to

$$J(t) = S(t)/S_0 = (1 - \lambda_1 t)(1 - \lambda_2 t).$$

Comparing this equality with (5.8) we express λ_1 and λ_2 via the auxiliary random values α and ρ:

$$\lambda_1 = -\frac{1}{R} - \sqrt{2\alpha} - \sqrt{\rho}, \quad \lambda_2 = -\frac{1}{R} - \sqrt{2\alpha} + \sqrt{\rho}.$$

From this, allowing for (5.9), we infer the joint probability density of the ray tube cross-section stretch rates along the mutually perpendicular principal axes of the stretching as

$$W(\lambda_1, \lambda_2) = \frac{(\lambda_2 - \lambda_1)}{\sqrt{\pi}\,(2\sigma_0 k_1)^3} \exp\left\{-\frac{1}{(4\sigma_0 k_1)^2}\left[3\left(\lambda_1 + \frac{1}{R}\right)^2 + 3\left(\lambda_2 + \frac{1}{R}\right)^2 - 2\left(\lambda_1 + \frac{1}{R}\right)\left(\lambda_2 + \frac{1}{R}\right)\right]\right\}. \tag{5.15}$$

From this it is easily seen that $\lambda_2 > \lambda_1$ and hence coincidence of the values λ_1 and λ_2 is ruled out. Therefore, the compression of the ray tube along the axes $0z_1$ and $0z_2$ progresses with different principal rates, even with the possibility of the tube being compressed along one axis while being stretched along the other. To sum up, we conclude that the intensity in the vicinity of a caustic has the same type of singularity as that beyond the one-dimensional phase screen.

As a final touch, it is instructive to note that the distribution analogous to (5.15) for the case of three-dimensional motion of noninteracting particles was elaborated in [80].

4.6 Concentration of a passive impurity in a flow with random velocity field

1. In what follows we shall present another example in which the utilisation of the Lagrangian method proves efficient. One of the fundamental problems in the theory of turbulence is the problem of particle diffusion in a medium with a random velocity field, whereas it is the Eulerian statistical properties of the field that are theoretically or experimentally known. Within the framework of the present discussion we confine ourselves to a one-dimensional model version of this problem.

Let $v(x, t)$ be the prescribed velocity field of a one-dimensional compressible gas. The concentration of a passive impurity in the gas, denoted by $\rho(x, t)$, satisfies the equation of continuity

$$\frac{\partial \rho}{\partial t} + \frac{\partial}{\partial x}(\rho v) = 0.$$

Examine first the gas statistical characteristics in the Lagrangian representation. The coordinate $X(y, t)$ and divergence $J^{\mathrm{L}}(y, t)$ (see (3.3.31)) of a fixed fluid particle satisfy the stochastic differential equations

$$\frac{\mathrm{d}X}{\mathrm{d}t} = v_0 + \tilde{v}(x, t), \quad \frac{\mathrm{d}J^{\mathrm{L}}}{\mathrm{d}t} = \frac{\partial \tilde{v}}{\partial X} J^{\mathrm{L}}, \quad X(y, 0) = y, \quad J^{\mathrm{L}}(y, 0) = 1. \qquad (6.1)$$

It is assumed here that $v(x, t) = v_0 + \tilde{v}(x, t)$, where v_0 is the average velocity of the particles while $\tilde{v}(x, t)$ is a fluctuation component of the velocity field with given statistical properties ($\langle \tilde{v}(x, t) \rangle = 0$).

It should be emphasised first of all that the equation for $J^{\mathrm{L}}(y, t)$ and a boundedness assumption for $\partial \tilde{v}(x, t)/\partial x$ lead to an inequality $J^{\mathrm{L}} > 0$, meaning that the particle motion in a given velocity field remains a single-stream one at all times.

2. The equations (6.1) generally appear as a quite complicated system of nonlinear stochastic equations. Under certain conditions, however, one can go over from stochastic differential equations to simpler deterministic ones for some statistical properties of the stochastic equation solutions. As the most typical (and frequently encountered in applications) examples of such deterministic equations, one can consider the Fokker–Planck equations (FPE) with respect to the solution probability density.

In the ensuing discussion we shall resort to the FPE and related equations when analysing random processes and waves. Thus, using (6.1), we propose a possible method to deduce such equations, and speculate concerning their applicability conditions. Getting ahead of the discussion, we remark that one out of two basic applicability conditions for FPE is that of two-scale separation. In the case of (6.1) it manifests itself as follows: within a characteristic period τ_v of the velocity field variation $\tilde{v}(x, t)$ these fluctuations only weakly influence the motion X and divergence J of a particle. This is why one of the most frequently applied techniques, when analysing the stochastic equations, is to replace the random influence, in our case that of the fluctuating velocity field $\tilde{v}(x, t)$, with one, delta-correlated in time, for which $\tau_v = 0$ (see, for example [61, 77, 78]). Further calculations here become harmonious and mathematically rigorous, lacking, however, both physical obviousness and, indeed, some vital effects which, when allowed for, show that he characteristic time τ_v finiteness must be retained, as a matter of principle. In view of this we propose a non-rigorous but, in our opinion, 'obvious' approximate derivation of equations for $W^{\mathrm{L}}_{XJ}(x, j; y, t)$, being the joint Lagrangian probability density of the fixed particle coordinate and divergence which accounts explicitly for the finiteness of τ_v. In order to deduce them one needs to know the covariance function of the velocity fluctuations

$$B_v(s, \tau) = \langle \tilde{v}(x, t)\, \tilde{v}(x + s, t + \tau) \rangle,$$

that we regard as given.

3. Let $\varphi(x, j)$ be an arbitrary twice continuously differentiable function of its arguments. Having differentiated $\varphi[X(y, t), J^{\mathrm{L}}(y, t)]$ with respect to time, using (6.1), and having averaged the resulting equality, one obtains

$$\frac{\mathrm{d}\langle \varphi \rangle}{\mathrm{d}t} = v_0 \left\langle \frac{\partial \varphi}{\partial X} \right\rangle + \left\langle \tilde{v} \frac{\partial \varphi}{\partial X} \right\rangle + \left\langle \frac{\partial \tilde{v}}{\partial X} J^{\mathrm{L}} \frac{\partial \varphi}{\partial J^{\mathrm{L}}} \right\rangle. \tag{6.2}$$

To calculate the average on the left-hand side and the first average on the right-hand side of this equality it is sufficient to know the required probability density $W^{\mathrm{L}}_{XJ}(x, j; y, t)$. In order to determine the two last averages, however, one has to employ more general joint probability densities for X, J^{L} and the field $\tilde{v}$. Therefore, to deduce an equation closed with respect to $W^{\mathrm{L}}_{XJ}(x, j; y, t)$ one needs, so to speak, to split the last averages, i.e. to express them via a combination of other averages, each being completely determined either by $W^{\mathrm{L}}_{XJ}(x, j; y, t)$ alone, or only by previously known statistical properties of the field $\tilde{v}(x, t)$. These averages are amenable to approximate splitting provided the two-scale requirement is met. We demonstrate it, and qualitatively formulate the two-scale condition below, using our particular example.

Let the field $\tilde{v}(x, t)$ possess a time τ_v of statistical dependence. The latter means that values $\tilde{v}(x'', \tau'')$ taken at times differing by $|t' - t''| > \tau_v$ are statistically independent. Hence it appears, in particular, that at $|\tau| > \tau_v$, $B_v(s, \tau) = 0$.

Prescribe a certain time instant t_0, and write the solutions of (6.1) for $t > t_0$ as

$$X(y, t) = X_0(y, t) + Z(t_0, t), \quad J^{\mathrm{L}}(y, t) = J^{\mathrm{L}}_0(y, t) + u(t_0, t).$$

Here X_0 and J^{L}_0 are the particle coordinate and divergence, respectively when for $t > t_0$ the medium velocity fluctuation is absent. It is clear that X_0 and J^{L}_0 satisfy the equations

$$\frac{\mathrm{d}X_0}{\mathrm{d}t} = v_0 + \tilde{v}(X_0, t)\, 1(t_0 - t), \quad \frac{\mathrm{d}J^{\mathrm{L}}_0}{\mathrm{d}t} = \frac{\partial \tilde{v}}{\partial X}\, 1(t_0 - t) J^{\mathrm{L}}_0,$$
$$X_0(y, 0) = y, \quad J^{\mathrm{L}}_0(y, 0) = 1. \tag{6.3}$$

The expressions for $z(t_0, t)$ and $u(t_0, t)$ are the coordinate and divergence perturbations, respectively, caused by the velocity field fluctuations $\tilde{v}(x, t)$ in the time interval (t_0, t). They satisfy the equations

$$\frac{\mathrm{d}z}{\mathrm{d}t} = \tilde{v}(X_0 + z, t), \quad \frac{\mathrm{d}u}{\mathrm{d}t} = \frac{\partial \tilde{v}(X_0 + z, t)}{\partial X_0} (J^{\mathrm{L}}_0 + u),$$
$$z(t_0, t_0) = u(t_0, t_0) = 0. \tag{6.4}$$

Introduce explicitly l_v, the characteristic spatial scale of the fluctuation velocity field variation. The above-outlined two-scale condition in this case is reduced to the requirement for the perturbations z and u to be so small within the time interval $t - t_0 \sim \tau_v$ that inequalities

$$z \ll l_v, \quad u \ll J_0^{\mathrm{L}}, \tag{6.5}$$

hold true. This requirement met, one can for $t - t_0 \leq \tau_v$ solve equations (6.4) by the method of successive approximations, and be confined to the first non-vanishing approximation to a high degree of accuracy,

$$z(t_0, t) = \int_{t_0}^{t} \tilde{v}[X_0(y, t'), t']\,\mathrm{d}t', \quad u(t_0, t) = \int_{t_0}^{t} J_0^{\mathrm{L}}(y, t') \frac{\partial \tilde{v}[X_0(y, t'), t']}{\partial X_0(y, t')}\,\mathrm{d}t'.$$

From equations (1.3) it follows that at $t' > t_0$

$$X_0(y, t') = X_0(y, t) - v_0(t - t'), \quad J_0^{\mathrm{L}}(y, t') = J_0^{\mathrm{L}}(y, t).$$

Putting these equalities into the previous one we eventually obtain

$$\begin{gathered} z(t_0, t) = \int_{t_0}^{t} \tilde{v}[X_0(y, t) - v_0(t - t'), t']\,\mathrm{d}t', \\ u(t_0, t) = J_0^{\mathrm{L}}(y, t) \int_{t_0}^{t} \frac{\partial \tilde{v}[X_0(y, t) - v_0(t - t'), t']}{\partial X_0(y, t)}\,\mathrm{d}t'. \end{gathered} \tag{6.6}$$

Let us return once again to the two-scale condition. Estimate roughly the values of z and u within the interval $t - t_0 \sim \tau_v$ as $z \sim \tau_v \sigma_v$, $u \sim \tau_v \sigma_v / l_v$, where σ_v is the standard deviation of the velocity fluctuations. Substituting these estimates into (6.5) yields the inequality

$$\varepsilon = \tau_v \sigma_v / l_v \ll 1, \tag{6.7}$$

which embodies qualitatively the two-scale condition in the case considered. It can be noted that a more careful analysis of the two-scale separation suggests majorant estimates of the proximity to the true solutions of equations (6.4) of the approximations (6.6). Without resorting to majorant estimations, we point out only that in the case of a Gaussian field $\tilde{v}(x, t)$ one can see that the two-scale condition is actually reduced to the requirement of the smallness of the dimensionless parameter (6.7).

Let us calculate the last two averages in (6.2). To begin with, consider the first of them in detail,

$$\left\langle \tilde{v}(X, t) \frac{\partial \varphi(X, J^{\mathrm{L}})}{\partial X} \right\rangle = \left\langle \tilde{v}(X_0 + z, t) \frac{\partial \varphi(X_0 + z, J^{\mathrm{L}} + u)}{\partial X_0} \right\rangle.$$

Let the characteristic variation scales of $\varphi(x, j)$ over x and j be not less than some given values of x_φ, j_φ. Then if

$$z \sim \varepsilon l_v \ll x_\varphi, \quad u \sim \varepsilon \ll j_\varphi \tag{6.8}$$

one can, when calculating the right-hand side of the latter equality, expand the functions being averaged in Taylor series in terms of z and u, and be confined during the expansion to the terms of not higher than the first order,

$$\left\langle \tilde{v}\frac{\partial\varphi}{\partial X}\right\rangle = \left\langle \tilde{v}(X_0, t)\,\frac{\partial\varphi(X_0, J_0^{\mathrm{L}})}{\partial X_0}\right\rangle + \left\langle \frac{\partial\tilde{v}(X_0, t)}{\partial X_0}\,z\frac{\partial\varphi(X_0, J_0^{\mathrm{L}})}{\partial X_0}\right\rangle$$
$$+ \left\langle \tilde{v}(X_0, t)\,z\frac{\partial^2\varphi(X_0, J_0^{\mathrm{L}})}{\partial X_0^2}\right\rangle + \left\langle \tilde{v}(X_0, t)\,\frac{\partial^2\varphi(X_0, J_0^{\mathrm{L}})}{\partial X_0\,\partial J_0^{\mathrm{L}}}\right\rangle. \tag{6.9}$$

Before we engage ourselves in discussing each of the averages involved here, it seems useful to underline that when deducing the FPE, the condition of dynamic causality, apart from the two-scale condition, plays, perhaps, even more of a principal role. In the case under consideration it is reduced to the requirement for $X(y, t)$ and $J^{\mathrm{L}}(y, t)$ to functionally depend on the field values $\tilde{v}(x', t')$ only over $t' < t$, and to be functionally independent of their values for $t' > t$. It is worth emphasising that the condition of dynamic causality with respect to the solutions of stochastic differential equations is provided by the formulation of Cauchy conditions, i.e. here the initial conditions at $t = 0$.

Further we assume that $t_0 = t - 2\tau_v$. Therefore, in view of the dynamic causality condition, $X_0(y, t)$ and $J_0^{\mathrm{L}}(y, t)$ functionally depend on the field values $\tilde{v}(x', t')$ within the interval $t' \in (0, t - 2\tau_v)$ alone. Furthermore, due to the probable statistical relations between $\tilde{v}(x', t')$ and $\tilde{v}(x'', t'')$, X_0 and J_0^{L} may depend statistically on the field values $\tilde{v}(x', t')$ over the interval $t' \in (t - 2\tau_v, t - \tau_v)$. However, X_0 and J_0^{L} depend neither functionally, nor statistically, on $\tilde{V}(x, t)$, and also there is no dependence of $\tilde{V}(x, t)$ on X_0 and J_0^{L}.

After these preliminary remarks, let us proceed to the consecutive splitting of each average on the right-hand side of (6.9). Utilising the general properties of statistical averages one can write the first of them as

$$\left\langle \tilde{v}(X_0, t)\,\frac{\partial\varphi(X_0, J_0^{\mathrm{L}})}{\partial X}\right\rangle = \left\langle \langle\tilde{v}(X_0, t)\rangle_{X_0 J_0^{\mathrm{L}}}\,\frac{\partial\varphi(X_0, J_0^{\mathrm{L}})}{\partial X_0}\right\rangle.$$

Here $\langle \dots \rangle_{X_0 J_0^{\mathrm{L}}}$ means averaging provided the values of X_0 and J_0^{L} are given, while the outer angle brackets refer to averaging over an ensemble of random X_0 and J_0^{L}. On account of the dynamic causality condition, claiming that statistical properties of $\tilde{v}(x, t)$ do not depend on X_0 and J_0^{L}, the conditional average can be replaced by an unconditional one,

$$\langle\tilde{v}(X_0, t)\rangle_{X_0 J_0^{\mathrm{L}}} = \langle\tilde{v}(x, t)\rangle\Big|_{x = X_0} = 0,$$

which is equal to zero in the present situation. Thus, the first average on the right-hand side of (6.9) vanishes.

Let us pass to the analysis of the second average in (6.9),

$$\left\langle \frac{\partial\tilde{v}(X_0, t)}{\partial X_0}\,z\frac{\partial\varphi(X_0, J_0^{\mathrm{L}})}{\partial X_0}\right\rangle = \left\langle \left\langle \frac{\partial\tilde{v}(X_0, t)}{\partial X_0}\,z\right\rangle_{X_0 J_0^{\mathrm{L}}}\frac{\partial\varphi(X_0, J_0^{\mathrm{L}})}{\partial X_0}\right\rangle.$$

Let us formulate in detail the conditional average involved here. Making use of (6.6) we obtain

$$\left\langle \frac{\partial \tilde{v}}{\partial X_0} z \right\rangle_{XJ_0^L} = \int_{t-2\tau_v}^{t} \left\langle \frac{\partial \tilde{v}(X_0, t)}{\partial X_0} \tilde{v}(X_0 - v_0(t - t'), t') \right\rangle_{X_0 J_0^L} dt'. \tag{6.10}$$

The average under the integral sign can be split differently, depending on the values of t'. For $t - 2\tau_v < t' < t - \tau_v$ the values of $\tilde{v}(x, t)$ and $\tilde{v}(x', t')$ prove statistically independent. Hence

$$\left\langle \frac{\partial \tilde{v}(X_0, t)}{\partial X_0} \tilde{v}(X_0 - v_0(t - t'), t') \right\rangle_{X_0 J_0^L}$$
$$= \left\langle \frac{\partial \tilde{v}(x, t)}{\partial x} \right\rangle \Bigg|_{x=X_0} \langle \tilde{v}(X_0 - v_0(t - t'), t') \rangle_{X_0 J_0^L} = 0,$$

as the first of the factors on the right-hand side equals zero. If, however, $t - \tau_v < t' < t$, then neither $\tilde{v}(x, t)$ nor $\tilde{v}(x', t')$ depends statistically on X_0 and J_0^L, and the conditional average can be replaced by an unconditional one

$$\left\langle \frac{\partial \tilde{v}(X_0, t)}{\partial X_0} \tilde{v}(X_0 - v_0(t - t'), t') \right\rangle_{X_0 J_0^L} = \left\langle \frac{\partial \tilde{v}(x, t)}{\partial x} \tilde{v}(x - v_0(t - t'), t') \right\rangle \Bigg|_{x=X_0}.$$

The latter average is expressible through the correlation function of the velocity fluctuations. Consequently, for $t - \tau_v < t' < t$ we have

$$\left\langle \frac{\partial \tilde{v}(X_0, t)}{\partial X_0} \tilde{v}(X_0 - v_0(t - t', t') \right\rangle_{X_0 J_0^L} = \frac{\partial B_v(s, \tau)}{\partial s} \Bigg|_{\substack{s=v_0(t - t') \\ \tau=t-t'}}.$$

Inasmuch as $B_v(s, t - t') \equiv 0$ for $t' < t - \tau_v$, the latter equality is valid at all t'' within the integration interval in (6.10). for the same reason the lower integration limit in (6.10) can be replaced by $-\infty$. Proceeding to a new integration variable $\tau = t - t'$ we obtain finally

$$\left\langle \frac{\partial \tilde{v}}{\partial X} z \right\rangle_{X_0 J_0^L} = -D' = \int_0^\infty \frac{\partial B(s, \tau)}{\partial s} \Bigg|_{s=v_0\tau} d\tau.$$

Correspondingly, the second average on the right-hand side in (6.9),

$$\left\langle \frac{\partial \tilde{v}(X_0, t)}{\partial X_0} z \frac{\partial \varphi(X_0, J_0^L)}{\partial X_0} \right\rangle = -D' \left\langle \frac{\partial \varphi(X_0, J_0^L)}{\partial X_0} \right\rangle,$$

has been broken into the product of the coefficient D' depending on the velocity fluctuation correlation function alone, and the average completely determined by the required probability density $W_{XJ}^L(x, j; y, t)$.

Analogous reasoning allows one to split the two remaining averages on the right-hand side in (6.9) as

$$\left\langle \tilde{v}(X_0, t) z \frac{\partial^2 \varphi(X_0, J_0^L)}{\partial X_0^2} \right\rangle = D \left\langle \frac{\partial^2 \varphi(X_0, J_0^L)}{\partial X_0^2} \right\rangle$$

and

$$\left\langle \tilde{v}(X_0, t)\, u \frac{\partial^2 j(X_0, J_0^{\mathrm{L}})}{\partial X_0^2\, \partial J_0^{\mathrm{L}}} \right\rangle = D' \left\langle J_0 \frac{\partial^2 \varphi(X_0, J_0^{\mathrm{L}})}{\partial X_0\, \partial J_0^{\mathrm{L}}} \right\rangle.$$

Here

$$D = \int_0^\infty B_v(v_0\tau, \tau)\, \mathrm{d}\tau.$$

Substituting all the averages split in this way into (6.9), we come to

$$\left\langle \tilde{v} \frac{\partial j}{\partial X} \right\rangle = -D' \left\langle \frac{\partial \varphi}{\partial X_0} \right\rangle + D \left\langle \frac{\partial^2 \varphi}{\partial X_0^2} \right\rangle + D' \left\langle J_0^{\mathrm{L}} \frac{\partial^2 \varphi(X_0, J_0^{\mathrm{L}})}{\partial X_0\, \partial J_0^{\mathrm{L}}} \right\rangle.$$

Transforming in the same way the final average in (6.2), we write

$$\left\langle \frac{\partial \tilde{v}}{\partial X} J^{\mathrm{L}} \frac{\partial \varphi}{\partial J} \right\rangle = D' \left\langle J_0^{\mathrm{L}} \frac{\partial^2 \varphi}{\partial X_0\, \partial J_0^{\mathrm{L}}} \right\rangle + D'' \left\langle J_0^{\mathrm{L}} \frac{\partial^2 \varphi}{\partial J_0^{\mathrm{L}^2}} \right\rangle,$$

where

$$D'' = -\int_0^\infty \left. \frac{\partial^2 B_v(s, \tau)}{\partial s^2} \right|_{s=v_0\tau} \mathrm{d}\tau,$$

Inserting both the last equalities into (6.2) yields

$$\frac{\mathrm{d}\langle \varphi \rangle}{\mathrm{d}t} = v_0 \left\langle \frac{\partial \varphi}{\partial X} \right\rangle - D' \left\langle \frac{\partial \varphi}{\partial X_0} \right\rangle + D \left\langle \frac{\partial^2 \varphi}{\partial X_0^2} \right\rangle + D'' \left\langle J_0^{\mathrm{L}^2} \frac{\partial^2 \varphi}{\partial J_0^{\mathrm{L}^2}} \right\rangle.$$

Perform a final approximation, i.e. in particular using the smoothness of function $\varphi(x, j)$ qualitatively expressed by inequalities (6.8), we neglect here the differences between X_0, J_0^{L} and X, J^{L}. This results in

$$\frac{\mathrm{d}}{\mathrm{d}t} \langle \varphi \rangle = (v_0 - D') \left\langle \frac{\partial \varphi}{\partial X} \right\rangle + D \left\langle \frac{\partial^2 \varphi}{\partial X^2} \right\rangle + D'' \left\langle J^{\mathrm{L}^2} \frac{\partial^2 \varphi}{\partial J^{\mathrm{L}^2}} \right\rangle.$$

Each average here is closed with respect to $W_{xJ}^{\mathrm{L}}(x, j; y, t)$ and is uniquely defined by this probability density. Representing the averages explicitly through $W_{xJ}^{\mathrm{L}}(x, j; y, t)$ one gets the equality

$$\int_{-\infty}^{\infty} \varphi(x, j) \frac{\partial W_{XJ}^{\mathrm{L}}}{\partial t}\, \mathrm{d}x\, \mathrm{d}j = (v_0 - D') \int_{-\infty}^{\infty} \frac{\partial \varphi(x, j)}{\partial x} W_{XJ}^{\mathrm{L}}\, \mathrm{d}x\, \mathrm{d}j$$

$$+ D \int_{-\infty}^{\infty} \frac{\partial^2 \varphi(x, j)}{\partial x^2} W_{XJ}^{\mathrm{L}}\, \mathrm{d}x\, \mathrm{d}j + D'' \int_{-\infty}^{\infty} \frac{\partial^2 \varphi(x, j)}{\partial j^2} j^2 W_{XJ}^{\mathrm{L}}\, \mathrm{d}x\, \mathrm{d}j.$$

Let us here get rid of the arbitrary function derivatives by transforming the integrals, integrating by parts and assuming that $\varphi(x, j) \to 0$ at $x, j \to \infty$. Uniting then all the integrals into a single one we obtain

$$\int_{-\infty}^{\infty} \varphi(x, j) \left\{ \frac{\partial W_{XJ}^{L}}{\partial t} + (v_0 - D') \frac{\partial W_{XJ}^{L}}{\partial x} - D \frac{\partial^2 W_{XJ}^{L}}{\partial x^2} - D'' \frac{\partial^2}{\partial j^2} (j^2 W_{XJ}^{L}) \right\} \mathrm{d}x \, \mathrm{d}j = 0.$$

Using the arbitrariness of $\varphi(x, j)$ we pass from an integral equality to the differential Fokker–Planck equation (FPE) with respect to $W_{xJ}^{L}(x, j; y, t)$,

$$\frac{\partial W_{XJ}^{L}}{\partial t} + (v_0 - D') \frac{\partial W_{XJ}^{L}}{\partial x} = D \frac{\partial^2 W_{XJ}^{L}}{\partial x^2} + D'' \frac{\partial^2}{\partial j^2} (j^2 W_{XJ}^{L}). \tag{6.11}$$

Recall that when deducing this equation we imposed the smoothness condition (6.8) on the arbitrary function $\varphi(x, j)$. Because of this, generally speaking, equation (6.11) results from the previous integral equality only if the characteristic variation scales of $W_{XJ}^{L}(x, j; y, t)$ with respect to x and j satisfy the same, if not more stringent, inequalities (6.8) than the characteristic scales of $\varphi(x, j)$. Nevertheless, when equation (6.11) is solved under the initial conditions resulting from (6.1), namely

$$W_{XJ}^{L}(x, j; y, 0) = \delta(x - y)\, \delta(j - 1),$$

we obtain an asymptotically correct expression for $W_{XJ}^{L}(x, j; y, t)$ at those times t when the characteristic scales of the solution with respect to x and j do satisfy inequalities of the type (6.8).

It seems noteworthy that the notion of the statistical dependence time τ_v introduced above was needed to rigorously split the averages of type (6.9), while in practice it is sufficient for them to be split that the statistical corrections between $\tilde{v}(x, t)$ and $\tilde{v}(x', t')$ quickly weaken when $|t - t'| > \tau_v$ and, in particular, that the correlation function of the velocity fluctuations $B_v(s, \tau)$ vanish quickly enough for $|\tau| > \tau_v$.

4. Let us embark upon a discussion of the statistical properties of the passive impurity concentration resulting from equation (6.11). The average concentration is given by a formula analogous to (3.3.12),

$$\langle \rho(x, t) \rangle = \int_{-\infty}^{\infty} \rho_0(y)\, W_{xJ}^{L}(x; y, t)\, \mathrm{d}y, \tag{6.12}$$

through the initial concentration $\rho_0(y)$ and the Lagrangian probability density $W^{L}(x, y, t)$ of a fixed particle. As follows from (6.11), (6.12), the latter satisfies the equation

$$\frac{\partial W_{X}^{L}}{\partial t} + (v_0 - D') \frac{\partial W_{X}^{L}}{\partial x} = D \frac{\partial^2 W_{X}^{L}}{\partial x^2}, \quad W_{x}^{L}(x; y, 0) = \delta(x - y).$$

The solution of this equation has the form of the Gaussian probability density

$$W_{X}^{L}(x; y, t) = \frac{1}{\sqrt{4\pi D t}} \exp\left\{ -\frac{[x - (v_0 - D')t]^2}{4Dt} \right\}.$$

It becomes evident, in particular, that the particle coordinate diffuses with the diffusion factor D, and drifts along the x-axis with a velocity $v_0 - D'$ that is slightly less than the average medium velocity v_0. The effect of the average particle velocity drop, as compared with the average velocity of the medium, is attributed to the particle diffusion asymmetry: the greater is the particle velocity the quicker it runs across the elementary inhomogeneity of the field, which leads to a shorter interaction time and to the particle diffusion retardation. It is this diffusion slow-down against the particle velocity rise that leads to the drop in the average particle drift velocity. A similar mechanism results in the drop of the diffusion factor D as v_0 increases. Let us illustrate it by an example of a Gaussian covariance function for the medium velocity,

$$B_v(s, \tau) = \sigma_v^2 \exp\left(-\frac{s^2}{2l_v^2} - \frac{\tau^2}{2\tau_v^2}\right).$$

The particle diffusion factor in this case is equal to

$$D = \int_0^\infty B_v(v_0\tau, \tau)\, d\tau = \sqrt{\frac{\pi}{2}}\, \frac{\sigma_v^2 \tau_v}{\sqrt{1 + (v_0\tau_0/l_v)^2}}.$$

The diffusion factor decrease with increase of v_0 can be interpreted as a result of an effective correlation time shortening for the field $\tilde{v}(x, t)$, to

$$\tau_v^{\text{eff}} = \frac{\tau_v}{\sqrt{1 + (v_0\, \tau_v/l_v)^2}}$$

i.e. the time for a particle travelling with velocity v_0 to traverse the elementary inhomogeneity of the field $\tilde{v}(x, t)$. It seems natural that, when passing to a medium with merely temporal velocity fluctuations, for which $l_v \to \infty$, this effect disappears and D' tends to zero.

5. The solution to equation (6.11) contains information on the probability distribution of the passive impurity concentration. Let us discuss its evolution in time for a particular case, of constant initial concentration ρ_0 = const. Making use of the relations between the Lagrangian and Eulerian statistics and of the expression $\rho = \rho_0/J^{\mathrm{E}}$ connecting the passive impurity concentration field with that of divergence, we come to the following form for the Eulerian density of a statistically homogeneous passive impurity concentration field,

$$W_\rho^{\mathrm{E}}(\rho; t) = \frac{\rho_0^2}{\rho^3}\, W_J^{\mathrm{L}}\left(\frac{\rho_0}{\rho}; t\right),$$

where $W_J^{\mathrm{L}}(j; t)$ is the Lagrangian probability density of the divergence, satisfying the equation resulting from (6.11),

$$\frac{\partial W_J^L}{\partial t} = D''\frac{\partial^2}{\partial j^2}(j^2 W_J^L), \quad W_J^L(j; 0) = \delta(j-1).$$

Although this equation lends itself well to a simple direct solution, we shall solve it utilising the method of stochastically equivalent equations. To this end we note that the probability density of the auxiliary random process $J(t)$ satisfying the stochastic differential equation

$$\frac{dJ}{dt} = D''J = \eta(t)J, \tag{6.15}$$

where $\eta(t)$ is the Gaussian delta-correlated process with correlation function

$$\langle \eta(t)\, \eta(t+\tau)\rangle = 2D''\delta(\tau),$$

satisfies exactly the same FPE, (6.14). The solution of the stochastic equation (6.15), meanwhile, has the form

$$J(t) = \exp\,[-D''t + \omega(t)],$$

where $\omega(t) = \int_0^t \eta(t')\,dt'$.

The probability density of process $J(t)$ is embodied through the Gaussian probability density

$$W_\omega(\omega; t) = \frac{1}{\sqrt{4\pi D''t}} \exp\left(-\frac{\omega^2}{4D''t}\right)$$

by an obvious equality,

$$W_J(j; t) = \int_{-\infty}^{\infty} \delta(j - e^{-D''t+\omega})\, W_\omega(\omega; t)\, d\omega = \frac{1}{2j\sqrt{\pi D''t}} \exp\left[-\frac{(\ln j + D''t)^2}{4D''t}\right].$$

Substituting it into (6.13) gives the required probability density of the passive impurity concentration,

$$W_\rho^E(\rho; t) = \rho_0 W(\rho/\rho_0; \tau), \tag{6.16}$$

where

$$W(\gamma; \tau) = \frac{1}{2\gamma\sqrt{\pi\tau}} \exp\left[-\frac{(\ln\gamma + \tau)^2}{4\tau}\right]$$

is the so-called log-normal probability density of the normalised concentration $\gamma = \rho/\rho_0$, depending on the dimensionless time $\tau = D''t$ characterising the developing degree of concentration fluctuations.

It is obvious from (6.16) that as τ evolves, the probability density maximum is shifted towards $\gamma < 1$, which corresponds to the appearance of increasingly wide zones of a medium with reduced concentration of passive impurity. Simultaneously, due to the concentration increase in the compression zones, the probability density acquires a tail slowly falling off for $\gamma > 1$ ($\langle\gamma\rangle = 1$ retained). Nevertheless, caustic singularities typical of hydrodynamic flows of noninteracting particles are not observed here, and hence all the passive impurity concentration moments remain bounded, although they grow with τ. At the same time, the shift of the most probable values towards the increasingly small values $\gamma > 1$ leads also to increase of inverse concentration moments. Let us clarify it by calculating $\langle\rho^2(x, t)\rangle_E$ and $\langle 1/\rho(x, t)\rangle_E$. Making use of the relations between the Lagrangian and Eulerian statistics, one can formulate these Eulerian averages through the Lagrangian averages of the divergences $J^L(y, t)$:

$$\langle\rho^2\rangle_E = \rho_0^2\,\langle 1/J^L\rangle, \quad \langle 1/\rho\rangle_E = \langle (J^L)^2\rangle/\rho_0 .$$

The averages on the right-hand sides of the equalities can be calculated by means of the familiar Lagrangian probability density $W_J^L(j; t)$.

It is easier, however, in this instance, to deduce and solve equations for them which directly follow from (6.14). So, multiplying (6.14) by j^2 and integrating with respect to j from 0 to ∞, we can write

$$\frac{d}{dt}\langle (J^L)^2\rangle = 2D''\langle (J^L)^2\rangle, \quad \langle (J^L)^2\rangle\Big|_{t=0} = 1.$$

Analogously

$$\frac{d}{dt}\langle 1/J^L\rangle = 2D''\langle 1/J^L\rangle, \quad \langle (J^L)^2\rangle\Big|_{t=0} = 1.$$

Hence

$$\left\langle\left(\frac{\rho}{\rho_0}\right)^2\right\rangle_E = \left\langle\frac{\rho_0}{\rho}\right\rangle_E = e^{2\tau}. \tag{6.17}$$

Lastly we write down the expression for the diffusion factor D'', the value of which governs the growth rate of the passive impurity concentration fluctuation. In the aforementioned particular case, of velocity fluctuations with Gaussian correlation function, it is equal to

$$D'' = \sqrt{\frac{\pi}{2}}\,\frac{\sigma_v^2\,\tau_v}{l_v^2\,[1 + (v_0\tau_v/l_v)^2]^{3/2}} .$$

Thus it is clear, for instance, that for concentration fluctuations to arise, the spatial–temporal nature of the velocity fluctuations is a matter of principle. Within the limits of purely temporal fluctuations, when $l_v \to \infty$, we have simply $D'' \to 0$.

4.7 Motion of noninteracting particles under the action of external forces

1. The problem of noninteracting particle flow behaviour in the field of a random force with prescribed Eulerian statistical properties is not less

important than that analysed in Section 4.6. To avoid cumbersome calculations we discuss, as earlier, the characteristic features of the particle motion caused by random forces, using an example of a one-dimensional flow such that its particles, along the x-axis, are acted upon by a force $f(x, t)$ depending on time and only one spatial coordinate x. Then the fixed particle coordinate and velocity (with masses assumed, for simplicity, to be equal to unity) satisfy a set of equations

$$\frac{\mathrm{d}X}{\mathrm{d}t} = V, \quad \frac{\mathrm{d}V}{\mathrm{d}t} = f(X, t), \quad X(y, 0) = y, \quad V(y, 0) = v_0(y), \tag{7.1}$$

where $v_0(x)$ is a given initial field of the flow particle velocity. Bearing in mind a further transition from Lagrangian statistics to Eulerian, we supplement (7.1) with the equations describing the flow divergence,

$$\begin{gathered}\frac{\mathrm{d}J^{\mathrm{L}}}{\mathrm{d}t} = \frac{\partial V(y, t)}{\partial y} = U, \quad \frac{\mathrm{d}V}{\mathrm{d}t} = J^{L}\frac{\partial f(X, t)}{\partial X}, \\ J^{\mathrm{L}}(y, 0) = 1, \quad U(y, 0) = v_0'(y).\end{gathered} \tag{7.2}$$

Additionally, we consider the random force $f(x, t)$ as stationary and statistically homogeneous, with zero average and given covariance function

$$B_f(s, \tau) = \langle f(x + s, t + \tau)\, f(x, t)\rangle.$$

2. As in Section 4.6, when meeting the two-scale conditions (the condition of dynamic causality is provided by specifying initial conditions in (7.1) and (7.2)) one can pass from the stochastic equations over to FPE. Let us derive in more detail an equation for $W^{\mathrm{L}}_{XV}(x, v; y, t)$ i.e. the joint Lagrangian probability density of coordinate $X(y, t)$ and velocity $V(y, t)$ of a fixed particle. For this purpose we differentiate a rather smooth arbitrary function $\varphi(X, V)$ with respect to t. Making use of equations (7.1) and averaging the resulting equality, we obtain

$$\frac{\mathrm{d}\langle\varphi\rangle}{\mathrm{d}t} = \left\langle V\frac{\partial\varphi}{\partial X}\right\rangle + \left\langle f(X, t)\,\frac{\partial\varphi}{\partial V}\right\rangle. \tag{7.3}$$

Transform the latter average, which is non-closed with respect to $W^{\mathrm{L}}_{XV}(x, v; y, t)$ Therefore, repeating the line of reasoning of Section 4.6, represent $X(y, t)$ and $V(y, t)$ as

$$\begin{gathered}X(y, t) = X_0(y, t) + z(t_0, t), \\ V(y, t) = V_0(y, t) + \omega(t_0, t), \\ X_0(y, t') = X_0(y, t) - V_0(y, t)\,(t - t'), \\ V_0(y, t') = V_0(y, t), \ (t_0 < t' < t).\end{gathered} \tag{7.4}$$

Here $z(t_0, t)$, $\omega(t_0, t)$ are the coordinate and velocity perturbations due to the action of random force $f(x, t)$ within the time interval (t_0, t) , satisfying equations similar to (6.4):

$$\frac{dz}{dt} = \omega, \quad \frac{d\omega}{dt} = f(X_0 + z, t), \quad \omega(t_0, t_0) = z(t_0, t_0) = 0.$$

Considering the two-scale condition as fulfilled, i.e. assuming that during time τ_f a random force causes only slight deviation of the particle: $z \ll l_f$, where τ_f and l_f are characteristic scales of $f(x, t)$, we restrict ourselves to the solution of these equations in the first non-vanishing approximation, in the method of successive approximations, namely

$$\omega(t_0, t') = \int_{t_0}^{t'} f(X_0(y, t) - V(y, t)(t - t''), t'')\, dt'',$$

$$z(t_0, t) = \int_{t_0}^{t} dt' \int_{t_0}^{t'} dt''\, f(X_0(y, t) - V_0(y, t)(t - t''), t'').$$

A rough estimate of z at $t - t_0 \sim \tau_f$ yields: $z \sim \sigma_f \tau_f^2$. Hence the two-scale condition in this case takes the form

$$\varepsilon = \frac{\sigma_f \tau_f^2}{l_f} \ll 1.$$

Substituting the equalities (2.4) into the last average in (7.3), and making assumptions similar to those set forth when deriving relations (6.9), one can write

$$\left\langle f\frac{\partial\varphi}{\partial V}\right\rangle = \left\langle f\frac{\partial\varphi}{\partial V_0}\right\rangle + \left\langle \frac{\partial f}{\partial X_0} z \frac{\partial\varphi}{\partial V_0}\right\rangle + \left\langle fz \frac{\partial^2\varphi}{\partial V_0\, \partial X_0}\right\rangle + \left\langle f\omega \frac{\partial^2\varphi}{\partial V_0^2}\right\rangle.$$

Splitting each of the averages on the right-hand side and neglecting the difference between X_0, V_0 and X, V we get

$$\left\langle f\frac{\partial\varphi}{\partial V}\right\rangle = \left\langle A(V)\frac{\partial^2\varphi}{\partial X\, \partial V}\right\rangle + \left\langle \frac{\partial}{\partial V}\left[D(V)\frac{\partial\varphi}{\partial V}\right]\right\rangle, \tag{7.5}$$

where

$$A(v) = \int_0^\infty \tau B_f(v\tau, \tau)\, d\tau, \quad D(v) = \int_0^\infty B_f(v\tau, \tau)\, d\tau. \tag{7.6}$$

Substituting (7.5) into (7.3) and using the arbitrariness of the function $\varphi(x, v)$ we go over to FPE with respect to the probability density $W_{XV}^{\mathrm{L}}(x, v; y, t)$:

$$\frac{\partial W_{XV}^{\mathrm{L}}}{\partial t} + v\frac{\partial W_{XV}^{\mathrm{L}}}{\partial x} = \frac{\partial}{\partial v}\left[A(v)\frac{\partial W_{XV}^{\mathrm{L}}}{\partial x}\right] + \frac{\partial}{\partial v}\left[D(v)\frac{\partial W_{XV}^{\mathrm{L}}}{\partial v}\right]. \tag{7.7}$$

Let us show the expressions for the coefficients included here, in particular when the force field potential

$$g(x, t) = \int^{x} f(x', t)\, \mathrm{d}x'$$

is statistically homogeneous, stationary and has Gaussian covariance

$$B_g(s, \tau) = \sigma_g^2 \exp\left(-\frac{s^2}{2l_f^2} - \frac{\tau^2}{2\tau_f^2}\right).$$

In this case the force correlation is equal to

$$B_f(s, \tau) = -\frac{\partial^2 B_g(s, \tau)}{\partial s^2} = \frac{\sigma_g^2}{l_f^2}\left(1 - \frac{s^2}{l_f^2}\right) \exp\left(-\frac{s^2}{2l_f^2} - \frac{\tau^2}{2\tau_f^2}\right).$$

Putting it into (7.6) one can find that

$$D(v) = \frac{D}{(1+\gamma^2)^{3/2}}, \quad A(v) = A\,\frac{1-\gamma^2}{(1+\gamma^2)^2}.$$

Here we use the notation $\gamma = v/v_f$, $v_f = l_f/\tau_f$,

$$A = \frac{\sigma_g^2}{v_f^2} = \varepsilon^2 v_f^2, \quad D = \sqrt{\frac{\pi}{2}}\,\frac{\sigma_g^2}{v_f^2 \tau_f} = \varepsilon^2 \sqrt{\frac{\pi}{2}}\,\frac{v_f^2}{\tau_f},$$

where

$$\varepsilon = \sigma_g / v_f^2 \ll 1$$

is a small parameter characterising the two-scale separation.

In this instance it seems expedient to change equation (7.7) into one with respect to $\tilde{w}(s, \gamma; s_0, \tau)$, i.e. the joint probability density of the dimensionless coordinate $s = \varepsilon^2 x/l_f (s_0 = \varepsilon^2{}_y/l_f)$ and velocity γ, that depends on the dimensionless time

$$\tau = \varepsilon^2 t/\tau_f \tag{7.8}$$

taken on a coarse scale, with a time unit appearing as τ_f/ε^2. This equation takes the form

$$\frac{\partial \tilde{w}}{\partial \tau} + \gamma \frac{\partial \tilde{w}}{\partial s} = \varepsilon^2 \frac{\partial}{\partial \gamma}\left[a(\gamma)\frac{\partial \tilde{w}}{\partial s}\right] + \frac{\partial}{\partial \gamma}\left[d(\gamma)\frac{\partial \tilde{w}}{\partial \gamma}\right]. \tag{7.9}$$

Here the dimensionless coefficients are derived as

$$a(\gamma) = \frac{1-\gamma^2}{(1+\gamma^2)^2}, \quad \mathrm{d}(\gamma) = \sqrt{\frac{\pi}{2}}\,\frac{1}{(1+\gamma^2)^{3/2}}. \tag{7.10}$$

3. Let us discuss now the main properties of equation (7.9) and of coefficients (7.10) entering into it. The velocity diffusion factor drop with γ is governed, as was noted in Section 4.6, by the fact that the particle velocity rise results in its quicker emergence from the elementary

inhomogeneity of the random force field. Hence we observe the shortening of the effective time of particle residence inside an elementary inhomogeneity, where an additional turbulent diffusion arises, and, consequently, the diminishing of the particle velocity diffusion factor, in proportion to the said time. For the same reason, the increase in γ is accompanied by a decrease of the coefficient $a(\gamma)$ which arose due to correlation between the random force value $f(x, t)$ and the particle trajectory perturbation z caused by the force. Moreover, when $\gamma > 1$, i.e. a particle covers distance $v\tau_f > l_f$ within time τ_f, the coefficient $a(\gamma)$ becomes negative. This is a consequence of the negative correlation, in this case, of the random force values at points with distance $s > l_f$ between them (i.e. owing to $B_f(s > l_f, \tau) < 0$).

Let us examine separately $\tilde{W}(\gamma; \tau)$, which is the probability density of the particle dimensionless velocity, which satisfies the equation resulting from (7.9),

$$\frac{\partial \tilde{w}}{\partial \tau} = \frac{\partial}{\partial \gamma}\left[d(\gamma)\,\frac{\partial \tilde{w}}{\partial \gamma}\right]. \tag{7.11}$$

It embodies the particle velocity diffusion as time evolves in a field of random forces. Should the initial velocity of a particle be zero, $\gamma_0 = 0$, and $\tau \ll 1$, implying that the particle velocity variance $\sigma_\gamma^2(\tau) \ll 1$, one can approximately move from (7.11) to an ordinary equation of diffusion,

$$\frac{\partial \tilde{w}}{\partial \tau} = d_0\frac{\partial^2 \tilde{w}}{\partial \gamma^2}, \tag{7.12}$$

where $d_0 = \sqrt{\pi/2}$. This equation implies in particular that the particle velocity variance, $\sigma_\gamma^2(\tau) = 2d_0\tau$, grows with time according to a linear law. With rise of the velocity γ, however, the diffusion factor $d(\gamma)$ diminishes, leading therefore to decrease of the velocity diffusion rate at large γ. Unfortunately, an exact analytical solution to (7.11) describing the consequences of this effect remains unknown as yet.

If, however, we only take interest in minor corrections to the abovementioned linear law of diffusion, it seems possible to replace equation (7.11) with a simpler (albeit more exact than (7.12)) one by means of expanding $d(\gamma)$ in Taylor series with respect to γ and only retaining the quadratic term,

$$\frac{\partial \tilde{w}}{\partial \tau} = d_0\frac{\partial}{\partial \gamma}\left[\left(1 - \frac{3}{2}\gamma^2\right)\frac{\partial \tilde{w}}{\partial \gamma}\right].$$

This equation implies, in particular, that $\langle\gamma^2\rangle$ satisfies the equation

$$\frac{\mathrm{d}\langle\gamma^2\rangle}{\mathrm{d}\tau} = 2d_0 - 9d_0\langle\gamma^2\rangle.$$

Solving this with a zero initial condition one can write

$$\langle\gamma^2\rangle = \sigma_\gamma^2(\tau) = \tfrac{2}{9}\,[1 - \exp(-9d_0\tau)]. \tag{7.13}$$

From this relation it follows, for instance, that as $\tau \to \infty$ the mean square of the velocity fluctuations is saturated at a level of 2/9. In fact of course, such saturation is not observed. There is reported only the growth of $\sigma_\gamma^2(\tau)$ more slowly than according to the linear law. Such slowing of rate is properly described by the solution (7.13) only if $\gamma \ll 1$, when one can expand the exponent in (7.13) in Taylor series and be restricted to the first two non-vanishing terms, giving

$$\langle\gamma^2\rangle = 2d_0\tau - 9d_0^2\tau^2.$$

4. Let us demonstrate, as a concluding remark, an equation for $W_{JU}^L(j, u; t)$ – the joint Lagrangian probability density of the divergence J^{L} and the auxiliary parameter U with respect to a fixed particle. As will be shown later the divergence fluctuations become important even when $\tau \ll 1$. Therefore, when deducing the FPE one can use here the approximation $\gamma = 0$. In this case the required equation acquires the form

$$\frac{\partial W_{JU}^{\mathrm{L}}}{\partial t} + u\frac{\partial W_{JU}^{\mathrm{L}}}{\partial j} = Cj^2\frac{\partial^2 W_{JU}^{\mathrm{L}}}{\partial u^2}, \quad C = -\int_0^\infty \frac{\partial^2 B_f(s,\tau)}{\partial s^2}\bigg|_{s=0} \mathrm{d}\tau = \sqrt[3]{\frac{\pi}{2}}\,\frac{\varepsilon^2}{\tau_f^2}.$$

Passing, in this equation, to a slow time (7.8), we obtain the equation for $\tilde{W}^L(j, \kappa; \tau)$ – the probability density of the dimensionless values j and $\kappa = u\tau_f/\varepsilon^2$:

$$\frac{\partial \tilde{W}^{\mathrm{L}}}{\partial \tau} + \kappa\frac{\partial \tilde{W}^{\mathrm{L}}}{\partial j} = \frac{3d_0}{\varepsilon^4}\,j^2\frac{\partial^2 \tilde{W}^{\mathrm{L}}}{\partial \kappa^2}. \tag{7.14}$$

The large coefficient $3d_0/\varepsilon^4$ multiplying the term on the right-hand side indicates definitely that the divergence fluctuations become large even when $\tau \ll 1$.

Provided the initial concentration of the hydrodynamic flow particles is constant and equal to ρ_0 and the initial velocities of all the particles are zero, equation (7.14) should be subject to the initial condition

$$\tilde{W}^{\mathrm{L}}(j, \kappa; 0) = \delta(j-1)\,\delta(\kappa), \tag{7.15}$$

while the Eulerian probability distribution of the flow particle concentration is connected with the divergence Lagrangian probability density

$$W_J^{\mathrm{E}}(j; t) = j\int_{-\infty}^{\infty} \tilde{W}^{\mathrm{L}}(j, \kappa; t)\,\mathrm{d}\kappa$$

through a familiar equality (6.13). Unfortunately, an exact analytical solution to (7.14) is still lacking. It is easy, however, to pass to closed equations of the divergence moments. The latter are related to the Eulerian inverse moments of a beam particle concentration through the equality

$$\langle (J^{\mathrm{L}})^n \rangle = \langle (J^{\mathrm{E}})^{(n-1)} \rangle = \rho_0^{(n-1)} \langle \rho^{(1-n)} \rangle_{\mathrm{E}}.$$

Calculate, for example, the first inverse moment of the particle concentration,

$$\rho_0 \left\langle \frac{1}{\rho} \right\rangle_E = \langle (J^{\mathrm{L}})^2 \rangle.$$

From equation (7.14) and the initial conditions (7.15) it follows that $\langle (J^{\mathrm{L}})^2 \rangle$ satisfies the equation

$$\frac{\mathrm{d}^3}{\mathrm{d}\theta^3} \langle (J^{\mathrm{L}})^2 \rangle = \langle (J^{\mathrm{L}})^2 \rangle,$$

with the initial conditions

$$\langle (J^L) \rangle \bigg|_{\theta=0} = 1, \quad \frac{\mathrm{d}}{\mathrm{d}\theta} \langle (J^{\mathrm{L}})^2 \rangle \bigg|_{\theta=0} = \frac{\mathrm{d}^2}{\mathrm{d}\theta^2} \langle (J^{\mathrm{L}})^2 \rangle \bigg|_{\theta=0} = 0.$$

Here we introduce the dimensionless time

$$\theta = \sqrt[3]{\frac{12 d_0}{\varepsilon^4}}\, \tau = \sqrt[3]{12 d_0 \varepsilon^2}\, \frac{t}{\tau_f}.$$

Solving this equation we have

$$\rho_0 \left\langle \frac{1}{\rho} \right\rangle_{\mathrm{E}} = \frac{1}{3} \mathrm{e}^{\theta} + \frac{2}{3} \mathrm{e}^{-\theta/2} \cos \frac{\sqrt{3}}{2} \theta.$$

Here, as in the case of (6.17), the growth of the first inverse moment ρ with τ is due to the development of increasingly wide regions wherein the particle concentration is low, as compared with the initial density ρ_0.

5 Statistical properties of discontinuous waves

Statistical properties of random waves satisfying the Burgers equation are studied in this chapter. It is shown that at large Reynolds numbers such waves have three characteristic stages of evolution. At the first stage the wave profile is distorted, but discontinuities appear very seldom and only lead to change of the spectrum short-wave asymptotic behaviour and to weak energy damping. At the second stage the field is represented by a sequence of sawtooth pulses, with random velocities and amplitudes of discontinuities. It is this stage, characterised by multiple discontinuity coalescence, that we generally call Burgers turbulence (BT). and finally, a wave may enter the linear damping regime, when nonlinear effects become insignificant.

Here we develop the qualitative theory of BT, which allows us to evaluate the external scale growth rate and the energy damping laws, depending on the initial spectrum type. It is rigorously demonstrated that under certain conditions the statistical properties of the waves in a non-dissipative nonlinear medium tend to self-preserving ones. An asymptotic theory of BT is inferred which incorporates the two-point probability distributions, correlation functions and energy spectra of a field. The theory developed is extended to nondispersive media with arbitrary nonlinearity.

5.1 Discontinuity influence on the nonlinear wave statistics on the initial stage

Later on in this section we shall discuss discontinuity effects on the statistical behaviour of nonlinear waves in a nondispersive medium at the initial stage when discontinuities are comparatively rare and their influence can be accounted for by the perturbation method.

1. Consider the behaviour of the nonlinear random waves $v(x, t)$ satisfying the BE

$$\frac{\partial v}{\partial t} + v\frac{\partial v}{\partial x} = \mu\frac{\partial^2 v}{\partial x^2}, \qquad v(x, 0) = v_0(x). \tag{1.1}$$

Let the initial field $v_0(x)$ be statistically homogeneous and have just one characteristic amplitude σ_0 and spatial scale l_0. Then, as was already discussed in Section 2.1, the times of nonlinearity t_n and of dissipation t_d and the initial Reynolds number, can be estimated as

$$t_n = l_0/\sigma_0, \quad t_d = l_0^2/\mu, \quad R_0 = \sigma_0 l_0/\mu.$$

Since the amplitude and the spatial scale of a wave vary with time, $\sigma = \sigma(t)$, $l = l(t)$, the Reynolds number describing the relative contribution of the nonlinear and dissipative effects to the wave evolution becomes time-dependent as well,

$$R(t) = \sigma(t)\ l(t)/\mu. \tag{1.2}$$

At large Reynolds numbers the effects of multiple harmonic interaction grow in importance and lead to an 'avalanche'-type increase of the harmonic number. Therefore, the utilisation of the spectral approach in this case is difficult, giving way to a more adequate technique based upon the BE solution analysis in the coordinate representation. The evolution of the field $v(x, t)$ as $\mu \to 0$, at a stage prior to discontinuity formation, can be described within the framework of RE (2.1.2), which has the following implicit solution

$$v(x, t) = v_0(x - v(x, t)t). \tag{1.3}$$

Statistical properties of Riemann waves were examined in Sections 4.1 and 4.2. However, the inevitable appearance of non-single-valued parts of the Riemann wave profile restricts the applicability of the implicit solution, when analysing nonlinear random wave statistics. It might seem that as long as the non-single-valued parts of the profile are comparatively infrequent, their effect on the statistical properties of a wave could be neglected. Nevertheless, discontinuities emerging on these portions lead to conceptually new qualitative effects, such as wave energy damping, initiation of specific short-wave asymptotic behaviour of the spectrum, and variation of the wave probability distribution.

2. First of all we estimate the average discontinuity number per unit length, $n(t)$. Discontinuity formation in a continuous wave is associated with multivaluedness of the Riemann solution profile which, in its turn, appears when the Lagrangian divergence field, i.e. Jacobian of (4.1.4), $J = 1 + v_0'(y)t$, changes sign. Therefore one can define $n(t)$ as an average number of drops of the function $v_0'(y)$ below the level $(-1/t)$, per unit length, $\lambda(-1/t)$ say. Under Gaussian statistics for $v_0(y)$, one obtains in analogy with (3.1.9)

$$\lambda(-1/t) = -\int_{-\infty}^{0} v_0'' W_0(-1/t, v_0'')\, dv_0'' = k_2 \exp(-1/2\tau^2)/2\pi k_1. \tag{1.4}$$

Here $W_0(v_0', v_0'')$ is the joint probability density of the first two derivatives of the initial field $v_0(x)$, $\tau = k_1\sigma_0 t = t/t_n$, and k_n is given by (4.2.8). With respect to the initial field characterised by a single scale $k_1 \sim 1/l_0$, $k_2 \sim 1/l_0^2$.

In the general case $n(t)$ and $\lambda(-1/t)$ are related through the inequality

$$n(t) \leq \lambda(-1/t), \tag{1.5}$$

inasmuch as two or more adjacent drops of $v_0'(y)$ below the level may be corresponded to by a single discontinuity. At the initial stage, however, when the discontinuity coalescence can be neglected the evaluation $n(t) = \lambda(-1/t)$ proves rather accurate. Equation (1.4) implies that the number of discontinuities builds up with the field initial amplitude σ_0 and its spectrum width. At the initial stage $t \ll t_n$, the discontinuity effects can be allowed for, using a method of perturbation with respect to the small parameter

$$n(t)l_0 \sim \exp(-1/2\tau^2)$$

equal to an average number of discontinuities on the characteristic spatial scale of the initial field.

3. Let us discuss the influence of discontinuities on the single-point probability density and on wave energy damping. In Section 4.1 it was shown that the probability density of the statistically homogeneous Riemann wave is conserved. The obvious explanation of this follows from (4.1.9), which describes the probability density through the limit of the relative residence length of the realisation $v(x, t)$ within the range $[v, v + \Delta v]$. The conservation of the probability density, as interpreted here, follows from the equality (4.1.11), valid prior to discontinuity formation in a Riemann wave. Following the discontinuity appearance one should omit, in (4.1.9), the terms corresponding to the intervals Δx_m cut off by the discontinuity (Fig. 5.1) and get

$$W_v^E(v; t) = \lim_{\substack{L\to\infty \\ \Delta v \to 0}} \frac{1}{L}\left[\sum_k \frac{\Delta x_k}{\Delta v} - \sum_m \frac{\Delta x_m}{\Delta v}\right] = W_0(v) + P(v; t). \tag{1.6}$$

The second sum here and the corresponding term $P(v; t)$ take into account the probability density variation due to discontinuity formation. Obviously

$$P(v; t) = \langle p(v; t) \rangle, \tag{1.7}$$

where

$$p(v; t) = \begin{cases} \dfrac{\partial x(x)}{\partial v}, & v \in [v_1, v_2], \\ 0, & v \notin [v_1, v_2], \end{cases} \tag{1.8}$$

and $x(v)$ is a branch of the Riemann solution $v(x, t)$ inverse function in the neighbourhood of which the discontinuity $[v_1, v_2]$ emerges (see Fig. 5.2),

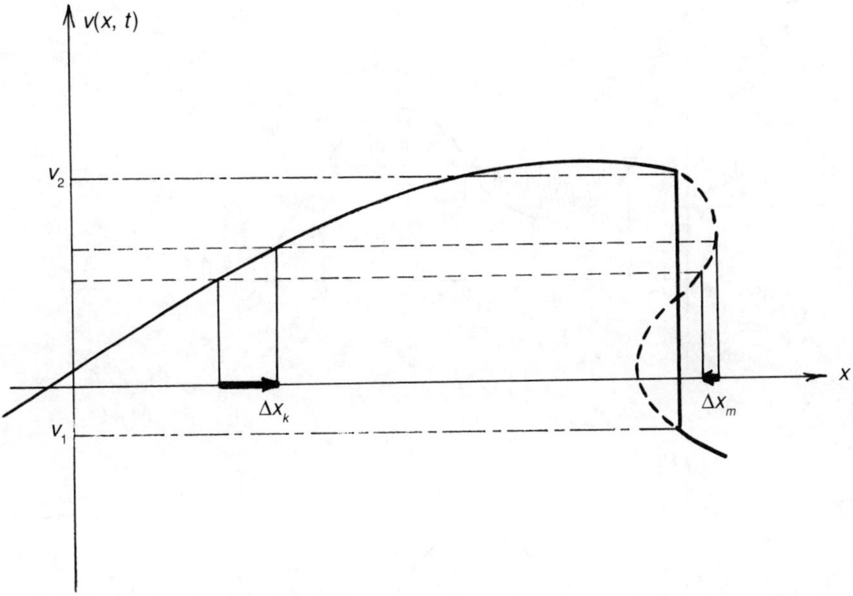

Fig. 5.1 Plots of the Riemann wave realisation and the corresponding discontinuous wave. Δx_m is the interval cut off by a discontinuity

$$x(v) = x_0(v) + vt. \tag{1.9}$$

Here $x_0(v)$ is the function inverse to $v_0(x)$.

The averaging in (1.7) is carried out over the number of discontinuities formed in a wave per unit length and over the Riemann solution form in the vicinity of the discontinuity. Generally speaking, in order to estimate this average one has to know the infinite-dimensional probability densities of the initial field $v_0(x)$. At the initial stage, however, when curve (1.8) has the shape of a cubic curve [83] in the vicinity of a discontinuity, it suffices only to know the finite-dimensional probability densities. Actually, at this stage the neighbourhood of function (1.9) corresponding to the discontinuity can be approximated with a high degree of accuracy by the expression

$$\begin{gathered} x = x_* + (x_0'(v_*) + t)\,(v - v_*) + \\ x_0^{(2)}(v_*)\,(v - v_*)^2/2! + x_0^{(3)}(v_*)\,(v - v_*)^3/3!, \\ x_* = x_0(v_*) + v_* t. \end{gathered} \tag{1.10}$$

The discontinuity location is determined from the condition that the areas cut off on either side by a discontinuity should be equal (see Fig. 5.2), and to the approximation (1.10) is in conformity with the point where $x_0^{(2)}(v_*) = 0$. If one writes the coefficients in (1.10) in terms of the initial field derivatives, $u = v_0'(x)$, $z = v_0^{(3)}(x)$ at the point of discontinuity formation (where $\eta = v^{(2)}(x) = 0$, $1 + ut < 0$, and $z > 0$) the following relation will result:

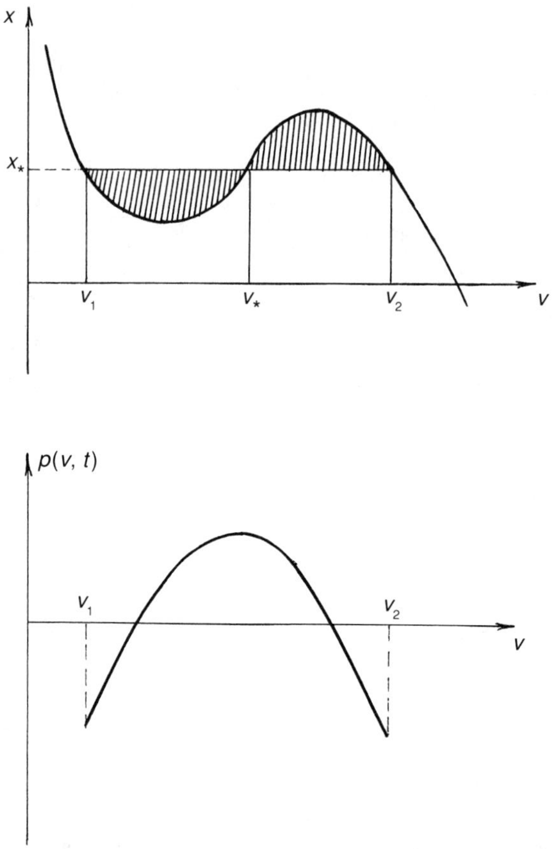

Fig. 5.2 Function inverse to the Riemann solution in the vicinity of a discontinuity, and the function describing the influence of an individual discontinuity on the wave probability density

$$x - x_* = (t + 1/u)(v - v_*) - z(v - v_*)^3/6u^2. \tag{1.11}$$

Consequently, the function (1.8) takes the form (see Fig. 5.2)

$$P(v; t) = \begin{cases} 1 + (t/u) - z(v - v_*)^2/2u^2, \ |v - v_*| < \Delta v/2, \\ 0, \ |v - v_*| \geq \Delta v/2, \end{cases} \tag{1.12}$$

where

$$\Delta v = 2\sqrt{6u^3(1 + ut)/z} \tag{1.13}$$

is the discontinuity amplitude. It is evident from (1.12) that an individual discontinuity leads to an increase of the probability density near the discontinuity centre (where the cut-off interval $\Delta x_m < 0$) and to its decrease at the edges (where $\Delta x_m > 0$).

In order to find the average of (1.17), one has to average (1.12) over the discontinuity number per unit length and over the parameters of curve $x(v)$ in the neighbourhood of the discontinuities, for which $1 + ut < 0$, $\eta = 0$, $z > 0$. Assuming the four-dimensional probability density $W_0(v, u, \eta, z)$ of the statistically homogeneous field $v_0(x)$ and its derivatives to be given, we obtain

$$P(v; t) = \int_{-\infty}^{\infty} dv_* \int_{-\infty}^{1/t} du \int_0^{\infty} dz\, P(v, t; v_*, u, z)\, z W_0(v_*, u, 0, z), \qquad (1.14)$$

where $P(\ldots)$ is determined by (1.12). From (1.12) and (1.14) it follows, in particular, that $\int_{-\infty}^{\infty} P(v; t)\, dv = 1$, i.e. that the probability density (1.6) satisfies the normalisation condition. It is also evident from (1.12) and (1.14) that the appearance of discontinuities in a wave leads to probability density growth at those values of v which are the most probable for discontinuity formation. Namely, for a wave with Gaussian initial statistics, $W_v^E(v; t)$ increases near $v = 0$ and diminishes at large values of $|v|$.

Due to discontinuity generation and the consequent energy absorption there, the average wave energy $\sigma^2(t) = \langle v^2(x, t)\rangle$ is attenuated. Utilising (1.12) and (1.14) with respect to $\Delta\sigma^2(t) = \sigma_0^2 - \sigma^2(t)$, one comes to

$$\Delta\sigma^2(t) = (16\sqrt{6}/5) \int_{-\infty}^{-1/t} du \int_0^{\infty} dz\, W_0(u, 0, z)\, (1 + ut)^{\frac{5}{2}}\, u^{\frac{7}{2}}/\sqrt{z}. \qquad (1.15)$$

Here $W_0(u, \eta, z)$ is the probability density of u, η and z. It can be noted that the average energy attenuation (1.15) does not depend directly on the probability properties of $v_0(x)$, since at the initial stage the energy absorption at each discontinuity does not depend on $v_0(x)$ but only on its derivatives.

For fields with Gaussian initial statistics the energy damping is determined by the first three moments k_n of the initial spectrum (see 3.2.8),

$$\begin{gathered}\langle u^2\rangle = k_1^2\sigma_0^2,\ \langle \eta^2\rangle = k_2^2\sigma_0^2,\ \langle z^2\rangle = k_3^2\sigma_0^2,\\ r_{v\eta} = -k_1^2/k_2,\ r_{uz} = -k_2^2/k_1k_3,\end{gathered} \qquad (1.16)$$

where $\sigma_0^2 = \langle v_0^2(x)\rangle - \langle v_0(x)\rangle^2$. With sufficiently rapidly decreasing spectra $G_0(k)$, for which $k_3 < \infty$, the energy absorption does not depend on the correlation coefficient value r_{uz} and is defined as follows [84]:

$$\Delta\sigma^2(t) = \sqrt{54/\pi}\, \sigma_0^2 r_0^2 \tau^3 \exp(-1/2\tau^2), \qquad (1.17)$$

where, as previously, $\tau = \sigma_0 k_1 t = t/t_n$, $r_0 = r_{v\eta}$. As could have been anticipated, the wave energy decrease associated with shock front formation is proportional to their average number per unit time $n(t) = \exp(-1/2\tau^2)$. It is worthwhile to remark also that, given the same initial energy, a wave with a wider initial spectrum decays more rapidly, since it is characterised by a faster discontinuity formation.

Now compare, as an example, the fading of wide-band noise with a covariance function $B_0(s) = \sigma_0^2 \exp(-s^2/s_0^2)$ and that of a quasi-harmonic wave with $B_0(s) = \sigma_0^2 \exp(-s^2/s_1^2) \cos k_* s$ ($k_* s \gg 1$). For the former case $k_1^2 = 2/s_0^2$, $r_0^2 = 1/3$, while for the latter $k_1^2 = k_*^2 + 2/s_1^2$, $r_0^2 \approx 1$. Given equal velocity gradient variance $|u|^2 = \sigma_0^2 k_1^2$, the attenuation of a quasi-harmonic wave is more pronounced, which is attributed to the large discontinuity values, on the average.

4. The strongest impact of a discontinuity is that with respect to the short-wave asymptotic behaviour of the spectrum, determined exactly by the field behaviour within the small-scale range. For the short-wave asymptotic behaviour, one can neglect interference between the spectra of individual chaotically distributed discontinuities. In analogy with (3.2.6) we have for the spectral density asymptotic behaviour

$$G_v(k;\, t) = n(t)\, \langle \Delta v^2(t) \rangle / 2\pi k^2, \tag{1.18}$$

where $\langle \Delta v^2(t) \rangle$ is the mean square of the discontinuity amplitudes at time t. Note that as $\nu \to 0$ equation (1.18) is of a universal nature and does not only hold at the initial stage. The power law $G_v \sim k^{-2}$ as applied to acoustic turbulence was qualitatively discussed in [85]. In practical applications it is useful to know how the short-wave asymptotic intensity $\langle \Delta v^2(t) \rangle$ varies with time. Utilising the expression (1.13) for the discontinuity amplitude and the assumptions for (1.17), one obtains, for the energy spectrum short-wave asymptotic behaviour,

$$G_v(k;\, t) = 6\sigma_0^2\, r_0\, k_1 \exp(-1/2\tau^2)/\pi k^2. \tag{1.19}$$

Comparing (1.19) with the spectral density asymptotic behaviour of a Riemann wave (4.2.9),

$$G_v(k;\, t) = \sigma_0^2\, k_1^2 \exp(-1/2\tau^2)/\sqrt{2\pi}\; k^3 \tau^3, \tag{1.20}$$

we are able to characterise the nonlinear wave spectrum evolution as follows. As $k \to \infty$ the spectrum asymptotic behaviour is associated with the appearance of discontinuities in the wave, and $G_v \sim k^{-2}$. For $k < k_*(t) \approx k_1/r_0 \tau^3$ the spectrum asymptotics are governed by the Riemann wave behaviour in the vicinity of discontinuities where $v \sim \sqrt{x - x_*}$, and hence, in accordance with (1.20), $G_v \sim k^{-3}$ (see Fig. 5.3). The power spectrum $G_v \sim k^{-3}$ for a Riemann wave was formally obtained in [86]. Due to the increase of the characteristic amplitude of a discontinuity, the transition region of the power asymptotics of the spectrum, around $k_*(\tau) \sim 1/l_0 \tau^3$, travels towards the large scales, i.e. towards lower spatial frequencies. Therefore, in the course of the wave evolution, discontinuities become more and more important in the determination of the field energy spectrum.

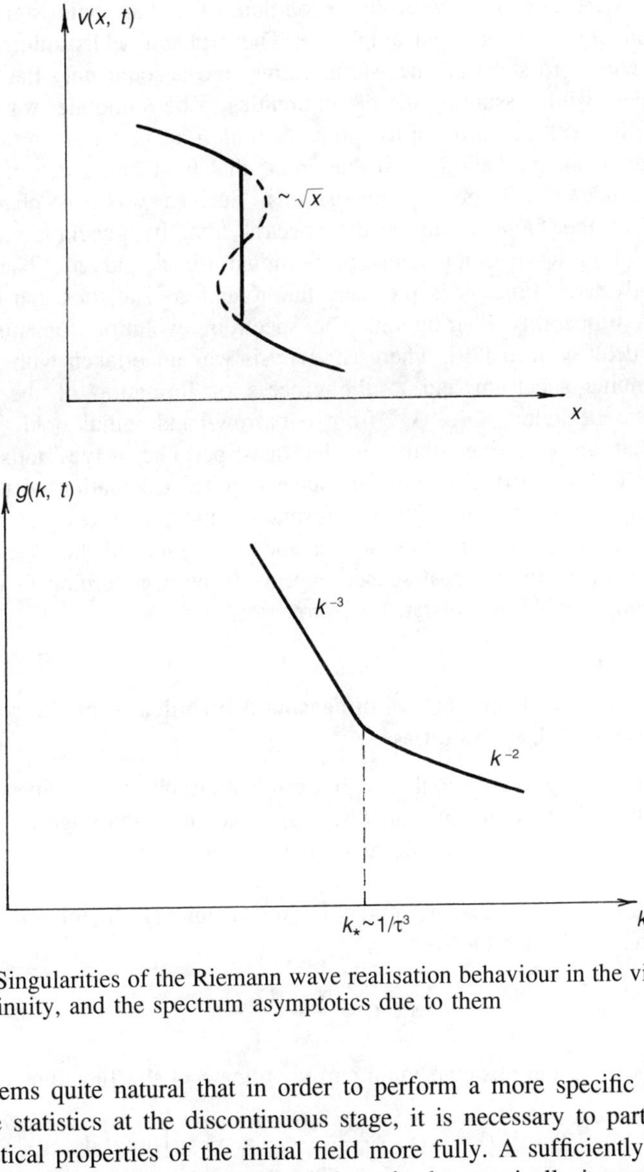

Fig. 5.3 Singularities of the Riemann wave realisation behaviour in the vicinity of a discontinuity, and the spectrum asymptotics due to them

5. It seems quite natural that in order to perform a more specific study of the wave statistics at the discontinuous stage, it is necessary to particularise the statistical properties of the initial field more fully. A sufficiently exhaustive analysis may be successfully carried out in the practically important case of a quasi-harmonic signal

$$v_0(x) = a(x) \sin (k_0 x + \varphi(x)) \tag{1.21}$$

with fluctuating amplitude $a(x)$ and phase $\varphi(x)$. Using the slowness of their change, one can initially assume that at each instant the evolution of a random wave coincides with the evolution of a harmonic signal of the corresponding frequency and amplitude. The probability distribution of such a wave with zero spectral line width, taking into account only the amplitude randomness while assuming the discontinuities to be immobile, was proposed in [87]. In [88] the probability properties of a quasi-harmonic field were investigated, the probability distribution of this field being determined both by amplitude and frequency fluctuations. Such a wave is characterised, because of the finite width of the spectral line, by generation of slowly decaying long-wave components proportional to φ' and φ''. Therefore, at sufficiently large times it is precisely the φ' and φ'' statistics that determine the field probability distribution. The spectrum evolution for this type of wave is dealt with in [89], where an analysis was undertaken with respect to the harmonic spectrum and to the process of formation of the universal asymptotic behaviour $G_v \sim k^{-2}$ from a narrow-band initial field. It can be stated that an effective study of the quasi-periodic wave statistics at a sufficiently developed discontinuous stage requires calculation of the proper time-span between the discontinuity formation and coalescence events. Later on in this chapter we shall present a detailed discussion of the case when the process of discontinuity coalescence appears to be a governing factor in the determination of the wave statistical properties.

5.2 Qualitative theory of one-dimensional turbulence, at the stage of developed discontinuities

1. Before going on to a thorough examination of the nonlinear random wave statistical properties in a nondispersive medium at the stage of developed discontinuities, a qualitative discussion of the wave evolution at this stage seems useful.

Consider first the case of infinite Reynolds number. In this situation the BE solution (1.1) has the form

$$v(x, t) = \frac{x - y(x, t)}{t}, \tag{2.1}$$

where $y(x, t)$ is the absolute minimum coordinate of the function

$$\Phi(x, y, t) = s_0(y) + \frac{(x - y)^2}{zt}, \quad s_0(y) = \int^y v_0(z)\, dz. \tag{2.2}$$

To perform a qualitative analysis of the field (2.1), it proves extremely handy to employ the graphical construction procedure described in Section 2.2, according to which $y(x, t)$ coincides with the coordinate of a point where the parabola $\alpha = H - (x - y)^2/2t$, pressed up to the initial action graph $s_0(y)$ from below, touches this graph (see Fig. 2.1). Relying on this procedure, let us single out the main stages of the BT development with time.

If the curvature of the parabola α, equal to $1/t$ at all values of y, exceeds the initial action curvature, equal to $s_0''(y) = -v_0'(y)$, i.e. if $1 + v_0'(y)\,t > 0$, then, with change of x, the parabola with slide along the curve $s_0(y)$ in such way that the coordinate of the contact point $y(x, t)$ appears to be a continuous function of x. In this instance the field $v(x, t)$ is continuous also, and is described by a Riemann solution (2.2.8). It is quite reasonable to take $\langle (v_0'(y))^2 \rangle^{\frac{1}{2}} \approx \sigma_0/l_0$ as an estimate of the initial action curvature. Then the initial stage mentioned takes place for $t < t_n \approx l_0/\sigma_0$.

For $t > t_n$ the parabola α sliding, with growth of x, along the curve $s_0(y)$, at some values x_k touches the initial action simultaneously at *two* points with *different* coordinates y_k^+, y_{k+1}^- (see Fig. 5.4). This implies that $y(x, t)$ is broken at points x_k, changing from y_k^+ to y_{k+1}^- in a jump. Accordingly, discontinuities are formed in the field $v(x, t)$ (2.1) as well. The coordinates x_k and velocities V_k of the discontinuities are given in (2.2.11), (2.2.12),

$$x_k = \frac{y_{k+1}^- + y_k^+}{2} + V_k t, \tag{2.3}$$

where V_k is the discontinuity propagation speed, equal to

$$V_k = \frac{s_0(y_{k+1}^-) - s_0(y_k^+)}{\eta_k}, \tag{2.4}$$

$$\eta_k = y_{k+1}^- - y_k^+.$$

Given $t \gg t_n$, the peak of parabola α changes smoothly within the scale of the initial action $s_0(y)$. Because of this, the coordinates of the tangency points of α and $s_0(y)$ (global minima $\Phi(x, y, t)$) are close to the coordinates of some local minima of the initial action $s_0(y)$, i.e. to some of the coordinates of the initial field zeros y_k: $v_0(y_k) = 0$, $v_0'(y_k) > 0$ (see Fig. 5.4).

Denote by $y_k(x, t)$ the values of function $y(x, t)$ in the interval between discontinuities (x_{k-1}, x_k) (Fig. 5.5). It is apparent that $y_k^- < y_k(x, t) < y_k^+$. In order to evaluate the deviation of $y_k(x, t)$ from the coordinate of the corresponding local minimum y_k of the initial action $s_0(y)$ we solve equation (2.2.1) by means of the perturbation method. The result can be written down as follows:

$$\Delta y_k = y_k(x, t) - y_k \approx |x - y_k|/t\, v_0'(y_k) \sim |x - y_k|\, t_n/t. \tag{2.5}$$

The field (2.1) can, in terms of new notation, be rewritten as

$$v(x, t) = \frac{x - y_k(x, t)}{t} = \frac{x - y_k + \Delta y_k}{t}. \tag{2.6}$$

From (2.5) one can see that when $t \gg t_n$ the field $v(x, t)$ between the discontinuities appears to be an almost linear function of x, inasmuch as $y_k(x, t)$ remains essentially unchanged when x grows from x_{k-1} up to the next discontinuity coordinate x_k. When passing through x_k, coordinate $y(x, t)$ changes

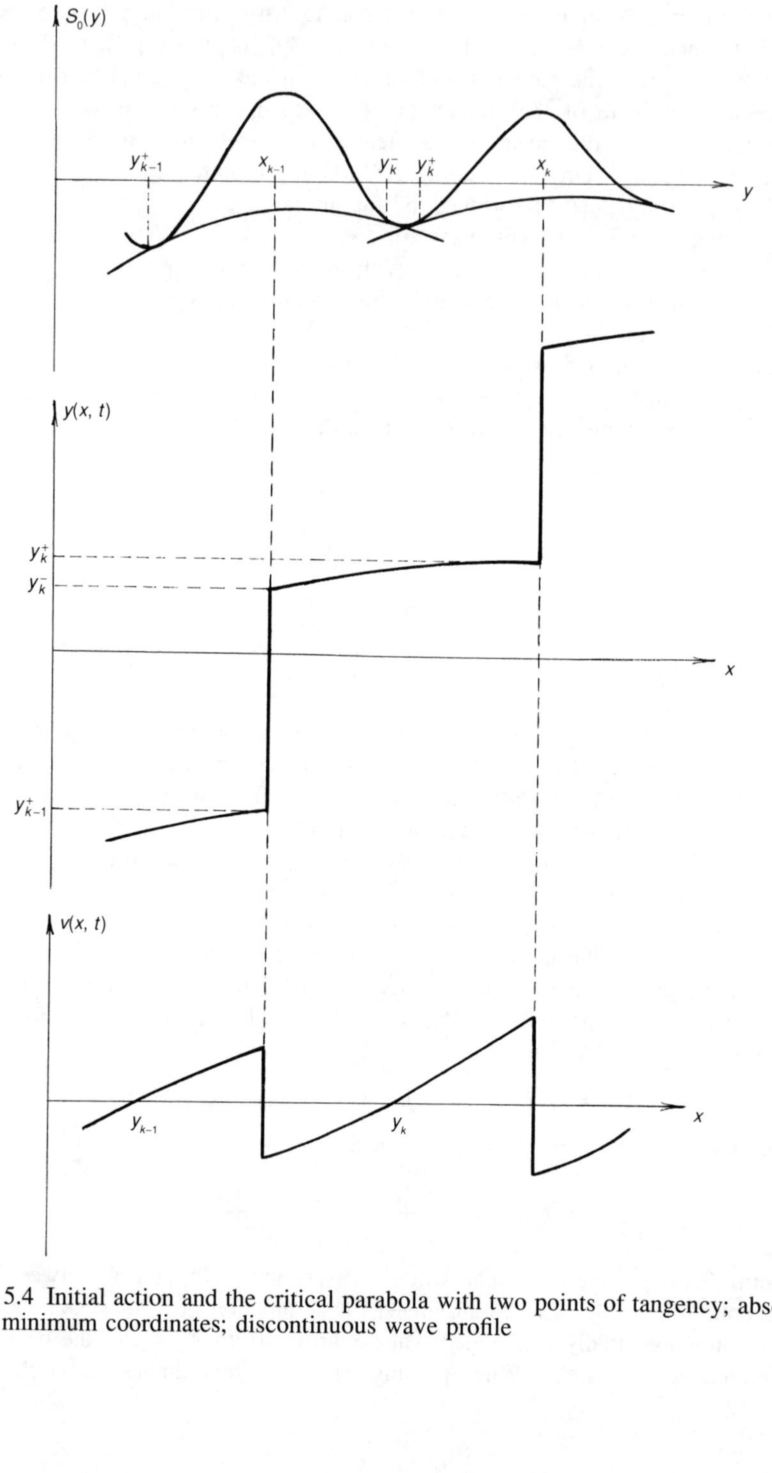

Fig. 5.4 Initial action and the critical parabola with two points of tangency; absolute minimum coordinates; discontinuous wave profile

abruptly from y_k to y_{k+1} which, in turn, remains, as a matter of fact, constant till the next discontinuity point x_{k+1}. Therefore, the structure of the field $v(x, t)$ at the stage $t \gg t_n$ is quite exhaustively described by a set of critical parabolas having two points of contact with the initial action $s_0(y)$. The co-ordinates of these parabola intersections virtually match the local minima of $s_0(y)$ and define 'zeros' y_k of the field $v(x, t)$, while the critical parabola centre coordinates characterise the positions of the discontinuities (see Fig. 5.5). Within the intervals between the discontinuities the field has a universal structure [7, 91]

$$v(x, t) = \frac{x - y_k}{t}, \quad x_{k-1} < x < x_k, \tag{2.7}$$

where y_k is a field 'zero', i.e. the intersection of the axis $v = 0$ by an inclined straight line or its continuation. Since, when $t \gg t_n$ the contact point coordinates are almost constant, the field $v(x, t)$ can be viewed as a sequence of triangular pulses all with the same slope $1/t$, together with discontinuities at $x = x_k$. The wave evolution at this stage, called the stage of developed discontinuities, may be portrayed by the following simple laws;

- the field gradient v_x between the discontinuities decreases monotonically, as $1/t$;
- 'zeros' of a sawtooth wave are immobile;
- discontinuities x_k of (2.3) travel with constant velocity V_k defined by (2.4), where $y_k^+ = y_k^- = y_k = \text{const}$;
- the discontinuity amplitudes are proportional to the distance between two adjacent zeros: $\Delta v_k = (y_{k+1} - y_k)/t$;
- at the shock front collision time, the number of 'zeros' and discontinuities is diminished by one, while two merged discontinuities make up a single one with an amplitude equal to the sum of the coalesced discontinuity amplitudes (for example, in the case of merging for the discontinuities with coordinates x_k and x_{k-1}, then the amplitude of the newly-formed discontinuity $\Delta\tilde{v}_k$ will be equal to

$$\Delta\tilde{v}_k = (y_{k+1} - y_{k-1})/t = (y_{k+1} - y_k + y_k - y_{k-1})/t = \Delta v_k + \Delta v_{k-1}.$$

2. For a wide range of initial conditions the field $v(x, t)$ at the stage of developed discontinuities may be characterised by a single external scale $l(t)$, which is equal to a characteristic distance between 'zeros' or discontinuities. With a random initial field $v_0(x)$ the discontinuity propagation velocities are, generally speaking, random as well. Due to the discontinuity-velocity discrepancies, their coalescence occurs, resulting in an increase of the external scale $l(t)$ with time. The growth of the external scale of field $v(x, t)$ can be observed directly from the graphical plotting: as t goes up, the parabola α becomes flatter, and the number of local minima $s_0(y)$ which can claim the right to be the absolute minimum of $\Phi(x, y, t)$ goes down.

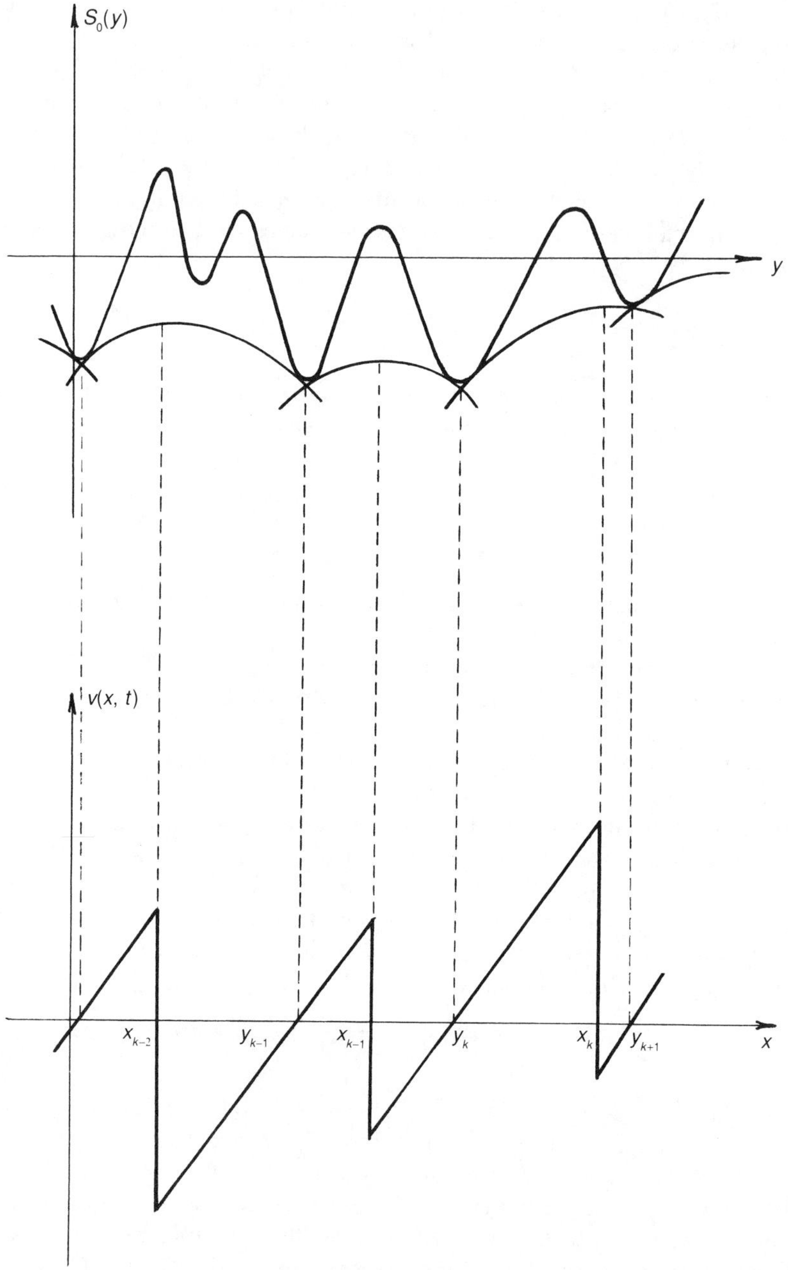

Fig. 5.5 A set of critical parabolas defining the discontinuity coordinates x_k and 'zeros' y_k of the field $v(x, t)$ at a stage of developed discontinuities

Let us present another obvious interpretation of the evolution of field $v(x, t)$, in terms of gas particles with absolutely inelastic collisions. Assume that at $t = 0$ the particle gas had initial velocities $v_0(x)$ and a uniform initial density $\rho_0 = 1$. In the course of time, however, the gas density concentrations and rarefactions $\rho(x, t) = 1/j(x, t)$ (3.3.4) arise, where $j(x, t)$ is the Jacobian of the Euler-to-Lagrangian coordinate transformation. As long as $j > 0$ the density is finite everywhere. At a time when j tends to zero, the light particles merge to form clusters, i.e. heavy particles. Therefore, discontinuity formation in the field $v(x, t)$ is accompanied by a coalescence of the initial light particles to form heavy ones. It is clear that the mass of a heavy particle is equal to the mass of the corresponding light particles 'stuck' to it, $m_k = (y_{k+1}^- - y_k^+)$, with momentum as follows:

$$P_k = m_k V_k = s_0(y_{k+1}^-) - s_0(y_k^+) = \int_{y_k^+}^{y_{k+1}^-} v_0(x)\, dx.$$

Here we took into account the fact that the particle initial density is $\rho_0 = 1$. It is clear from this, and from (2.4) as well, that the expressions for the discontinuity velocities V_k result from the law of conservation of momentum. Thus, at a discontinuous stage one can interpret the BE solution as an evolution of gas consisting of two types of particles. The momentum transfer from the light particles to the heavy ones and the collision of the heavy particles occur in this case under the laws of absolutely inelastic collisions. In terms of particles, in particular, it is easy to understand the velocity decrease in the unipolar pulse discontinuity motion (2.3.4). The discontinuity is equivalent to a heavy particle which travels in a field of immobile light particles, absorbs their mass, and consequently slows down. The interpretation of unipolar-pulse interaction with a random field in terms of particles is given in [130], where it is shown that such an interaction results in an additional drop in the discontinuity velocity.

When $t \gg t_n$ the total gas mass is, in fact, concentrated in the heavy particles with masses $m_k = y_{k+1} - y_k$ and velocities

$$V_k = \frac{P_k}{m_k} = \frac{s_0(y_{k+1}) - s_0(y_k)}{y_{k+1} - y_k}.$$

Owing to the velocity differences, the discontinuities merge. The amplitudes $\Delta v_k = m_k/t$ and velocities of the merged discontinuities are controlled by the laws of conservation of mass and momentum,

$$\tilde{m}_k = m_k + m_{k+1} = (y_{k+1} - y_k) + (y_k - y_{k-1}), \tag{2.8}$$

$$\tilde{V}_k = \frac{m_k V_k + m_{k-1} V_{k-1}}{m_k + m_{k-1}} = \frac{s_0(y_{k+1}) - s_0(y_{k-1})}{y_{k+1} - y_{k-1}}. \tag{2.9}$$

In terms of particles the increase of the external scale $l(t)$ corresponds to the growth of the characteristic mass of the heavy particles due to inelastic collisions. The trajectories of the discontinuities are depicted in Fig. 5.6.

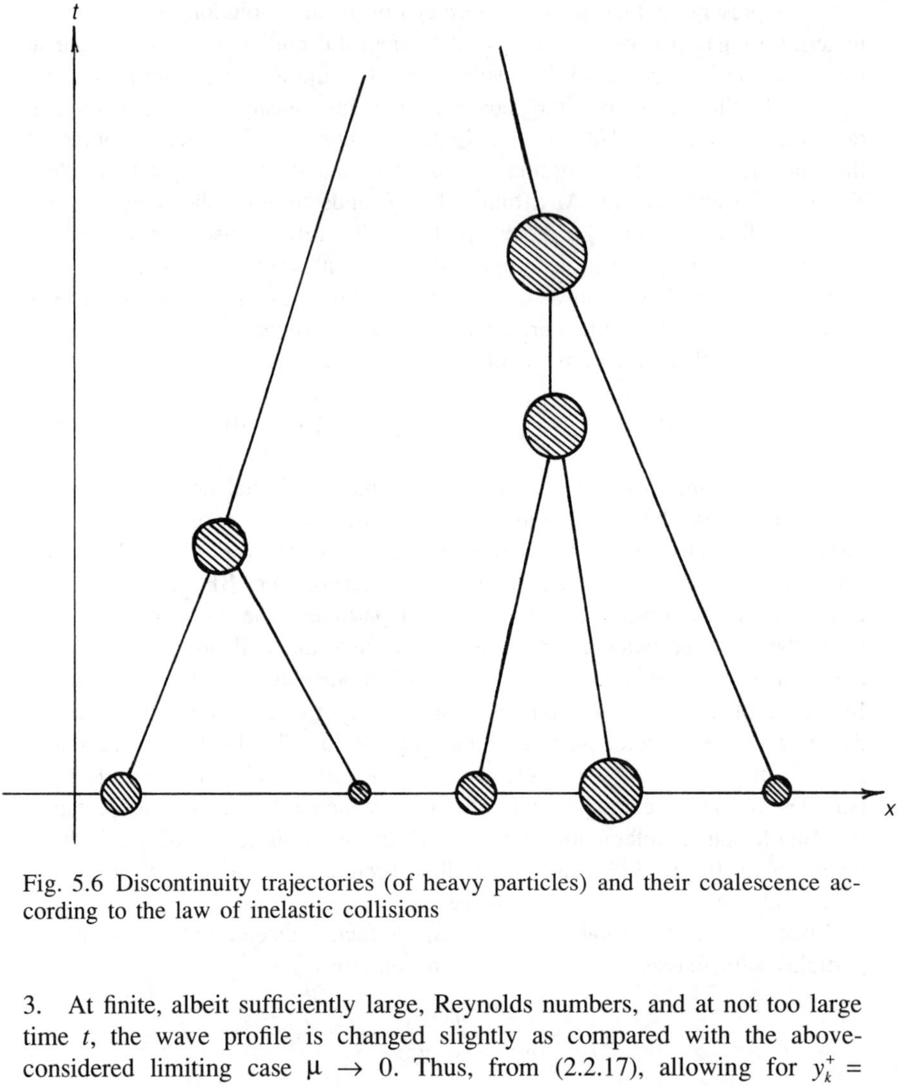

Fig. 5.6 Discontinuity trajectories (of heavy particles) and their coalescence according to the law of inelastic collisions

3. At finite, albeit sufficiently large, Reynolds numbers, and at not too large time t, the wave profile is changed slightly as compared with the above-considered limiting case $\mu \to 0$. Thus, from (2.2.17), allowing for $y_k^+ = y_k^- = y_k$, we have

$$v(x, t) = \frac{1}{t}\left(x - \frac{y_{k+1} - y_k}{2}\right) - \frac{\eta_k}{2t} \tanh\left[\frac{\eta_k(x - x_k)}{4\mu t}\right]. \tag{2.10}$$

$$\eta_k = y_{k+1} - y_k .$$

Consequently, for finite Reynolds number the field $v(x, t)$ is characterised, along with the external scale $l(t) \sim \eta_k$, by the internal scale $\delta = \mu t/l(t)$ as well, which is equal to the characteristic width of a shock front. Equation (2.10) implies that the width of a shock front is enlarged with time, and, thus, the criterion for strongly nonlinear behaviour can be formulated via the smallness of the shock front width, i.e. that of the internal scale $\delta(t)$ as compared with the external one $l(t)$ (being a characteristic distance between discontinuities).

At the strongly nonlinear stage, $\delta(t) \ll l(t)$, the field energy decrease takes place within narrow neighbourhoods of the shock fronts, and does not depend on the viscosity coefficient value provided it is sufficiently small [91]. Actually, if one casts the BE (1.1) in a divergence form

$$\frac{\partial v}{\partial t} = \frac{\partial}{\partial x}\left\{-\frac{1}{2}v^2 + \mu\frac{\partial v}{\partial x}\right\}, \tag{2.11}$$

we find for the energy damping rate

$$\frac{\mathrm{d}\mathcal{E}}{\mathrm{d}t} = \frac{\mathrm{d}}{\mathrm{d}t}\frac{1}{2L}\int_{-L}^{+L} v^2(x, t)\mathrm{d}x = -\frac{\mu}{L}\int_{-L}^{+L}\left(\frac{\partial v(x, t)}{\partial x}\right)^2 \mathrm{d}x \tag{2.12}$$

Here it is assumed that $v(\pm L, t) = 0$. Using the expression (2.10) for the field and taking into consideration that at low μ the value of $\mu(v_x)^2$, determining the energy dissipation rate, differs substantially from zero only in ε-neighbourhoods of the shock fronts ($\delta \ll \varepsilon \ll l$), we obtain

$$\mu\int_{-L}^{L}\left(\frac{\partial v}{\partial v}\right)^2 \mathrm{d}x \approx \sum_{k=1}^{N}\mu\int_{x_k-\varepsilon}^{x_k+\varepsilon}\left(\frac{\partial v}{\partial x}\right)^2 \mathrm{d}x = \sum_{k=1}^{N}\frac{\eta_k^4}{64\mu^2 t^4}\int_{x_k-\varepsilon}^{x_k+\varepsilon}\mathrm{sech}^4\left[\frac{\eta_k}{4\mu t}(x - x_k)\right]\mathrm{d}x$$

$$\approx \frac{1}{16t^3}\int_{-\infty}^{\infty}\mathrm{sech}^4 s\,\mathrm{d}s\sum_{k=1}^{N}\eta_k^3 = \frac{1}{12t^3}\sum_{k=1}^{N}\eta_k^3.$$

Here N is the number of discontinuities of a sawtooth wave within the interval $2L$. Therefore, at the stage of developed discontinuities and sawtooth waves, the wave decay rate

$$\frac{\mathrm{d}\mathcal{E}}{\mathrm{d}t} = -\frac{1}{12t^3}\frac{1}{L}\sum_{k=1}^{N}\eta_k^3 = -\frac{1}{6}\frac{1}{2L}\sum_{k=1}^{N}(\Delta v_k)^3 \tag{2.13}$$

does not depend on the dissipation factor value μ, being defined by the distance between 'zeros' η_k, i.e. by the amplitude of the sawtooth wave discontinuities $\Delta v_k = \eta_k/t$.

As far as a random field is concerned, averaging of (2.13) and allowance for $\langle\eta\rangle = \langle N\rangle/2L$ yield for the average-over-a-statistical-ensemble energy density $E = \langle\mathcal{E}\rangle$

$$\frac{\mathrm{d}E}{\mathrm{d}t} = -\frac{\langle\eta^3\rangle}{6\langle\eta\rangle t^3}. \tag{2.14}$$

This relation clearly testifies to the fact that the law of random wave energy damping is defined by the first and the third statistical moments of the inter-'zero' distance for the sawtooth wave.

On the basis of the relations as obtained, it is instructive now to discuss the possible ways of random field evolution at the developed-discontinuity stage. first of all it should be noted that a situation is possible in which the discontinuities are immobile. It is, for example, the case for a periodic wave and, in particular, for a harmonic initial field. With respect to the initial field of a more general type $v_0(x) = a_0(\cos\psi(x))_x$ where $\psi(x)$ is an arbitrary monotonic function, it is clear from (2.4) that for $t \gg t_n$ the discontinuities are immobile also. In all these cases the wave shape for $t \gg t_n$ is invariant, while its amplitude decays as $1/t$. The field energy in this instance decays in time according to (2.14) as per the following law:

$$E(t) = \langle\eta^3\rangle/12\langle\eta\rangle t^2 = l_*^2/12t^2, \tag{2.15}$$

where l_* is a certain characteristic spatial scale of the field: $l_*^2 = \langle\eta^3\rangle/\langle\eta\rangle$. In (2.15) it is supposed that the average field equals zero. Therefore, provided the discontinuities are immobile the spatial scale of the field remains unchanged, its energy attenuates as t^{-2}, does not depend on the initial field amplitude, and is determined by its spatial scale l_* alone. In particular, as regards the harmonic initial field, l_* in equation (2.15) is the initial field period.

4. In a common case, however, due to different random velocities of the discontinuities the latter merge, leading to a building up of the characteristic spatial scale and, consequently, to a change of the energy decay law. Naturally, of course, the law of energy decay is defined by the discontinuity coalescence rate which, in its turn, depends on the initial field realisation structure. In order to evaluate the energy decay rate, we assume in (2.15) that $\langle\eta^3\rangle \approx \langle\eta\rangle^3 = l^3(t)$ where $l(t)$ is a field external scale, related to the average discontinuity number per unit length through the equality

$$n(t) = 1/l(t). \tag{2.16}$$

Then (2.14) is transformed into

$$\frac{dE}{dt} = -\frac{l^2(t)}{6t^3} = -\frac{1}{6n^2(t)\,t^3}, \tag{2.17}$$

i.e. to get an estimate of the energy damping (decay) one needs to know, to a given approximation, $l(t)$ or $n(t)$, only. Obviously, the decrease of $n(t)$ due to the discontinuity coalescence over the time interval Δt is proportional to $n(t)$ itself, as well as to the ratio of the possible discontinuity convergence $V\Delta t$ over this time interval to the characteristic distance $l(t)$ between discontinuities:

$$\Delta n \approx n(V\Delta t/l). \tag{2.18}$$

for $t \gg t_n$, when the characteristic inter-discontinuity distance $l(t) \gg l_0$ (the initial scale of field $v_0(x)$) we can take the root mean square velocity of an individual discontinuity $V = \sqrt{\langle V_k^2 \rangle}$ as an estimate of V. Dividing both sides of (2.18) by Δt, assuming that $\Delta t \to 0$ and allowing for the equality (2.16) one can write

$$\frac{dn}{dt} = -n^2 V. \tag{2.19}$$

Evaluate the characteristic convergence rate V by making use of (2.4) and replacing the random inter-'zero' distance in that expression with an external scale l as follows:

$$V^2 \approx \langle V_k^2 \rangle \approx \frac{d_s(l)}{l^2} = \frac{1}{l^2} \int_0^l (l - z)\, B_0(z)\, dz. \tag{2.20}$$

Here $B_0(z)$ is a covariance function of the initial field, whereas

$$d_s(z) = \langle [s_0(x + z) - s_0(x)]^2 \rangle \tag{2.21}$$

is the structure function of the initial action. Taking into account (2.16) and (2.20) we go over from (2.19) to a closed kinetic equation with respect to $l(t)$:

$$\frac{dl}{dt} = \frac{\sqrt{d_s(l)}}{l}. \tag{2.22}$$

It is implied in the above relation that when $l(t) \gg l_0$ the law of the external scale growth is determined by the initial action structure function $d_s(z)$ behaviour for $z \gg l_0$ or, which is the same, by the action spectral density type, $G_s(k) = G_0(k)/k^2$, in the low spatial frequency range [90].

Consider two cases: $G_0(0) = G_0 \neq 0$ and $G_0(k) \sim k^n$ as $k \to 0$, $n \geq 1$ (the latter we take $G^0 = 0$). Then, for the structure function $d_s(z)$ for $z \gg l_0$ we have

$$d_s(z) = \begin{cases} \sigma_0^2 l_0\, |z|, & G^0 \neq 0, \\ \sigma_0^2 l_0, & G^0 = 0, \end{cases} \tag{2.23}$$

where σ_0 is a characteristic amplitude of the initial field. The following estimates of the external scale growth result then from (2.22) and (2.23):

$$l(t) = \begin{cases} l_0 (t/t_n)^{\frac{2}{3}}, & G^0 \neq 0, \\ l_0 (t/t_n)^{\frac{1}{2}}, & G^0 = 0. \end{cases} \tag{2.24}$$

Here, as previously, $t_n = l_0/\sigma_0$. The type of dependence on t of the average discontinuity number per unit time (2.16), allowing for the initial stage (see (1.4) and (1.5)), is depicted in Fig. 5.7.

Inserting (2.24) into (2.17) we obtain the following expressions for the average energy densities:

$$E(t) = \begin{cases} (\sigma_0^2 l_0/t)^{\frac{2}{3}} = \sigma_0^2 (t_n/t)^{\frac{2}{3}}, & G^0 \neq 0, \\ \sigma_0 l_0/t = \sigma_0^2 (t_n/t), & G^0 = 0. \end{cases} \tag{2.25}$$

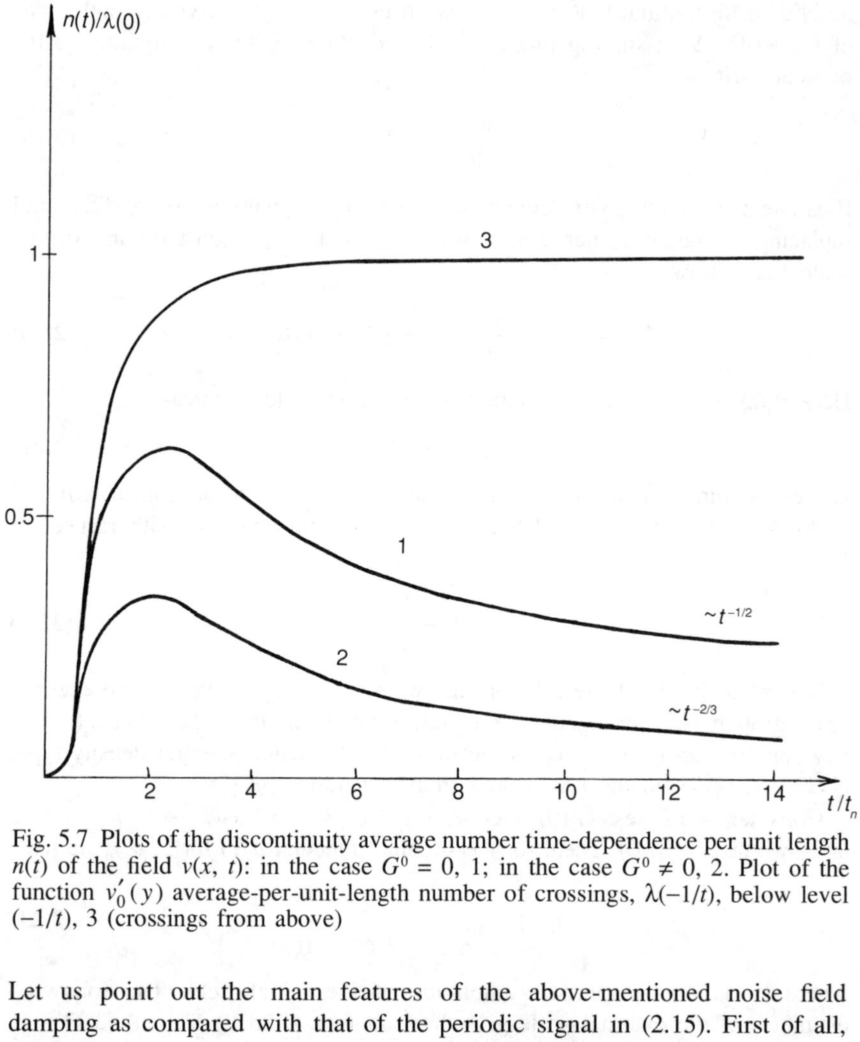

Fig. 5.7 Plots of the discontinuity average number time-dependence per unit length $n(t)$ of the field $v(x, t)$: in the case $G^0 = 0$, 1; in the case $G^0 \neq 0$, 2. Plot of the function $v_0'(y)$ average-per-unit-length number of crossings, $\lambda(-1/t)$, below level $(-1/t)$, 3 (crossings from above)

Let us point out the main features of the above-mentioned noise field damping as compared with that of the periodic signal in (2.15). First of all, it follows from (2.25) that the noise field decay is slower than that of a periodic signal. This relates to the external scale growth due to the discontinuity coalescence. Should the periodic field energy (2.15) for $t \gg t_n$ not depend on the initial amplitude and be proportional to l_0^2, the energy of the noise field will nevertheless increase with the initial energy $E_0 = \sigma_0^2$. This can be explained by the fact that the discontinuity coalescence rate depends on the initial energy and grows with σ_0. It can be added that at the stage of developed discontinuities, when the wave shape is universal and appears as a sequence of triangular pulses with the same slope $1/t$ and characteristic inter-discontinuity distance $l(t)$, the energy density can be expressed, from dimensionality considerations, as

$$E(t) = \langle v^2 \rangle \approx l^2(t)/t^2. \tag{2.26}$$

Substituting the external scale expression (2.24) here, one has, as before, (2.25) for $E(t)$.

The discontinuities emerging in the field lead to the formation of universal asymptotic behaviour of the spectrum in the range of high spatial frequencies. To find it,it seems helpful to initially consider the field gradient $u(x, t) = v_x\,(x, t)$, for which, when $t \gg t_n$ (2.7) implies that

$$u(x, t) = \frac{1}{t} - \sum_m \frac{\eta_m}{t}\,\delta(x - x_m). \tag{2.27}$$

The appropriate Fourier transform is equal to

$$C_u(k, t) = \frac{1}{t}\left[\frac{1}{2\pi}\sum_m \eta_m e^{-ikx_m} - \delta(k)\right]. \tag{2.28}$$

In order to reveal the energy spectrum $G_u(k; t)$ one can utilise a well-known relation with respect to the spectra of statistically homogeneous fields [78],

$$\langle C(k, t)\, C(k_2', t)\rangle = G(k; t)\,\delta(k - k').$$

Then, applying the obvious equality $G_v(k;\ t) = G_u(k;\ t)/k^2$ to the energy spectrum of the sawtooth field we have

$$G_v(k; t) = \frac{n(t)}{2\pi k^2 t^2}\left[\langle\eta^2\rangle + 2\sum_{p=1}^{\infty}\langle\eta_m\eta_{m+p}\,\cos k\Delta_p\rangle - \delta(k)\right]. \tag{2.29}$$

Here $\Delta_p = x_{m+p} - x_m$ is the distance between the discontinuities. The first term in (2.29) characterises an averaged spectrum of an individual discontinuity, while the remaining sum describes the interference between the different discontinuity spectra. Owing to the discontinuity location fluctuations, one can neglect their interference when $kl \gg 1$ and, as a result, we have from (2.29) the universal power law

$$G_v(k; t) = \frac{n(t)\,\langle\eta^2\rangle}{2\pi k^2 t^2}, \tag{2.30}$$

for the spectrum high-frequency asymptotic behaviour.

5. At finite Reynolds numbers, along with the external scale $l(t)$, the random field has an internal one $\delta(t)$, equal to the shock front thickness and connected with the external scale through the relation $\delta = \mu t/l$. Given the internal scale, one is capable of evaluating the applicability range of the discontinuous wave idealisation. It is supposed to remain valid as long as $l(t) \gg \sigma(t)$. The ratio l/σ produces the time-varying Reynolds number

$$R(t) = l(t)/\delta(t) = l^2(t)/\mu t. \tag{2.31}$$

It can easily be observed that (2.31) coincides with the definition of the Reynolds number introduced above in (1.2), inasmuch as the characteristic field amplitude at the discontinuous stage is equal to $\sigma(t) = l(t)/t$. Now for a periodic disturbance, we have $l = l_0$, and because of the shock wave broadening the Reynolds number decreases with time, $R \sim t^{-1}$. However, in the case of a noise field, the Reynolds number variation is governed by a competition between two effects: the external scale $l(t)$ growth because of the discontinuity coalescence and the internal scale increase because of the shock front broadening. Substituting the above-obtained external scale values into (2.31), we write

$$R(t) = \begin{cases} R_0 (t/t_n)^{\frac{1}{3}}, & G^0 \neq 0, \\ R_0, & G^0 = 0. \end{cases} \tag{2.32}$$

Therefore, we have arrived, for a noise field, at a surprising result: the increase of the characteristic inter-discontinuity distance attributed to their coalescence leads for $G^0 \neq 0$ to growth of the effective Reynolds number. Because of this, the idealisation of discontinuous waves becomes more correct with t, and the linear stage is never reached in this case at all. As to $G^0 = 0$, (2.32) also implies the absence of the linear stage. Later on, however, namely in Section 5.4, it will be demonstrated that when $G^0 = 0$ a wave does none the less enter the linear stage, though this process will be extremely lengthy, $t_l \approx t_n \exp(R_0^2)$.

Provided $R(t) \gg 1$, it follows then that the influence of the internal scale on the spectrum will be only pronounced for the short-wave spectrum components. Indeed, out of (2.10) we have for the field gradient

$$u(x,t) = \frac{1}{t} - \sum_m \frac{\eta_m^2}{8\mu t^2} \operatorname{sech}^2 \left[\frac{(x - x_m)}{4\mu t} \eta_m \right]. \tag{2.33}$$

The high-frequency asymptotic behaviour subject to this condition will be determined by an averaged spectrum for an individual discontinuity, and in analogy with (2.30), when $kl \gg 1$, for the asymptotic behaviour the following expression might be written:

$$G_v(k;t) = \frac{2\pi n(t)}{k^2} \left\langle \frac{k^2}{\sinh^2(2\pi k \mu t/\eta)} \right\rangle. \tag{2.34}$$

In order to obtain qualitative assessments we can assume $\eta \approx l$, which results in spectrum decay by a universal power law k^{-2} as in (2.30), within the inertial interval $l^{-1} \ll k \ll \delta^{-1}$. In the viscous region $k \gg \delta^{-1}$ the wave spectrum vanishes exponentially,

$$G_v(k;t) \sim n(t) \exp(-\delta k), \quad \delta = \mu t/l.$$

Rigorous analysis carried out in Section 5.4 shows that allowance for internal scale fluctuations actually leads to a slower wave spectrum decrease in the viscous region.

6. As was demonstrated above, the BT evolution at the stage of developed discontinuities is equivalent to particle gas behaviour with absolutely inelastic collisions. The equation (2.19) for the average discontinuity number presents a qualitative description of the particle quantity decrease due to coalescence. A more rigorous approach is to describe the particle gas with the aid of multi-particle distribution functions [91]. It has to be underlined, however, that the particle gas equivalent to the BT discontinuities differs from the arbitrary gas of inelastically interacting particles in some aspects. In fact when $t \gg t_n$ it is clear from (2.3) that the propagation velocity of a discontinuity and the field zeros y_k are interrelated through

$$V_k = \frac{dx_k}{dt} = \frac{1}{t}\left[x_k - \frac{1}{2}(y_k + y_{k+1})\right]. \tag{2.35}$$

Taking into account that the mass of the kth particle amounts to $m_k = y_{k+1} - y_k$ we have from (2.35) for the neighbouring particle convergence speed $\tilde{V}_k = V_k - V_{k+1}$ and the distance between them $\lambda_k = x_{k+1} - x_k$,

$$\tilde{V}_k = -\frac{d\lambda_k}{dt} = \frac{1}{t}\left[\frac{m_k + m_{k+1}}{2} - \lambda_k\right]. \tag{2.36}$$

Thus, at a time of collision ($\lambda_k = 0$) the particle convergence speed is uniquely determined by their masses, which allows us to formulate a kinetic equation for the multi-particle distribution functions without explicit introduction of the particle velocity.

The particle gas is exhaustively described by an average particle number per unit length $n(t)$ and by a set of multi-point distribution functions. The one-point functions $f(m; t)$ and $g(\lambda; t)$ offer the particle distribution over masses and distances between the adjacent particles. The joint distribution function of $N + 1$ successive particles

$$W_N(\lambda^1, \lambda^2, \ldots, \lambda^N, m^0, m^1, \ldots, m^N; t)$$

describes the statistical properties of the sequence of distances λ^k and masses m^k with respect to the particles within the given set. The basic concepts used to deduce the kinetic equations for the distribution functions will be shown in accordance with reference [91]. We restrict our analysis to the time variation of the number of particles $n(t)\, f(m; t)\, dm$ with masses within the interval $(m, m + dm)$. Provided $\tilde{V}$ is the convergence speed of adjacent particles with masses m' and m'', the time interval dt will see $W_1(0; m', m''; t) \times n(t)\tilde{V}\, dt\, dm'\, dm''$ such collisions where, allowing for (2.36), $\tilde{V} = (m' + m'')/2t$. Inasmuch as the variation of number of the particles with mass m is equal to the difference between the quantity of particles 'born' from the more lightweight ones ($m' + m'' = m$) and those 'killed' as a result of their coalescence with the adjacent particles ($m' + m = m''$), one can write down the following equation:

$$\frac{\mathrm{d}}{\mathrm{d}t}[n(t)\, f(m; t)] = n(t)\left\{\int_0^m \mathrm{w}_1(0; m', m - m'; t)\, \frac{m}{2t}\, \mathrm{d}m' \right. \tag{2.37}$$

$$\left. - \int_0^\infty [w_1(0; m, m'; t) + w_1(0; m', m; t)]\, \frac{m + m'}{2t}\, \mathrm{d}m' \right\}.$$

As a consequence of (2.37) we obtain an equation for the particle number,

$$\frac{\mathrm{d}n}{\mathrm{d}t} = -\frac{n}{2t}\int_0^\infty \mathrm{d}m \int_0^\infty (m + m')\, w_1(0; m, m'; t)\, \mathrm{d}m' \tag{2.38}$$

and an equation for $f(m; t)$. In analogy we can formulate equations for W_N as well, but each such equation, however, involves the distribution functions of higher order, W_{N+1}.

This 'chain' can be stopped by a series of rather strong assumptions. If it is assumed that all the distributions are self-preserving and described by a single parameter $l(t)$, a characteristic (external) scale of the turbulence, then in particular,

$$w_1(\lambda; m, m'; t) = \tilde{w}_1(\lambda/l; m/l, m'/l)/l^3.$$

Inserting this into (2.38) we come to

$$\int_0^\infty \mathrm{d}m \int_0^\infty \mathrm{d}m'(m + m')\, w_1(0; m, m'; t) = \text{const} = 2\alpha, \tag{2.39}$$

and then for the particle number $n(t)$ a closed equation can be written, namely

$$\frac{\mathrm{d}n}{\mathrm{d}t} = -\frac{\alpha}{t}\eta, \tag{2.40}$$

where α/t characterises the collision number per unit time, and α is an arbitrary constant. For $l(t)$ the obvious relation $n(t)l(t) = n(t_0)l(t_0)$ holds, allowing for which (2.40) gives the following power law with respect to $l(t)$,

$$l(t) = l_0(t/t_0)^\alpha. \tag{2.41}$$

In order to find the forms of the mass and interval probability distributions, additional assumptions have to be made, for example, with respect to the statistical independence of the adjacent particle masses. Therefore, we managed to obtain a single-parameter family of distributions depending on a parameter α. Index α in (2.44) cannot be assessed within the framework of kinetic equations. This task can be accomplished rigorously only for the single case of BT with a non-zero spectrum magnitude $G(k = 0) = G^0 \neq 0$. As G^0 = const appears as a BT invariant (4.2.4), while, on the other hand, the self-preservation hypothesis leads to

$$G^0 \sim l^3/t^2 \sim t^{3\alpha-2},$$

then from $G^0 \neq 0$ we immediately have $\alpha = 2/3$. To define α at $G^0 = 0$, it was supposed that as physically realisable one can imagine a distribution with maximum entropy H of the inter-particle distance,

$$H = -\int_0^\infty g(\lambda; t) \ln g(\lambda; t)\, d\lambda,$$

which results in $\alpha = 1/2$. Consequently the method of kinetic equations produces essentially the same result as the qualitative theory (2.24) developed here does. A more consistent examination of BT confirms that the kinetic theory based on the equation set truncation gives an incorrect distribution of the particle masses.

To deduce closed equations for the probability distributions one can, along with the self-preserving hypothesis, resort to other hypotheses as well. Thus, in [131], in order to describe the random discontinuity transformation, a procedure was adopted which led to a Boltzmann-type kinetic equation. The advantage of this approach, as compared with the self-preservation assumption, is that it allows us to consider within the framework of the Boltzmann kinetic equation the evolution of initial distribution functions of arbitrary type.

5.3 Self-preservation of random waves in nonlinear dissipative media

1. As is evident from the qualitative theory developed above, the BT behaviour for $t \gg t_n$ is governed by the special form of the initial field $v_0(x)$ within the range of low spatial frequencies. This can be associated with the fact that nonlinear excitation of the difference harmonics and strong dissipation of the small-scale components lead to a relative increase of the large-scale harmonic contribution to the field energetics.

The initial field spectrum is supposed to have the form

$$G_0(k) = \alpha_n^2 \, |k|^n b_0(k), \quad b_0(0) = 1, \tag{3.1}$$

where $b_0(k)$ is a function with characteristic scale $k_* = 1/l_0$ which vanishes quite rapidly for $k > k_*$. Let us investigate the field asymptotic behaviour dependence on the exponent n in (3.1).

We draw attention first to the case of a low initial Reynolds number ($R_0 = \sigma_0 l_0/\mu \ll 1$). Then, at the initial stage nonlinear effects can be neglected and the field will be described by a linear equation of diffusion, while the wave spectrum is supposed to change with time as

$$G_v(k; t) = G_0(k) \exp(-2k^2 t). \tag{3.2}$$

Due to the small-scale component damping the field behaviour at $t \gg t_1 \approx 1/\mu\, k_*^2$ will be only defined by the large-scale part of the initial spectrum

$$G_v(k; t) = \alpha_n^2 \, |k|^n \exp(-2\mu k^2 t) = \alpha_n^2 \, |k|^n \exp\left(-\frac{k^2 l^2(t)}{2}\right). \tag{3.3}$$

This implies that the spatial scale $l(t)$ growth and energy decay $\sigma^2(t) = 2\int_0^\infty G_v(k; t)\, dk$ caused by small-scale harmonic dissipation obey the laws

$$l(t) = \sqrt{2\mu t}\,, \quad \sigma^2(t) = \alpha_n^2\, \Gamma\left(\frac{n+1}{2}\right) (2\mu t)^{-(n+1)/2}, \tag{3.4}$$

and the current Reynolds number varies as

$$R(t) = \sigma(t) l(t)/\mu \approx \frac{\alpha_n}{\mu} (2\mu t)^{(1-n)/4}. \tag{3.5}$$

It immediately follows from (3.5) that the field evolutions for $n < 1$ and $n > 1$ are qualitatively different. If the large scale components of the initial spectrum are significantly suppressed ($n > 1$), then $R(t)$ diminishes with time only when the initial Reynolds number is high. If, however, $n < 1$, then the relative contribution of the large-scale components rises, together with the Reynolds number, according to (3.5). Consequently, the field eventually enters the nonlinear regime in this case.

2. Let us discuss now another limiting case, $R_0 \gg 1$, and demonstrate that for $n < 1$ the developing-with-time sawtooth field proves structurally stable, i.e. it will never enter the linear mode of dissipation. For this purpose, it needs to be shown that for all t the effective Reynolds number (2.31) satisfies the inequality $R(t) \gg 1$. Variation of $R(t)$ at the stage of sawtooth waves is governed by the law of the BT external scale $l(t)$ growth, which can be assessed from (2.22). For the situation under consideration the initial action structure function is, as per (3.1.23), equal to

$$d_s(z) = 4\alpha_n^2 \int_0^\infty (1 - \cos kz)\, k^{n-2}\, b_0(k)\, dk.$$

from this it follows, in particular, that for $n < 1$ and $z \gg l_0$ the asymptotic formula

$$d_s(z) = \beta_n^2\, |z|^{1-n} \quad (n < 1),$$
$$\beta_n^2 = 2\pi\, \alpha_n^2/[\Gamma(2-n) \sin(\pi(1-n)/2)], \tag{3.6}$$

holds. Substituting it into (2.22), one obtains for the turbulence external scale

$$l(t) = (\beta_n t)^{2/(3+n)}. \tag{3.7}$$

Before proceeding to a further analysis of turbulence, it seems instructive to outline one more approach to estimate the external scale $l(t)$ [92, 132] supported by the solutions (2.1) and (2.2). It is quite obvious that the absolute minimum coordinate $y(x, t)$ of the function $\Phi(x, y, t)$ falls within the domain $|x - y| < l(t)$ in which the local minima of the function $s_0(y)$ are not raised too much by the addition of the parabolic term $(x - y)^2/2t$ to $s_0(y)$. It seems rather natural to evaluate the dimensions of this region under the condition that the initial action increments $\Delta s = s_0(y + l) - s_0(y)$ and those of the parabola α are of the same order: $|\Delta s| \sim l^2/t$. Having estimated Δs with the help of the initial action structure function $\langle(\Delta s)^2\rangle = d_s(l)$, we are led to the following equation:

$$\frac{l^2(t)}{t} = \sqrt{d_s(l)}, \tag{3.8}$$

the solution of which also results in (3.7).

The substitution of (3.7) into (2.31) yields

$$R(t) = \beta^{4/(3+n)}\, t^{(1-n)/(n+3)}/\mu. \tag{3.9}$$

Thus, for $n < 1$ the Reynolds number increases at the nonlinear stage as well, meaning that the sawtooth wave regime proves structurally stable. For $n \geq 1$, as the Section 5.2 estimates show, the Reynolds number is constant.

3. Time has come to show more strictly that when $n < 1$, a strongly nonlinear regime of sawtooth waves is actually set up, independently of the R_0-value, and that the field statistical properties become self-preserving. To this end let us resort to an exact solution of the BE, (2.1.13).

Introduce new variables into (2.1.13) and (2.1.14),

$$x = \xi l(t), \quad y = \eta l(t), \tag{3.10}$$

where the external scale $l(t)$ is given by (3.7). In terms of the new variables, allowing for the expressions (2.31) and (3.9) with respect to the Reynolds number, the equalities (2.1.13) and (2.1.14) will take the form

$$v(x, t) = \frac{l(t)}{t} \frac{\int_{-\infty}^{\infty} (\xi - \eta) \exp[-R(t)\, \tilde{\Phi}(\xi, \eta, t)]\, d\eta}{\int_{-\infty}^{\infty} \exp[-R(t)\, \tilde{\Phi}(\xi, \eta, t)]\, d\eta}, \tag{3.11}$$

$$\tilde{\Phi}(\xi, \eta, t) = \frac{(\eta - \xi)^2}{2} + \tilde{s}_0(\eta, t), \quad \tilde{s}_0(\eta, t) = \frac{t}{l^2}\, s_0(\eta l). \tag{3.12}$$

Now we calculate a structure function of the dimensionless initial action $\tilde{s}_0(\eta, t)$ present in the above relation. It should be noted that the structure function of $s_0(y)$ can be represented as [79]

$$d_s(z) = \beta_n^2\, a(z)\, |z|^{(1-n)}, \tag{3.13}$$

where $a(z \gg l_0) = a(\infty) = 1$. Hence, taking into consideration (3.12), one can define the structure function in the form

$$d_{\tilde{s}}(\tilde{z}, t) = \langle[\tilde{s}_0(\eta + \tilde{z}, t) - \tilde{s}_0(\eta, t)]^2\rangle = t^2 d_s(\tilde{z}l)/l^4 = a(\tilde{z}l)\, |\tilde{z}|^{(1-n)}. \tag{3.14}$$

A remark is due here as to the fact that, as soon as the Reynolds number (3.9) increases so as to render the inequality $R(t) \gg 1$ valid, the integral (3.11) will be only contributed to by a point η where $\tilde{\Phi}$ attains an absolute minimum, and the solution (3.11) can be substituted by its asymptotic expression

$$v(x, t) = \frac{l(t)}{t}(\xi - \eta(\xi, t)) = \frac{l(t)}{t}\tilde{v}(\xi, t). \tag{3.15}$$

Here $\eta(\xi, t)$ stands for the coordinate of an absolute minimum of $\tilde{\Phi}(\xi, \eta, t)$. If the inequality $t \gg t_n = l_0/\sigma_0$ holds as well, then the coordinates of the absolute minima of $\tilde{s}_0$ coincide with the coordinates of the local minima of $\tilde{s}_0$, the function $\eta(\xi, t)$ is piecewise constant as viewed within the scale $l(t)$, and the field $\tilde{v}(\xi, t)$ itself is interpreted as a sequence of sawtooth pulses with shock front width proportional to $1/R(t)$. Within the ordinary scale l_0 the shock front width of the field $v(x, t)$ will enlarge according to $\delta \sim t^{(n+1)(n+3)}$ whereas the shock relative width $\delta/l = 1/R(t)$ decreases, which permits us to speak about a strongly nonlinear field evolution mode. Therefore, the field becomes locally self-preserving and its shape within the inter-discontinuity intervals becomes universal and does not depend on the initial perturbation type.

4. Let us emphasise now that not only does the wave profile become self-preserving, but so do its statistical properties. At $R(t) \gg 1$ the field $v(x, t)$ has the form (3.15). At times when $l(t) \gg l_0$ the structure function (3.14) can be replaced by

$$d_{\tilde{s}}(\tilde{z}, t) = |\tilde{z}|^{(1-n)}, \tag{3.16}$$

the function not depending on time and possessing no spatial scales of its own. In this case the statistical properties of the absolute minima coordinates $\eta(\xi, t)$ do not vary with time. The latter phenomenon means precisely that the field $v(x, t)$ statistical properties determined by time-independent statistics of $\eta(\xi, t)$ are rendered self-preserving according to (3.15).

It can be seen in particular from (3.15) and (3.16), that the correlation function and the turbulence energy spectrum at a self-preserving stage can be seen in the forms:

$$\begin{aligned}
B_v(z; t) &= \frac{l^2(t)}{t^2}\langle\eta^2\rangle R\left(\frac{z}{l(t)}\right),\\
\sigma^2(t) &= \frac{l^2(t)}{t^2}\langle\eta^2\rangle \sim t^{-2(n+1)/(n+3)},\\
G_v(k; t) &= \frac{l^3(t)}{t^2}\langle\eta^2\rangle \tilde{G}(kl(t)),\\
\tilde{G}(\kappa) &= \frac{1}{2\pi}\int_{-\infty}^{\infty} R(z)\, e^{i\kappa z}\, dz.
\end{aligned} \tag{3.17}$$

The field probability density appears self-preserving also,

$$W(v;\, t) = (t/l)\tilde{W}_v(vt/l(t)). \tag{3.18}$$

The dimensionless functions R, $\tilde{G}$, $\tilde{W}_v$ and a constant $\langle\eta^2\rangle$ in (3.17) and (3.18) might be obtained, for example, by means of the statistical handling of numerical simulation results for the dimensionless field $\tilde{v}(\xi) = \xi - \eta$, where η is the coordinate of an absolute minimum of the random function $\tilde{s}(\xi)$ with structure function (3.16). Certain properties of these functions can be assessed, making use of some general BT characteristics. first of all it should be noted that the field realisation at a self-preserving stage involves discontinuities leading to a universal short-wave asymptotic behaviour of the energy spectrum, $\tilde{G}(\kappa) = \lambda_n\, \kappa^{-2}$. The divergence form of the BE (2.11) evidently shows that nonlinearity and dissipation cannot (for $n < 2$) change the behaviour of the energy spectrum when $k \to 0$: $G_v(k;\, t) = G_0(k)$ (see also (2.2.6)). Therefore, in this case we have $\tilde{G}(\kappa) = \gamma_n \kappa^n (\kappa \to 0)$ with respect to a dimensionless spectrum.

The external scale growth law (3.7) can be inferred from the condition of the spectrum conservation within the large-scale range. As a matter of fact, using the spectrum self-preservation (3.17) under the condition that $G_v(k;\, t) = G_0(k)$ as $k \to 0$, one can formulate the following:

$$\alpha_n^2\, b_0(0)k^n = l^3\gamma_n k^n\, \langle\eta^2\rangle/t^2, \tag{3.19}$$

which yields the same growth law for $l(t)$ as that in (3.7). It is worth saying that in [23] the hypotheses of BT self-preservation and energy spectrum conservation within the small spatial frequency range were put forward, yielding, naturally, the growth law (3.7) with respect to $l(t)$. As is clear from the above, however, this hypothesis only holds for $n < 1$. When $n \geq 2$, because of the parametric generation of low-frequency components, the spectrum as $k \to 0$ is not preserved, though acquiring a universal form $G_v(k;\, t) \sim k^2$. Later on, in subsection 5, it will be shown that for $n \geq 2$ an incomplete self-preservation is observed (i.e. self-preservation over a finite but very large time interval) whereas the external scale growth law does not depend on $n = l(t) \sim \sqrt{t}$. The case of $1 < n < 2$ is considered to be intermediate between $n < 1$ and $n > 2$. For $1 < n < 2$ the field spectrum as $k \to 0$ behaves as $G_v = \alpha_n^2 k^n b_0(0)$, though due to the low-frequency component generation it is not energy-containing and is quite rapidly replaced by $G_v \sim k^2$; hence the external scale grows as $l(t) \sim \sqrt{t}$.

As far as BT with initial spectrum exponent $n < 1$ is concerned, one can develop a simple qualitative step-by-step model of the nonlinear mode successive decay [13].

5. The case of $n = 1$ in (3.1) was carefully examined by Burgers [7], who suggested that under this condition $s_0(y)$ should be approximated by a Wiener

process with $d_s(z) = 2J|z|$ (3.6) ($J = \beta_0^2/2 = \pi\,\alpha_0^2$). Burgers succeeded, in this approximation, in calculating the statistical properties of the amplitudes and the distances between the field discontinuities, expressing his final results through a complex set of analytical functions and constants which only permit numerical evaluation. It is a natural thing that all BT statistical properties are self-preserving in the Wiener approximation, and the correlation function and one-point probability density of BT can be formulated as in (3.17) and (3.18) where $l(t) = \sqrt[3]{2Jt^2}$, while the invariant form of the functions $R(\tilde{z})$ and $\tilde{w}(u)$ and the constant $\langle\eta^2\rangle$ can be deduced by utilising the theory of Markov processes, or else, numerically, by a method of statistical testing [92, 132]. So, in [132] it was observed that $\langle\eta^2\rangle = 0.7$ while $R(\tilde{z}) = 1 - 0.87\,|\tilde{z}|$ for $|\tilde{z}| \ll 1$. Consequently, in (3.17) the asymptotic behaviour of the dimensionless spectrum reads as $\tilde{G}(\kappa) = 0.87/\pi\,\kappa^2 = 0.28/\kappa^2$.

5.4 Asymptotic analysis of nonlinear random waves at large Reynolds numbers

1. Let us delve into the statistical properties of BT provided the initial spectrum behaves as $G_0(k) \sim k^n$ at $k \to 0$ where $n \geq 2$. We shall determine the one- and two-point probability densities, the spatial spectral density and the covariance function of BT.

We are confined here to the analysis of BT characteristics as $\nu \to 0$. In this case the field $v(x, t)$ is given by (2.1), which implies that the problem of statistical property evaluation for the field $v(x, t)$ is reduced to that of finding the statistical properties of the coordinate $y(x, t)$ of the function $\varphi(x, y, t)$ absolute minimum, (2.2). Consider the field $v(x, t)$ for $t \gg t_n$, when the turbulence external scale $l(t)$ of (2.24) is far in excess of the initial correlation length l_0. Subject to this condition, the parabola in (2.2) proves to be a smooth function as compared with the initial action $s_0(y)$, which means that a large number of local minima $s_0(y)$ compete to become the absolute minimum of $\Phi(x, y, t)$ and one can, when examining statistical properties of $y(x, t)$, use the asymptotic formulae of the theory of random process overshoots (see, for example, [93, 94]).

In the following discussion we shall assume $v_0(x)$ to be a Gaussian statistically homogeneous field with a zero average value. Then with $n \geq 2$ the initial action $s_0(y)$ appears also as a statistically homogeneous Gaussian field, with variance

$$\langle s_0^2\rangle = \sigma_*^2 = \sigma_0^2\, l_0^2 = \int_{-\infty}^{\infty} G_0(k)\, k^{-2}\, \mathrm{d}k,$$

$$\sigma_0^2 = \int_{-\infty}^{\infty} G_0(k)\, \mathrm{d}k.$$

Later on, for the sake of greater convenience, we shall interpret $y(x, t)$ as an absolute maximum coordinate of the function

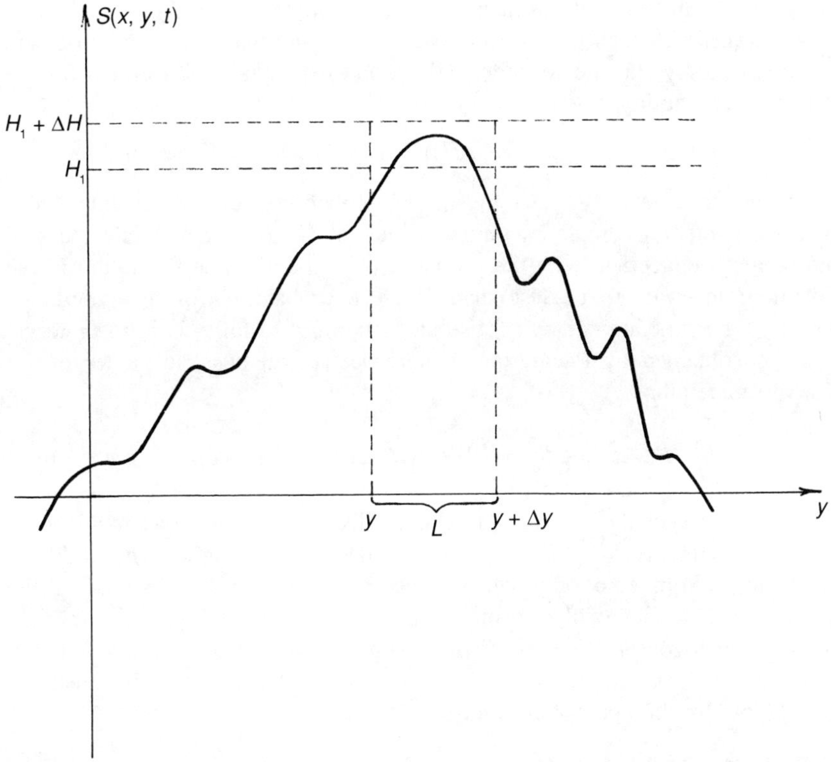

Fig. 5.8 Illustration related to the calculation of the absolute maximum coordinate one-point probability density with respect to y of the function $S(x, y, t)$

$$s(x, y, t) = -\Phi(x, y, t) = -s_0(y) - \frac{(x-y)^2}{2t}. \tag{4.1}$$

Introduce now the integral distribution function and the probability density of an absolute maximum H of function $s(x, y, t)$ in a prescribed region L:

$$F_{max}(H; L) = P(s(x, y, t) < H; y \in L),$$
$$W_{max}(H; L) = \frac{\partial}{\partial H} F_{max}(H; L). \tag{4.2}$$

The probability for the absolute maximum to be contained between H_1 and $H_1 + \Delta H$ and for its coordinate $y(x, t)$ to fall within the interval L: $(y, y + \Delta y)$ is equal to that for the absolute maximum $H \in (H_1, H_1 + \Delta H)$ to lie in L and for the absolute maxima in the remaining regions $\bar{L} = (-\infty, \infty)\backslash L$ to be less than H in magnitude (see Fig. 5.8).

If $l(t) \gg l_0$, which is assumed valid, one can choose the interval L length

Δy to be physically infinitesimal, so that inequalities $l_0 \ll \Delta y \ll l(t)$ hold. Provided these inequalities are fulfilled, the absolute maxima values in L and $\bar{L}$ are essentially statistically independent, and the asymptotic formula for the following probability holds:

$$P(y \in L, H \in (H_1, H_1 + \Delta H)) = W_{max}(H; L)\, \Delta H F_{max}(H; L),$$

where the first factor $W_{max}\,\Delta H$ means the probability for the absolute maximum over interval L to fall within the interval H_1, $H_1 + \Delta H$, while the second factor is concerned with the probability for the absolute maximum in the remaining intervals to be less than H_1. The probability for the coordinate $y(x, t)$ of the function $s(x, y, t)$ absolute maximum to fall within the interval L is then obtained by integrating with respect to all possible values of the absolute maximum,

$$P(y(x, t) \in L) = \int_{-\infty}^{\infty} W_{max}(H; L)\, F_{max}(H; \bar{L})\, \mathrm{d}H. \tag{4.3}$$

Now let us reveal the absolute maximum distribution functions involved in the above equality. As will be shown later, the integrand here differs significantly from zero only when $H \gg \sigma_*$. Hence, when calculating the distribution functions of the absolute maxima one can employ the asymptotic formulae analogous to (3.2.16), formulating the integral distribution function of the absolute maximum via the average number of given level intersections from below by the appropriate random function,

$$F_{max}(H; L) = \mathrm{e}^{-\langle M(H;L)\rangle} \tag{4.4}$$

Calculate the average number of the H-level intersections from below by the function $s(x, y, t)$ within the interval L. According to (3.2.14) it is equal to

$$\langle M(H; L)\rangle = \int_L \mathrm{d}y \int_0^{\infty} v W_{s,v}(H, v; x - y)\, \mathrm{d}v. \tag{4.5}$$

Here $W_{s,v}(s, v; y)$ is the joint probability density of functions s of (4.1) and

$$v(x, y, t) = S_y(x, y, t) = -v_0(y) + \frac{x - y}{t},$$

which, for the case of the initial action $s_0(y)$, is defined by an expression of type (3.1.19). Therefore, we have from (4.5)

$$\langle M(H; L)\rangle = \frac{1}{2\pi l_0^2} \int_L \mathrm{d}y \exp\left[-\left(H - \frac{(x - y)^2}{2t}\right)^2 \Big/ 2\sigma_*^2\right]$$
$$\times\left[\exp\left(-\frac{(x - y)^2}{2\sigma_0^2 t^2}\right) + \frac{(x - y)}{\sigma_0 t}\, g\left(\frac{x - y}{\sigma_0 t}\right)\right]. \tag{4.6}$$

Here $l_0 = \sigma_*/\sigma_0$, and $g(z)$ is determined by (A.16). Provided $H \gg \sigma_*$ the asymptotic formula (4.4) holds, and (4.6) can be replaced by a simpler approximate equality

$$\langle M(H; L)\rangle = \frac{1}{2\pi l_0} \exp\left(-\frac{H^2}{2\sigma_*^2}\right) \int_L \exp\left[-\frac{H(x-y)^2}{2\sigma_*^2\, t}\right] dy. \tag{4.7}$$

Inserting it into (4.4), and (4.4) into (4.3), we see after integration by parts that

$$P(y(x, t) \in L) = \int_{-\infty}^{\infty} \langle M(H; L)\rangle\, W_\infty(H)\, dH. \tag{4.8}$$

Here

$$W_\infty(H) = F'_\infty(H)$$

is the probability density, while $F_\infty(H) = F_{max}(H; (-\infty, \infty))$ is the integral distribution function of the absolute maximum of a function $s(x, y, t)$, expressed asymptotically for $t \gg t_n$ as

$$F_\infty(H) = \exp\left[-\sqrt{\frac{\sigma_*\tau}{H}} \exp\left(-\frac{H^2}{2\sigma_*^2}\right)\right], \tag{4.9}$$
$$\tau = \sigma_0 t/2\pi l_0 \approx t/t_n, \quad t_n = l_0/\sigma_0.$$

This implies that when $t \gg t_n$ (for $\tau \gg 1$) the probability density $W_\infty(H)$ is concentrated at $H >> \sigma_*$.

As was shown in Chapter 3, the values of the statistically homogeneous random function absolute maxima over quite a lengthy interval L obey the double exponential distribution (3.2.21). Formula (4.9) describes the integral distribution function of the statistically inhomogeneous function (4.1) absolute maximum over an infinite interval. Due to the presence of a decreasing parabola $-(x - y)^2/2t$ in (4.1), however, a certain *effective* interval length $l(t)$ is observed in this case, which can contain absolute maxima of the function $s(x, y, t)$. As in Chapter 2, one can portray H as

$$H = H_0(1 + \xi\sigma_*^2/H_0^2), \tag{4.10}$$

where H_0 is the solution to a transcendental equation

$$\sqrt{\frac{\tau\sigma_*}{H_0}} \exp\left(-\frac{H_0^2}{2\sigma_*^2}\right) = 1, \tag{4.11}$$

$$H_0 \approx \sigma_* \sqrt{\mu - \frac{\mu \ln \mu}{2\mu + 1}} \approx \sigma_* \sqrt{\ln \tau}, \tag{4.12}$$
$$\mu = \ln \tau, \quad \tau = t/t_n.$$

On substituting (4.10) into (4.7), we see that when $t \gg t_n$ the integral distribution function of the absolute maxima dimensionless values H is asymptotically described by distribution (3.2.21), while the absolute maxima values are concentrated in a narrow region

$$\frac{\Delta H}{H_0} \approx \frac{\sigma_*^2}{H_0^2} = \frac{1}{\ln \tau} \ll 1, \tag{4.13}$$

near $H_0 = \sigma_* \sqrt{\ln \tau}$. Therefore, as was stated earlier, the integration of (4.3) and (4.8) is, as a matter of fact, performed over a narrow region $\Delta H \ll H_0$ near $H_0 \gg \sigma_*$. Using this fact once again, substituting (4.9) and (4.7) into (4.8), and passing over to integration with respect to (4.10), for the absolute maximum coordinate probability density

$$W_y(y; x, t) = \lim_{\Delta y \to 0} \frac{1}{\Delta y} P(y; L),$$

one obtains the following asymptotic relation:

$$W_y(y; x, t) = \frac{1}{\sqrt{2\pi l^2}} \exp\left[-\frac{(x-y)^2}{2l^2}\right]. \tag{4.14}$$

Here

$$l(t) = \sqrt{\frac{\sigma_*^2 t}{H_0}} = \frac{l_0}{\sqrt[4]{\ln \tau}} \sqrt{\frac{t}{t_n}} \tag{4.15}$$

is the characteristic interval length along the y-axis within which absolute maxima of function (4.1) might fall. The length $l(t)$ can be also interpreted as a characteristic distance between the field $v(x, t)$ zeros, and be used as the BT external scale definition. Obviously, (4.15) coincides, to an accuracy of a logarithmic correction, with a semi-qualitative estimate of an external scale for the case $G^0 = 0$, (2.24). However, although it may seem that this logarithmic correction only slightly affects the external scale growth law, later on we shall see that its presence nevertheless proves vital for the description of statistics of nonlinear random waves with finite Reynolds number.

It follows from (2.1) and (4.14) that the one-point probability density of the field $v(x, t)$ for $t \gg t_n$ is Gaussian as well,

$$W_u(v; t) = \frac{1}{\sqrt{2\pi}\, b(t)} \exp\left(-\frac{v^2}{2b^2(t)}\right), \tag{4.16}$$

$$b^2(t) = \langle v^2(x, t)\rangle = \frac{l^2(t)}{t^2} = \frac{\sigma_0 l_0}{t\sqrt{\ln \tau}}. \tag{4.17}$$

The Gaussian nature of the one-point BT distribution seems to be quite typical, and appears to be valid not only in the one-dimensional model problem under consideration here [18]. This can be attributed to the fact that the field values at any given point are governed by the combined action of a great variety of causes, which allows us to treat Gaussianity as a consequence of the central limit theorem of probability theory. In our case the field $v(x, t)$ is determined for $t \gg t_n$ by the initial field $v_0(x)$ values taken from a region with length $l(t)$ far in excess of the initial correlation length l_0, and besides, if $G_0 = 0$ all values of $v_0(x)$ falling within this region make approximately the same contribution to determination of the field $v(x, t)$.

2. With respect to BT, in the situation considered here one can find the two-point probability densities as well. This permits us to exhaustively describe all the two-point statistical properties of BT, e.g. its spectral–correlation characteristics.

Before proceeding to the asymptotic analysis, it may be helpful to handle some exact features of the two-point probability density. Denote

$$v_1 = v(-s, t), \quad v_2 = v(s, t), \quad y_1 = y(-s, t), \quad y_2 = y(s, t), \tag{4.18}$$

where $2s$ is the distance between the observation points $-s$ and s. Since y_1, y_2 are the coordinates of the absolute maxima, the following inequalities hold:

$$S(-s, y, t) \leq S(-s, y_1, t), \quad S(s, y, t) \leq S(s, y_2, t).$$

Substituting in their left-hand sides y_2 and y_1 respectively, and performing the summation yields

$$y_2 \geq y_1, \quad v_2 - v_1 \leq 2s/t. \tag{4.19}$$

Evidently, $y(x, t)$ is a non-decreasing function of x, and the field $v(x, t)$ growth rate with x is restricted by the inequality $v'_x(x, t) < 1/t$. As $t \to \infty$, as we saw in Section 5.2, $y(x, t)$ is a step function of x with jumps at points x_k (see Fig. 5.9), while the field $v(x, t)$ appears as a sequence of teeth with the same slope $1/t$ (see Fig. 5.5). Therefore, as $t \to \infty$ two situation may arise:

(1) interval $x \in (-s, s)$ has no discontinuities and $y(-s, t) = y(s, t)$;
(2) one or more discontinuities fall within this interval.

Thus, the two-point probability densities of the absolute maxima and field coordinates can be represented as

$$\begin{aligned} W_y(y_1, y_2; 2s, t) &= \delta(y_1 - y_2)W_y^{\mathrm{d}} + W_y^{c}, \\ W_v(v_1, v_2; 2s, t) &= \delta(v_2 - v_1 - 2s/t)W_y^{\mathrm{d}} + W_y^{c}. \end{aligned} \tag{4.20}$$

Here the first and the second terms on the right-hand side refer to the statistical properties of $y(x, t)$ and $v(x, t)$ with the field discontinuities within the interval $(-s, s)$ absent or present, respectively. Moreover, in accord with (4.19), $W_y^{\mathrm{c}} \equiv 0$ if $y_2 < y_1$ and $W_v^{\mathrm{c}} = 0$ if $v_2 - v_1 > 2s/t$.

At sufficiently large $t \gg t_n$, the absolute maxima coordinates y_k differ from

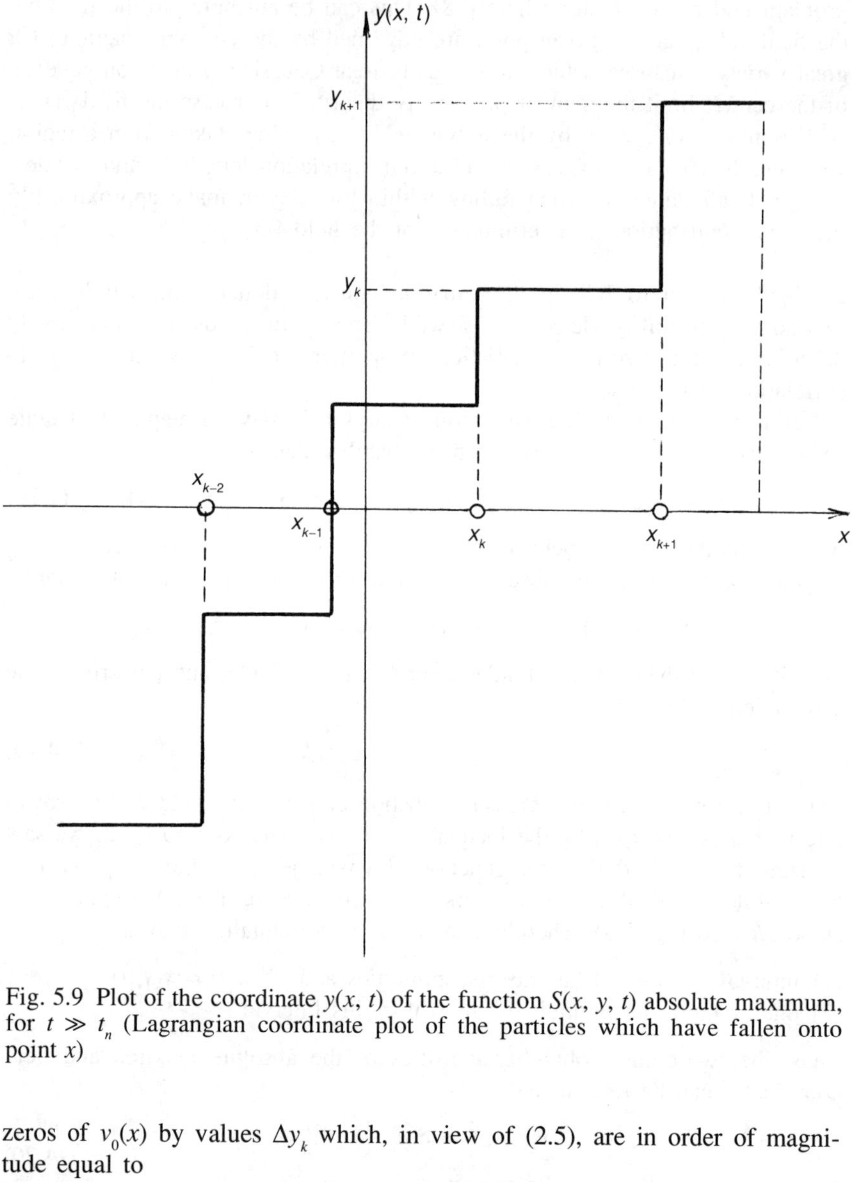

Fig. 5.9 Plot of the coordinate $y(x, t)$ of the function $S(x, y, t)$ absolute maximum, for $t \gg t_n$ (Lagrangian coordinate plot of the particles which have fallen onto point x)

zeros of $v_0(x)$ by values Δy_k which, in view of (2.5), are in order of magnitude equal to

$$\Delta y_k \approx |x - y_k| \, t_n/t \approx l_0\sqrt{t_n/t}. \tag{4.21}$$

The Δy_k differing from zero leads to deviation of the field $v(x, t)$ growth law (2.1) from the linear within the intervals between the discontinuities. As is clear from (4.21), Δy_k is far less than the initial scale l_0 and, consequently, in its turn, is much less than the BT external scale $l(t) \gg l_0$. The non-zero Δy_k

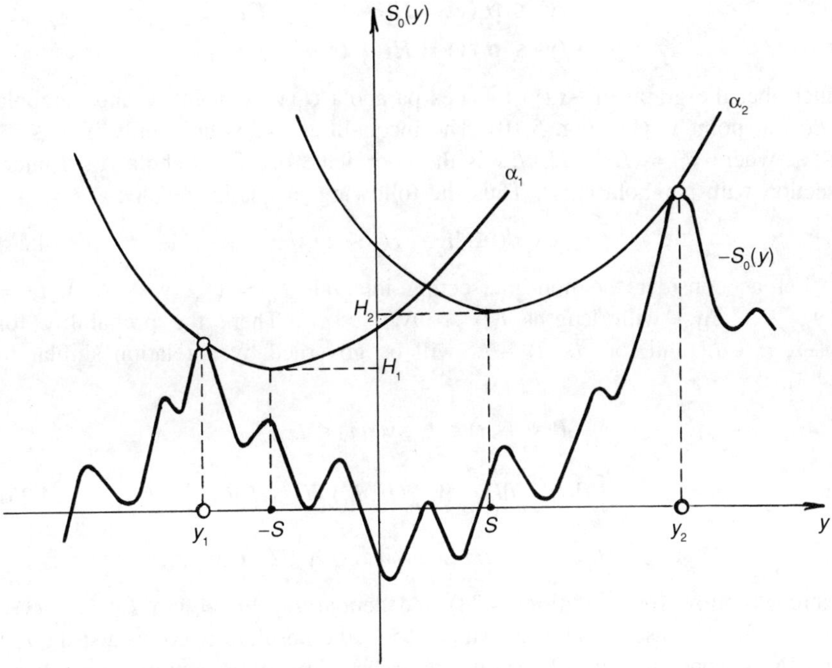

Fig. 5.10 Illustration related to the calculation of the absolute maximum coordinates $y(-s, t)$ and $y(s, t)$ two-point probability density

imply that the terms in (4.20) corresponding to the no-discontinuity case within the interval of length $2s$ are not, strictly speaking, considered to be delta functions. They are blurred. Such blurring, however, is of the order of $l_0\sqrt{t_n/t}$ (4.21) and much less compared to the characteristic width of $l(t)$, equal to $l_0\sqrt{t/t_n}$.

Now let us proceed to calculate the two-point probability densities (4.20). To begin with, separate out the continuous part of the absolute maxima coordinate probability density $W_y^c(y_1, y_2; 2s, t)$, i.e. assume that there are discontinuities within the interval $(-s, s)$, and $y_2 \neq y_1$. Let H_1 and H_2 be the values of the absolute maxima $S(-s, y, t)$ and $S(s, y, t)$ with coordinates equal to $y(-s, t) = y_1$ and $y(s, t) = y_2$, respectively. In other words, let the inequalities

$$\begin{aligned} S(-s, y, t) &\leq S(-s, y_1, t) = H_1, \\ S(s, y, t) &\leq S(s, y_2, t) = H_2 \end{aligned} \tag{4.22}$$

hold. they give birth to a vital-in-future restriction on H_1 and H_2. In fact (4.22) and (4.1) imply that

$$
\begin{aligned}
-s_0(y) \le \alpha_1(y) &= H_1 + (s+y)^2/2t, \\
-s_0(y) \le \alpha_2(y) &= H_2 + (s-y)^2/2t,
\end{aligned} \tag{4.23}
$$

such that the graph of $-s_0(y)$ touches parabola $\alpha_1(y)$ at point y_1 and parabola $\alpha_2(y)$ at point y_2 (see Fig. 5.10). The inequalities (4.23) hold only if $y_1 \le y^* \le y_2$, where $y^* = (H_2 - H_1)t/2s$ is the coordinate of the parabola $\alpha_1(y)$ intersection with parabola $\alpha_2(y)$. Thus the following inequality results:

$$
2y_1 s/t \le H_2 - H_1 \le 2y_2 s/t. \tag{4.24}
$$

Let us consider the non-intersecting intervals $L_1 = (y_1, y_1 + \Delta y_1)$, $L_2 = (y_2, y_2 + \Delta y_2)$ with lengths $l(t) \gg \Delta y_{1,2} \gg l_0$. Then, the probability for $y(-s, t) \in L_1$ and for $y(s, t) \in L_2$ will be governed by a relation similar to (4.3),

$$
\begin{aligned}
&P(y(-s,t) \in L_1,\ y(s,t) \in L_2) \\
&= \int_{-\infty}^{\infty} dH_1 \int_{H_1+2y_1 s/2}^{H_1+2y_2 s/t} dH_2\, W_{\max}(H; L_1)\, W_{\max}(H; L_2) \\
&\times F_{\max}(H; \overline{L}_1)\, F_{\max}(H; \overline{L}_2) \approx W_y^c(y_1, y_2; 2s, t)\Delta y_1\, \Delta y_2
\end{aligned} \tag{4.25}
$$

Here we allow for condition (4.24) and denote $\overline{L}_1 = (-\infty, y^*)\backslash L_1$, $\overline{L}_2 = (y^*, \infty)\backslash L_2$. In the case of interest, when $t \gg t_n$ the absolute maxima distribution functions appearing in (4.25) are described by an asymptotic expression (4.4); substituting the latter into (4.25) we obtain

$$
\begin{aligned}
W_y^c(y_1, y_2; 2s, t) &= \frac{1}{(2\pi l_0)^2} \exp\left[-\frac{H_0^2}{\sigma_*^2} - \frac{(y_1+s)^2}{2l^2} - \frac{(y_2-s)^2}{2l^2}\right] \\
&\times \int_{-\infty}^{\infty} d\xi_1 \int_{\xi_1+2y_1 s/l^2}^{\xi_2+2y_2 s/l^2} d\xi_2\, \exp\left[-\xi_1 - \xi_2 - e^{-\xi_1}\frac{1}{\sqrt{2\pi}}\, g\left(\frac{s}{l} + \frac{l(\xi_1-\xi_2)}{2s}\right)\right. \\
&\left. - e^{-\xi_2}\frac{1}{\sqrt{2\pi}}\, g\left(\frac{s}{l} - \frac{l(\xi_1-\xi_2)}{2s}\right)\right].
\end{aligned} \tag{4.26}
$$

Here the definition of the turbulence external scale $l(t)$ (4.15) was employed. In terms of dimensionless variables $\rho = s/l$, $z_{1,2} = y_{1,2}/l$, and taking into account that according to (4.11) and (4.15)

$$
\frac{1}{2\pi\, l_0^2} \exp\left(-\frac{H_0^2}{\sigma_*^2}\right) = \frac{1}{l^2(t)}, \tag{4.27}
$$

after replacing in (4.26) the integration variables by $\xi = \xi_1$ and $z = (\xi_1 - \xi_2)/l$ and calculating the integral over ξ, one arrives at the final expression for the continuous portion of the probability density of the dimensionless random values $z_1 = y(-s, t)/l(t)$, $z_2 = y(s, t)/l(t)$:

$$W_z^c(z_1, z_2; 2\rho) = l^2 W_y^c(z_1 l, z_2 l; 2\rho l) = \exp\left[-\frac{1}{2}(z_1+\rho)^2 - \frac{1}{2}(z_2-\rho)^2\right] \tag{4.28}$$
$$\times \int_{z_1}^{z_2} \frac{2\rho \, \mathrm{d}z}{[g(\rho+z)\exp(\rho z) + g(\rho - z)\exp(-\rho z)]^2}.$$

Recall that $g(x) = \int_{-\infty}^{x} \exp(-z^2/2)\, \mathrm{d}z$ from (A.19).

The discrete part W_z is to be found by making use of the compatibility condition for the one- and two-point probability densities. Prior to this let us write down an equality analogous to (4.20) for the joint probability density of z_1 and z_2:

$$W_z(z_1, z_2; 2\rho) = W_z^d(z_1; 2\rho)\,\delta(z_1 - z_2) + W_z^c(z_1, z_2; 2\rho). \tag{4.29}$$

Integrating it over z_2 within infinite limits we have

$$W_z(z_1) = W_z^d(z_1) + \int_{z_1}^{\infty} W_z^c(z_1, z_2; 2\rho)\, \mathrm{d}z_2. \tag{4.30}$$

In this expression $W_z(z_1)$ on the left-hand side is the one-point probability density of the dimensionless random value $z = y(-s, t)/l(t)$ equal, according to (4.14), to

$$W_z(z_1) = \frac{1}{\sqrt{2\pi}} \exp\left[-\frac{(z_1+\rho)^2}{2}\right]. \tag{4.31}$$

The equality (4.30) defines uniquely the form of the probability density discrete part W_z^d through known W_z and W_z^c. In order to recover an explicit expression for W_z^d one needs to calculate the integral in (4.30). Taking into consideration that

$$\exp[-\tfrac{1}{2}(z_2-\rho)^2]\, \mathrm{d}z_2 = -\mathrm{d}g(\rho - z_2),$$

we have

$$\int_{z_1}^{\infty} W_z^e \, \mathrm{d}z_2 = -\exp\left[-\frac{(z_1+\rho)^2}{2}\right] \int_{z_1}^{\infty} \mathrm{d}g(\rho - z_2) \int_{z_1}^{z_2} \frac{2\rho\, \mathrm{d}z}{g(\rho+z)\,\mathrm{e}^{\rho z} + g(\rho-z)\,\mathrm{e}^{-\rho z}}$$
$$= \exp\left[-\frac{1}{2}(z_1+\rho)^2\right] \int_{z_1}^{\infty} \frac{2\rho g(\rho - z)\, \mathrm{d}z}{[g(\rho+z)\,\mathrm{e}^{\rho z} + g(\rho-z)\,\mathrm{e}^{-\rho z}]^2}.$$

Recognising that the integrand here is equal to

$$\frac{2\rho g(\rho - z)}{[g(\rho+z)\,\mathrm{e}^{\rho z} + g(\rho-z)\,\mathrm{e}^{-\rho z}]^2} = \frac{\mathrm{d}}{\mathrm{d}z}\left(\frac{1}{g(\rho+z) + g(\rho-z)\,\mathrm{e}^{-2\rho z}}\right)$$

we eventually get

$$\int_{z_1}^{\infty} W_z^c \, \mathrm{d}z_2 = \frac{1}{\sqrt{2\pi}}\, \mathrm{e}^{-\frac{1}{2}(z_1+\rho)^2} - \frac{\exp(-z_1^2/2 - \rho^2/2)}{g(\rho+z_1)\,\mathrm{e}^{\rho z_1} + g(\rho - z_1)\,\mathrm{e}^{-\rho z_1}}.$$

Hence, it follows from (4.30) and (4.31) that

$$W_z^{\mathrm{d}}(z) = \frac{\exp(-z^2/2 - \rho^2/2)}{g(\rho + z)\,\mathrm{e}^{\rho z} + g(\rho - z)\,\mathrm{e}^{-\rho z}}.$$

Recalling that the continuous part of the two-point probability density of z_1 and z_2 is described by (4.29), one is capable of formulating a full expression concerning the two-point probability density of the absolute maxima coordinates,

$$W_z(z_1, z_2; 2\rho) = \frac{\delta(z_1 - z_2)\exp(-z_1^2/2 - \rho^2/2)}{g(\rho + z_1)\,\mathrm{e}^{\rho z_1} + g(\rho - z_2)\,\mathrm{e}^{-\rho z_2}}$$

$$+ \exp\left[-\frac{1}{2}(z_1 + \rho)^2 - \frac{1}{2}(z_2 - \rho)^2\right] \tag{4.32}$$

$$\times \int_{z_1}^{z_2} \frac{2r\,\mathrm{d}z}{[g(\rho + z)\,\mathrm{e}^{\rho z} + g(\rho - z)\,\mathrm{e}^{-\rho z}]^2}.$$

Passing from $y(-s, t)$, $y(s, t)$ to the field values $v(-s, t)$ and $v(s, t)$ and introducing the dimensionless normalised fields

$$q_1 = v(-s, t)/b(t), \quad q_2 = v(s, t)/b(t), \tag{4.33}$$

one derives from (4.33)

$$W_q(q_1, q_2; 2\rho) = W_z(-q_1 - \rho, -q_2 + \rho; 2\rho)$$

$$= \frac{\delta(q_2 - q_2 - 2\rho)}{g(-q_1)\exp(q_1^2/2) + g(q_2)\exp(q_2^2/2)} \tag{4.34}$$

$$+ \exp\left(-\frac{q_1^2}{2} - \frac{q_2^2}{2}\right)\int_{-\rho-q_1}^{\rho-q_2} \frac{2\rho\,\mathrm{d}z}{[g(\rho + z)\,\mathrm{e}^{\rho z} + g(\rho - z)\,\mathrm{e}^{-\rho z}]^2}.$$

As is evident from (4.33) and (4.34) the two-point probability density is self-preserving, and only depends on a single spatial scale, i.e. the turbulence external scale $l(t)$ of (4.15). As distinct from the one-point distribution (4.11), the two-point one is essentially non-Gaussian. The first term in (4.34) corresponds to lack of field discontinuities over the interval of length $2s = 2\rho l$. It is normalised to

$$P(2\rho) = \int_{-\infty}^{\infty} \frac{\mathrm{d}z}{g(\rho + z)\exp[(\rho + z)^2/2] + g(\rho - z)\exp[(\rho - z)^2/2]}, \tag{4.35}$$

being the probability of discontinuity absence within the interval of length $2\rho l$.

Consider the limit behaviour of the two-point probability density (4.34) as $\rho \to 0$ and $\rho \to \infty$. Observing that for $\rho \gg 1$, $g(\rho + z) \approx g(\rho - z) \approx \sqrt{2\pi}$ and $g(-\rho) + g(\rho) = \sqrt{2\pi}$, one can show from (4.34) that

$$W_q(q_1, q_2; 2\rho) = \begin{cases} \delta(q_1 - q_2)\dfrac{1}{\sqrt{2\pi}}\exp\left(-\dfrac{q_1^2}{2}\right), & \rho \to 0, \\ \dfrac{1}{2\pi}\exp\left(-\dfrac{q_1^2}{2} - \dfrac{q_2^2}{2}\right), & \rho \to \infty, \end{cases} \tag{4.36}$$

i.e. the distributions in these limiting cases tend to the Gaussian ones. This can be attributed to the fact that as $\rho \to 0$, $q_2 \to q_1$ and the two-point distribution degenerates into the one-point Gaussian (4.16) as well, while as $\rho \to \infty$ the two-point distribution is broken down into the product of one-point distributions. At $\rho \to 1$ the distribution (4.34) is essentially non-Gaussian (see Fig. 5.11).

3. The distribution (4.34) determines all the two-point moment functions of BT. When calculating them it seems more appropriate, however, to use the probability distribution of the absolute maxima coordinates (4.32), bearing in mind that dimensionless values of BT variables q_i are related to z_i through

$$q_1 = -\rho - z_1, \quad q_2 = \rho - z_2. \tag{4.37}$$

Let us reveal, for example, the correlation function of BT using (4.37) and the probability distribution (4.32). In analogy with the probability distribution, we break the correlation function down into two summands,

$$R(2\rho) = \langle q_1 q_2 \rangle = \langle (z_1 + \rho)(z_2 - \rho) \rangle = R_1 + R_2, \tag{4.38}$$

where

$$R_1 = \int_{-\infty}^{\infty} \frac{z^2 - \rho^2}{A(\rho, z)}\, dz,$$

$$R_2 = \int_{-\infty}^{\infty} dz_1 (z_1 + \rho)\exp\left[-\frac{(z_1 + \rho)^2}{2}\right] \int_{z_1}^{\infty} dz_2 (z_2 - \rho)$$

$$\times \exp\left[-\frac{(z_2 - \rho)^2}{2}\right] \int_{z_1}^{z_2} \frac{2\rho \exp(z^2 + \rho^2)\, dz}{A(\rho, z)}.$$

Here we took into consideration that $z_1 \geq z_1$ and introduced the notation

$$A(\rho, z) = g(\rho + z)\exp\left[\frac{(\rho + z)^2}{2}\right] + g(\rho - z)\exp\left[\frac{(\rho - z)^2}{2}\right].$$

Getting rid, in the expression for R_2, of the integrals over z_2 and z_1 by means of integration by parts, we formulate the following equation for the correlation function,

$$R(2\rho) = R_1 + R_2 = \int_{-\infty}^{\infty} \left[\frac{z^2 - \rho^2}{A(\rho, z)} - \frac{2\rho}{A^2(\rho, z)}\right] dz.$$

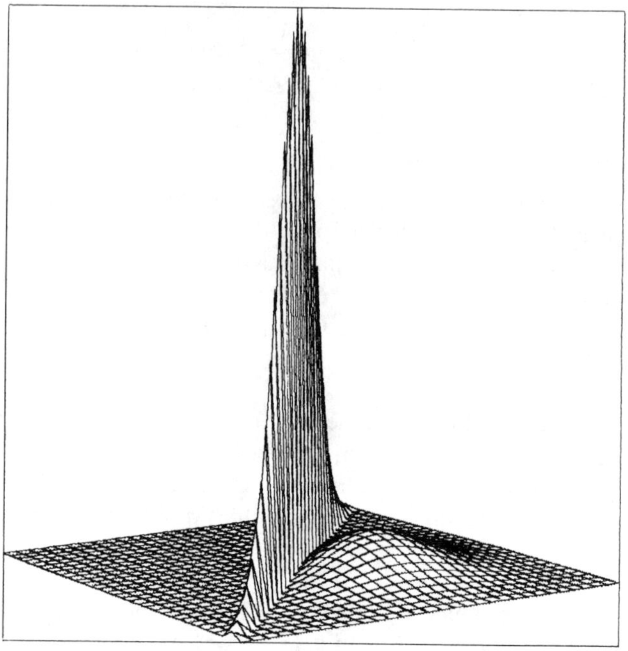

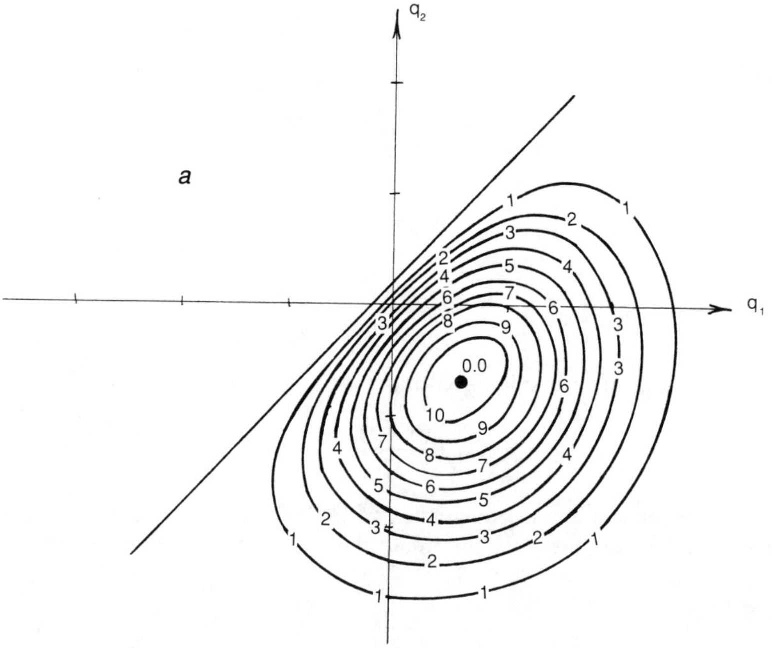

Fig. 5.11 Two-point probability distribution of the velocity field, and its level lines $\rho = 0.1$, (a); $\rho = 0.6$, (b); $\rho = 1$, (c)

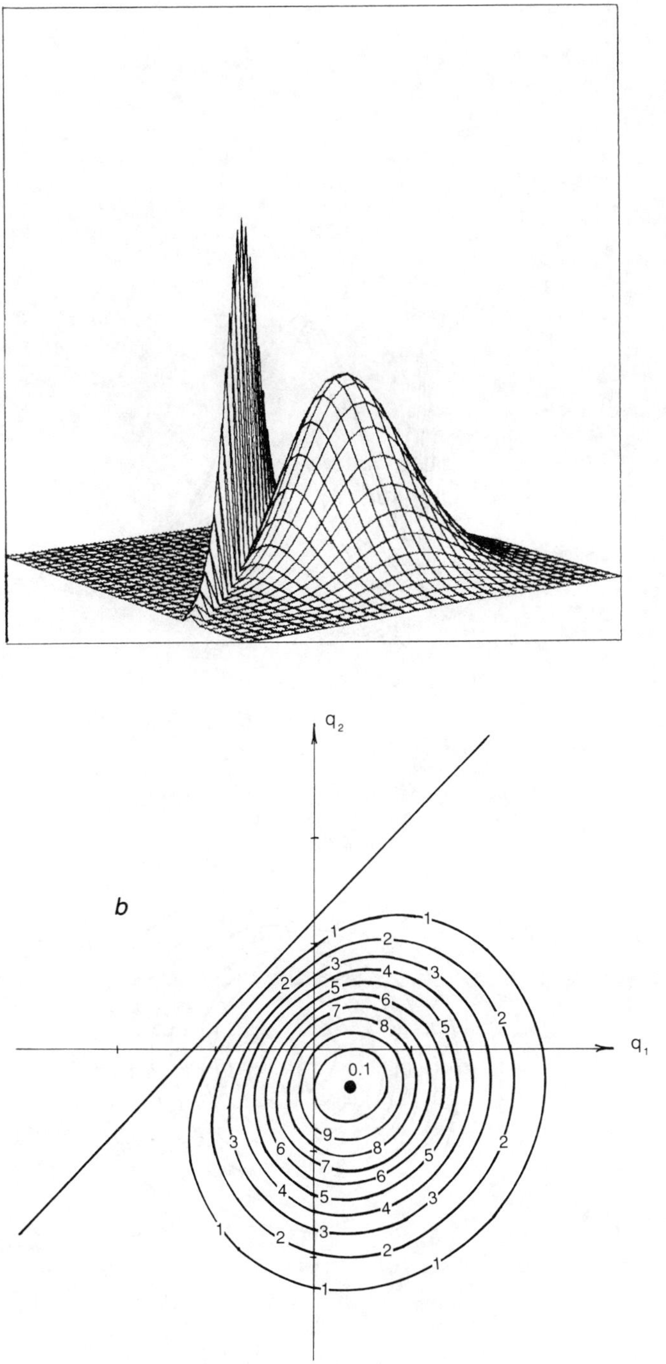
q2
q1
b
0.1

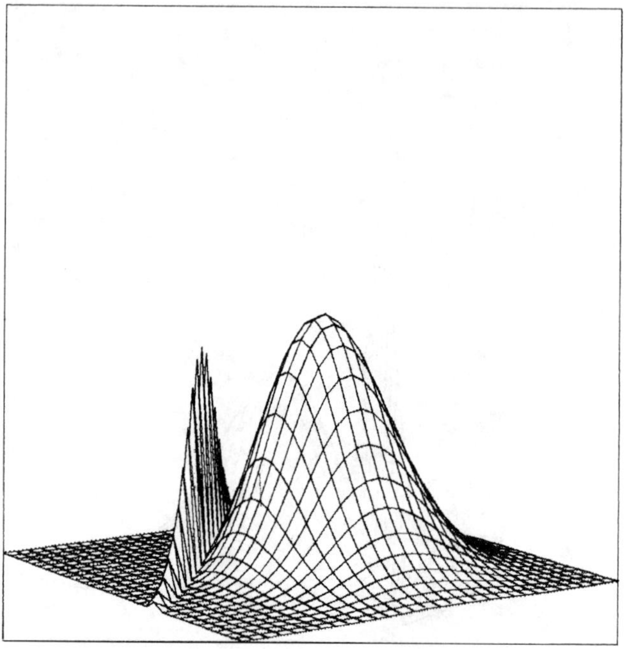

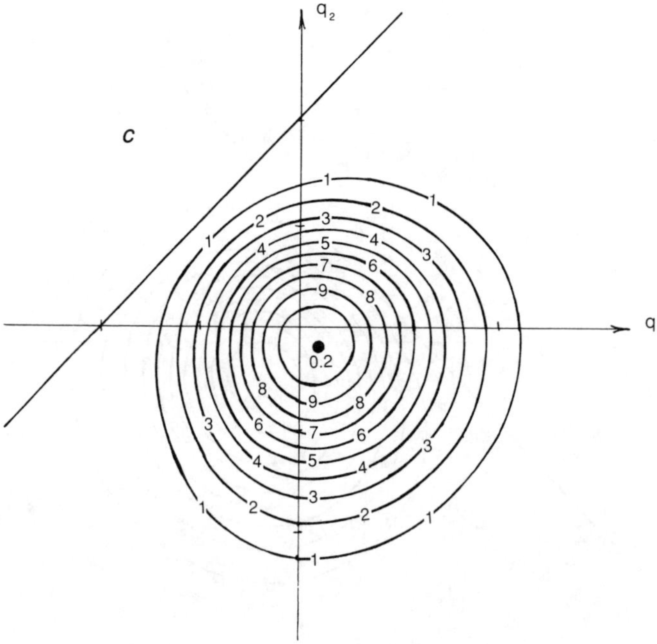
q2
q1
c
1
2
3
4
5
6
7
8
9
0.2

Then using the identity

$$\frac{z^2 - \rho^2}{A(\rho, z)} - \frac{2\rho}{A^2(\rho, z)} = \left(\rho\frac{\partial}{\partial\rho} - z\frac{\partial}{\partial z}\right)\frac{1}{A(\rho, z)}$$

we shall get

$$R(2\rho) = \rho\frac{d}{d\rho}\int_{-\infty}^{\infty}\frac{dz}{A(\rho, z)} - \int_{-\infty}^{\infty} z\frac{\partial}{\partial z}\left(\frac{1}{A(\rho, z)}\right) dz.$$

Taking the last integral by parts we finally find

$$R(2\rho) = \langle q_1 q_2\rangle = \frac{d}{d\rho}[\rho P(2\rho)]. \tag{4.39}$$

Here $P(2\rho)$ of (4.35) is the probability for the interval of length $2\rho l$ to have no discontinuities. Obviously, (4.39) satisfies the condition

$$\int_{-\infty}^{\infty} R(2\rho)\, d\rho = 0,$$

meaning that BT spectral density equals zero at zero frequency. The plot of $R(2\rho)$ is depicted in Fig. 5.12.

For the BT dimensionless spectral density one can deduce the following expression from (4.39):

$$\tilde{G}(\kappa) = \frac{1}{\pi}\int_0^{\infty} R(z)\cos(z\kappa)\, dz = \frac{\kappa}{\pi}\int_0^{\infty} zP(z)\sin(z\kappa)\, dz. \tag{4.40}$$

Inasmuch as the correlation function and the BT spectrum are expressed by means of the probability density (4.35), its asymptotic behaviours can be represented as

$$P(2\rho) = \begin{cases} 1 - \dfrac{2|\rho|}{\sqrt{\pi}}, & |\rho| \ll 1, \\ \sqrt{\dfrac{\pi}{8}}\dfrac{1}{\rho}\exp\left(-\dfrac{\rho^2}{2}\right), & |\rho| \gg 1. \end{cases} \tag{4.41}$$

Utilising (4.39) and (4.41), the appropriate asymptotic behaviours of the correlation function can be readily derived, as

$$R(2\rho) = \begin{cases} 1 - \dfrac{4|\rho|}{\sqrt{\pi}}, & |\rho| \ll 1, \\ -\sqrt{\dfrac{\pi}{8}}\,\rho\exp\left(-\dfrac{\rho^2}{2}\right), & |\rho| \gg 1. \end{cases} \tag{4.42}$$

As we demonstrated in Chapter 2, non-analyticity of the correlation function at $\rho = 0$ resulting from (4.42), and related to the presence of discontinuities in BT, leads to a power law of the BT spectral density decrease,

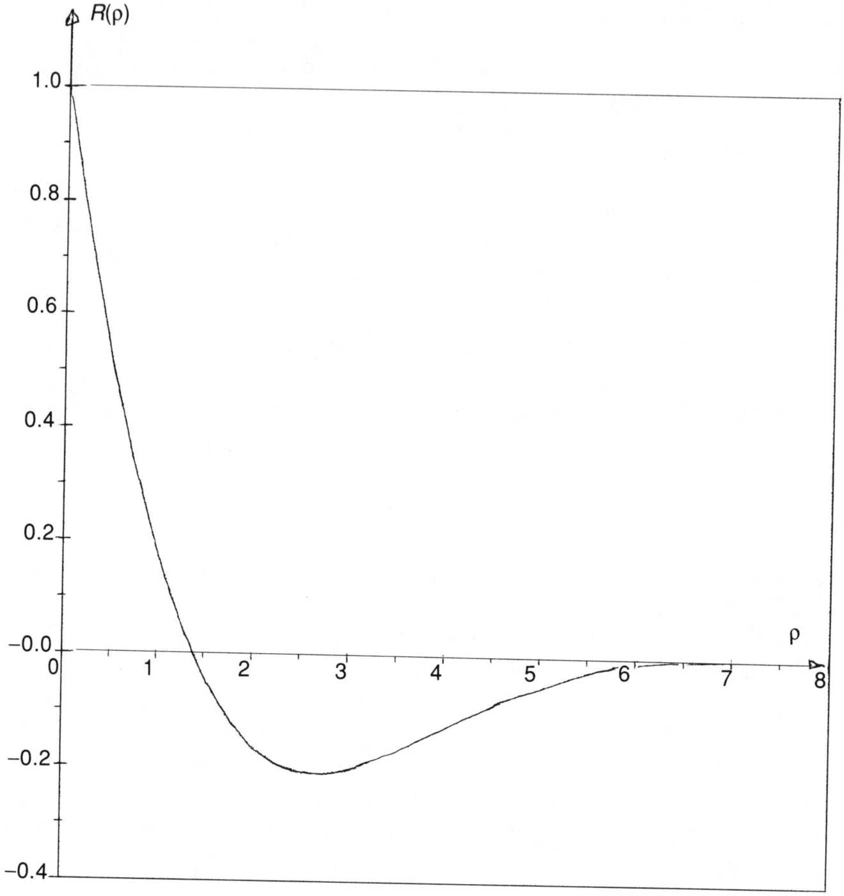

Fig. 5.12 Normalised correlation function of Burgers turbulence at the self-preserving stage

$$\tilde{G}(\kappa) \approx \frac{2}{\pi\sqrt{\pi}\,\kappa^2} \approx \frac{0.36}{\kappa^2}, \quad \kappa \gg 1. \tag{4.43}$$

With respect to the spectrum within the low-frequency range, (4.40) also yields a universal asymptotic behaviour,

$$\tilde{G}(\kappa) = \mu^2\kappa^2, \quad \mu^2 = \frac{8}{\pi}\int_0^\infty \rho^2\, P(2\rho)\, d\rho \approx 1.08. \tag{4.44}$$

Its emergence is associated with the energy pumping to the low-frequency spectrum range. The plots of $\tilde{G}(\kappa)$ are presented in Fig. 5.13. They demonstrate that, with the exception of a comparatively small frequency range in the vicinity of $\kappa_* \approx 0.9$ where the spectrum acquires a maximum value, its behaviour is well described by power laws (4.43) and (4.44). Allowing for the relation $G_v(k; t) = l^3\tilde{G}(kl)/t^2$, we assess the BT spectral density at large and low kl as

$$G_v(k;t) = \begin{cases} \dfrac{2l(t)}{k^2\pi\sqrt{\pi}\, t^2} \sim \dfrac{1}{k^2 t\sqrt{t}}, & kl \gg 1, \\ \dfrac{\mu^2 k^2 l^5}{t^2} \sim k^2\sqrt{t}, & kl \ll 1. \end{cases} \tag{4.45}$$

As has been already stated, the power asymptotic behaviour in the high-frequency range is related to the discontinuity formation, whereas the appearance of the power law in the low-frequency range is concerned with the low frequency spectral component parametric generation. A spectrum maximum is therefore shifted towards low k as $1/l(t)$, which is attributed to growth of the turbulence external scale due to discontinuity coalescence.

4. The theory developed here stems essentially from the assumption of Gaussianity of the initial field $v_0(x)$. This requirement, however, appears in fact to be not that strict. At the stage of developed discontinuities, when the characteristic turbulence length $l(t)$ is far in excess of the initial correlation radius l_0, one can for the BT analysis apply an approximate procedure set forth in [92]. Divide the y-axis in (4.1) into intervals I_n with lengths Δx. Provided $l_0 \ll \Delta x < l$, the actions on adjacent intervals can be considered as statistically independent, and the characteristic scale of the field $v(x, t)$ statistical property variation proves much more than the discretisation length Δx. Let S_n be the maximum value of the function S of (4.1) over the nth interval. Then, considering the sequence of random numbers S_n one can, by using standard procedures, reveal the probability for the absolute maximum to fall within the kth interval. This direction will be followed in Chapter 6 as regards the potential turbulence described by a three-dimensional BE. In [92] this approach was developed in order to find the probability distributions of the discontinuity amplitudes and velocities. The application of the multi-point distribution functions of the discontinuity allowed one to derive the BT correlation functions and spectra. The expressions so obtained, being properly normalised, coincide with the equations (4.39) and (4.40) given above. So, the correlation function $\tilde{B}(\tilde{r})$ offered in [92] is connected with the correlation factor $R(z)$ (4.39) through $R(z) = \pi\tilde{B}(z/\sqrt{\pi})$, while the corresponding spectra are $\tilde{G}(\kappa) = \pi\sqrt{\pi}\, E(\kappa\sqrt{\pi})$. The differences only refer to the external scale $l(t)$ dependence of the initial action $s_0(y)$ characteristics. If, within the current consideration, the external scale $l(t)$ is defined uniquely under Gaussian statistics of $v_0(x)$ by the first two moments of the initial spectrum $G_0(k)$ then, in the framework of a discrete model it is necessary to make assumptions as to the probability distribution form of the function s_0 absolute maximum over the interval Δx. Let h be a value of the maximum and let h have the following asymptotic behaviour:

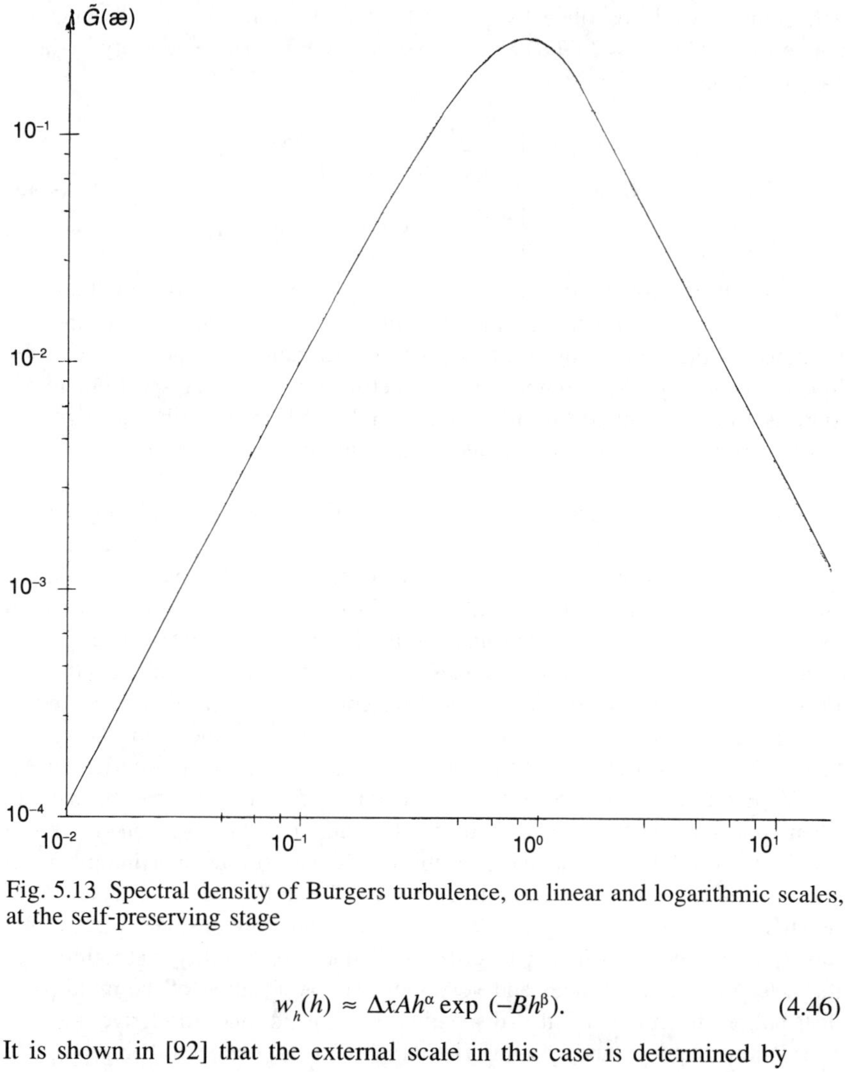

Fig. 5.13 Spectral density of Burgers turbulence, on linear and logarithmic scales, at the self-preserving stage

$$w_h(h) \approx \Delta x A h^{\alpha} \exp\,(-Bh^{\beta}). \tag{4.46}$$

It is shown in [92] that the external scale in this case is determined by

$$l(t) = \sqrt{\frac{\pi t}{\beta B^{\frac{1}{\beta}}}} \left[\ln \left\{ A \sqrt{\frac{\pi t}{2}}\, \beta^{-\frac{3}{2}}\, \beta^{-(2\alpha+3)/2\beta} \right\} \right]^{(1-\beta)/2\beta} . \tag{4.47}$$

Thus the external scale growth law $l(t)$ does not, in fact, depend on the parameters of the probability distribution (4.46). One only requires $w_h(h)$ to fall off rapidly enough as $h \to \infty$. It can be noted that at $\beta = 1$, $\alpha = 0$ the power law $l \sim \sqrt{t}$ appears to be exact. Out of four parameters in the probability distribution there are only two, B and β, that critically affect the value of $l(t)$, whereas the dependence on A and α proves logarithmically weak.

Finally, it can be remarked that numerous contributions are known wherein

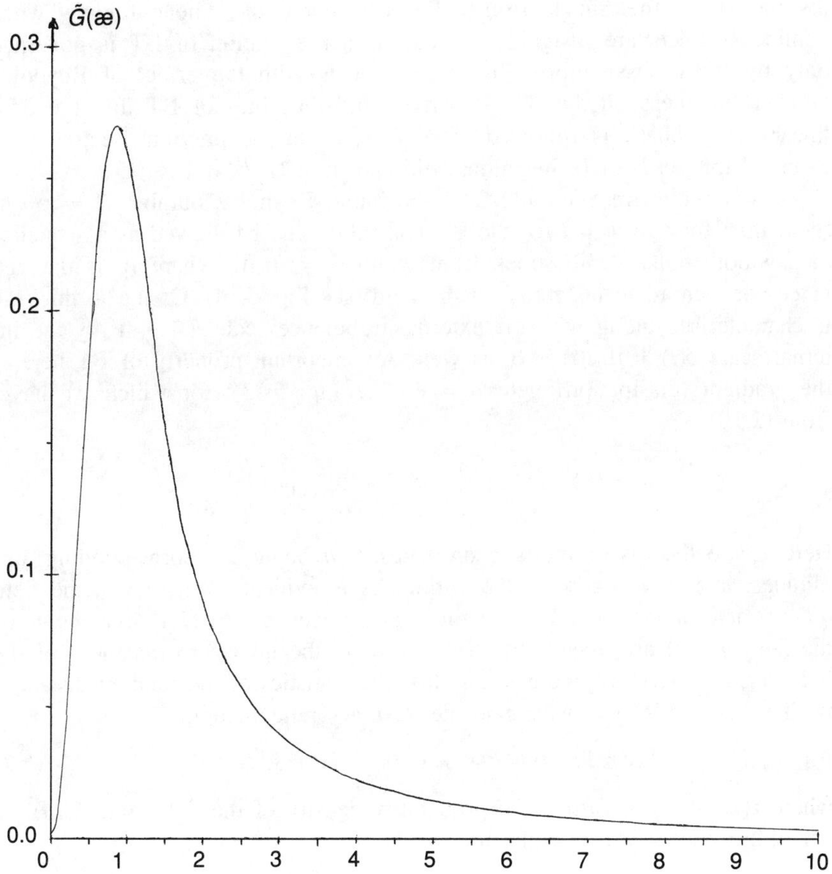

a numerical simulation of BT was performed for various types of initial spectrum (see, for example, [47, 92, 133–135]). The results obtained confirm the existence of two qualitatively different growth laws of the BT external scale, $l \sim t^{1/2}$ and $l \sim t^{2/3}$. It is worth emphasising that the universal self-preserving power laws of the spectrum are observed to be initially established within the high-frequency range, the low-frequency one coming next. As concerns intense acoustical noise, the initial stage of the self-preserving regime formation was observed experimentally in [136, 137].

5.5 Turbulence at finite Reynolds numbers; final stage of decay

1. As far as finite Reynolds numbers are concerned, when $\nu \neq 0$ the BT shock fronts have a non-zero thickness and as internal turbulence scale $\delta(t)$. In this case, due to the increase of the chock front thickness and the energy

dissipation in the shock fronts, BT can enter the linear regime when nonlinear effects are insignificant, and further evolution of BT is governed only by linear dissipation. This section deals with the effect of Reynolds number finiteness on the BT properties, and the entry of BT into the final linear decay phase is followed. Therefore, as in the previous section it is assumed that as $k \to 0$ the initial field spectrum $G_0(k) \sim k^n$, $n \geq 2$.

As was shown in Section 5.2, if the initial Reynolds number $R_0 = \sigma_0 l_0/\mu \gg 1$, then for $t \gg t_n = l_0/\sigma_0$ the BT realisation can be viewed as a sequence of sawtooth pulses with shock front width $\delta_k = \mu t/\eta_k$, where η_k is the distance between adjacent 'zeros' of the field (see Fig. 5.14). Owing to this, BT is characterised along with the external turbulence scale $l(t) \sim \eta_k$ by the internal scale $\delta(t) = \mu t/l(t) \sim \delta_k$ as well. An important property of BT here is the gradient (dissipation) field $q = v_x$ (see Fig. 5.13), for which we have, from (2.10),

$$q(x, t) = \frac{\partial v(x, t)}{\partial x} = \frac{1}{t} - \frac{1}{t} \sum_k \frac{\eta_k}{2\delta_k} \operatorname{sech}^2 \left(\frac{x - x_k}{\delta_k} \right). \tag{5.1}$$

Here x_k are the discontinuity coordinates, η_k/t being the corresponding amplitudes, and $\delta_k = 4\mu t/\eta_k$ is the width. As is evident from (2.12), the field $q^2(x, t)$ determines the BT dissipation rate. According to (5.1) overshoots of the field $q(x, t)$ are essentially determined by the inter-zero distances of the field $v(x, t)$. Therefore, we examine first the statistics of the random distances η_k. To this end let us investigate the auxiliary random field

$$\Delta(x; 2s, t) = z(x + s, t) - z(x - s, t) \geq 0, \tag{5.2}$$

where $z(x, t) = y(x, t)/l(t)$. The probability density of the field $\Delta(x; 2s, t)$, as per (4.29) and (4.35), is equal to

$$W_\Delta(\Delta; 2\rho) = P(2\rho)\, \delta(\Delta) + \int_{-\infty}^{\infty} W_z^c(z, z + \Delta; 2\rho)\, dz,$$
$$\rho = s/l(t). \tag{5.3}$$

Write another representation of $W_\Delta(\Delta; 2\rho)$, noting that if not a single discontinuity x_k falls within the interval $(x - s, x + s)$ then $\Delta(x; 2s, t) = 0$; provided there is one discontinuity in this interval $\Delta(x; 2s, t) = \eta_k/l(t) = \theta_k$, i.e. $\Delta = \theta_k$ is equal to the dimensionless discontinuity amplitude or to the distance between the adjacent zeros. Generally, if the interval $(x - s, x + s)$ includes N discontinuities we have

$$\Delta(x; 2s, t) = \sum_{n=1}^{N} \eta_{k+n-1}/l(t) = \sum_{n=1}^{N} \theta_{k+n-1}.$$

Therefore, using the total probability formula one can write

$$W_\Delta(\Delta; 2\rho) = P(2\rho; 0)\, \delta(\Delta) + \sum_{N=1}^{\infty} P(2\rho; N)\, W_\Delta(\Delta|N). \tag{5.4}$$

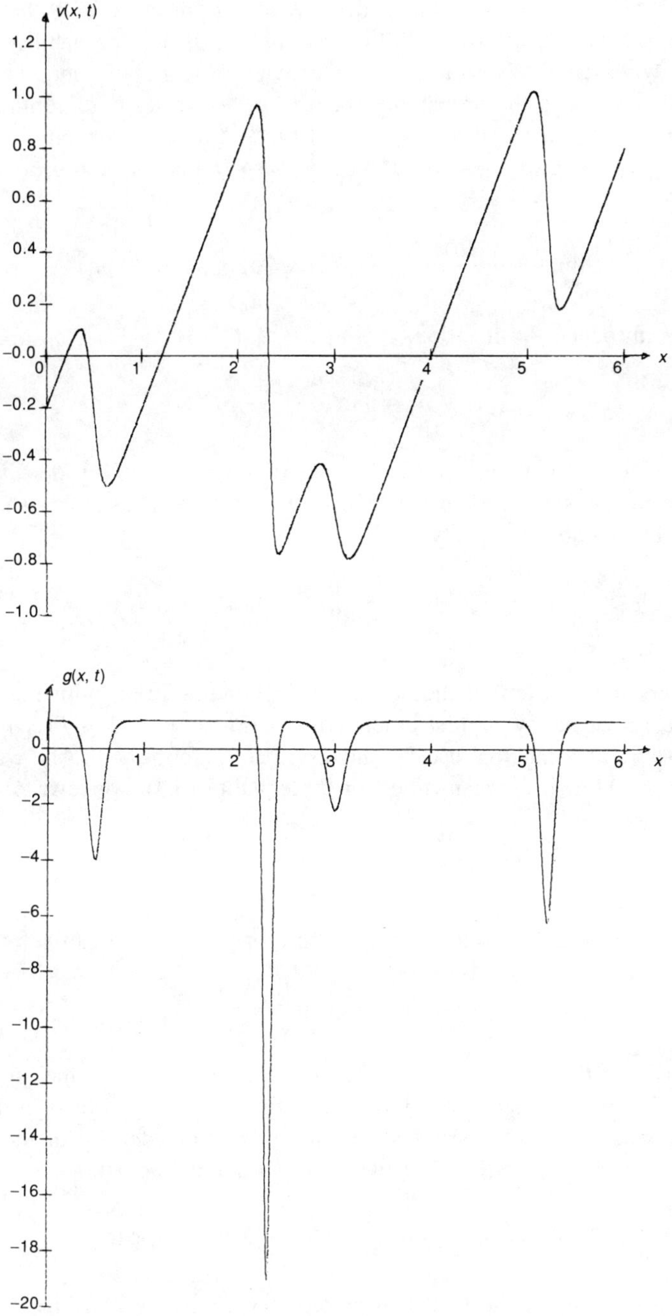

Fig. 5.14 Realisation of the Burgers turbulence velocity field and the corresponding dissipation field at $t = 1$ and $4\mu t = 0.1$

Here $W_\Delta(\Delta | N)$ is the probability density of the distance sum between N adjacent 'zeros', $P(2\rho;\ 0) = P(2\rho)$, see (4.35), is the probability for the interval with length $2s = 2\rho l$ not to contain a single discontinuity, while $P(2\rho, N)$ formulates the probability for it to have exactly N discontinuities of the field $v(x, t)$. Note that as $\rho \to 0$ and for $N \geq 1$, the probability $P(2\rho;\ N)$ tends to zero as $P(2\rho;\ N) \sim \rho^N$. Hence as $\rho \to 0$ and for $\Delta \neq 0$, (5.4) yields the equality

$$\lim_{\rho\to 0} \frac{W_\Delta(\Delta;\ 2\rho)}{2\rho} = \frac{1}{2}\frac{\partial}{\partial\rho} W_\Delta(\Delta;\ 2\rho)\Big|_{\rho=0} = W_0(\Delta) n_*. \tag{5.5}$$

Here we introduce the notation

$$n_* = \lim_{\rho\to 0} \frac{P(2\rho;\ 1)}{2\rho}, \tag{5.6}$$

and take into account that $W_\Delta(\Delta | 1) = W_\theta(\Delta)$ is the probability density of the dimensionless distances between 'zeros' $\theta_k = \eta_k / l(t)$. Thus, (5.5) shows that the sought probability density is

$$W_\theta(\Delta) = \frac{1}{2n_*}\frac{\partial}{\partial\rho} W_\Delta(\Delta;\ 2\rho)\Bigg|_{\rho=0}. \tag{5.7}$$

We remark that in view of the normalisation condition the following equality holds, to the accuracy of first-order terms ρ, $P(2\rho;\ 1) = 1 - P(2\rho)$. Putting it into (5.6) and making use of the asymptotic formula (4.41) we obtain $n_* = 1/\sqrt{\pi}$. Allowing, meanwhile, for the equality (5.3), we rewrite (5.7) as

$$W_\theta(\Delta) = \frac{\sqrt{\pi}}{2}\frac{\partial}{\partial\rho} \int_{-\infty}^{\infty} W_z^c(z, z + \Delta;\ 2\rho)\, dz\Bigg|_{\rho=0}.$$

Inserting here (4.29) and performing simple operations one can see that

$$W_\theta(\Delta) = \frac{\Delta}{2} \exp\left(-\frac{\Delta^2}{4}\right). \tag{5.8}$$

Let us clarify in more detail the geometrical sense of the probability density (5.8) of the inter-'zero' distances of the field $v(x, t)$ formal in this way. In this connection it can be noted that the probability density of the ergodic random field $\Delta(x;\ 2s, t)$ of (5.2) may be also defined as a limit

$$W_\Delta(\Delta;\ 2\rho) = \lim_{L\to\infty} \frac{1}{2L} \int_{-L}^{L} \delta[\Delta - \Delta(x;\ 2s, t)]\, dx. \tag{5.9}$$

for $s \ll l(t)$ ($\rho \ll 1$) the field realisations $\Delta(x;\ 2s, t)$ are considered as a sequence of rectangular pulses with centres matching the discontinuity points x_k, amplitudes $\theta_k = \eta_k / t$ and the same duration $2s = 2\rho l(t)$. Accordingly, equality (5.9) as $\rho \to 0$ tends asymptotically to

$$W_\Delta(\Delta; 2\rho) = 2\rho l(t) \lim_{L\to\infty} \frac{N(2L)}{2L} \frac{1}{N(2L)} \sum_{k=1}^{N(2L)} \delta(\Delta - \theta_k). \tag{5.10}$$

Here $N(2L)$ is the number of discontinuities falling within the interval $(-L, L)$. Recall that at sufficiently large L

$$\frac{1}{N(2L)} \sum_{k=1}^{N(2L)} \delta(\Delta - \theta_k) = W_\theta^L(\Delta),$$

where, in accordance with Section 3.3, $W_\theta^L(\Delta)$ is the Lagrangian probability density of the interval lengths between the adjacent 'zeros'. Consequently, the equality (5.10) tends asymptotically as $\rho \to 0$ to

$$W_\theta(\Delta; 2\rho) = 2\rho l(t)\, W_\theta^L(\Delta), \tag{5.11}$$

where

$$n(t) = \lim_{L\to\infty} \frac{N(2L)}{2L}$$

is the average frequency of discontinuity appearance in the field $v(x, t)$. Comparing (5.11) with the equality (5.7), we find out that the revealed probability density (5.8) proves Lagrangian, while the average frequency has the form

$$n(t) = n_*/l(t) = 1/\sqrt{\pi}\; l(t). \tag{5.12}$$

Apart from the Lagrangian probability density of the interval lengths between the field $v(x, t)$ zeros, it seems useful to know the Lagrangian probability density of the dimensionless interval lengths between discontinuities $\mu_k = (x_{k+1} - x_k)/l(t)$ as well. To obtain it we note that

$$-\frac{dP(\Delta)}{d\Delta} = W_+(\Delta),$$

where $W_+(\Delta)$ is the probability density of a distance from a given point x to the nearest discontinuity lying to the right of this point. According to (3.3.25), $W_+(\Delta)$ is expressed through the appropriate Lagrangian probability density of the inter-discontinuity interval lengths, denoted as $W_\mu^L(\Delta)$, using the equality

$$W_+(\Delta) = -\frac{dP(\Delta)}{d\Delta} = \frac{1}{\langle\mu\rangle_L} \int_\Delta^\infty W_\mu^L(\Delta')\, d\Delta'.$$

Hence we have

$$W_\eta^L(\Delta) = \langle\mu\rangle_L \frac{d^2 P(\Delta)}{d\Delta^2}, \tag{5.13}$$

where $P(\Delta)$ is given by (4.35). Out of the normalising condition or $W_\mu^L(\Delta)$ and the equalities (5.13) and (4.41) it follows, in particular, that

$$1 = -\langle\mu\rangle_L \frac{dP(\Delta)}{d\Delta} = \frac{\langle\mu\rangle_L}{\sqrt{\pi}}.$$

Consequently $\langle\mu\rangle_L = \sqrt{\pi}$, and the Lagrangian average distance between the discontinuities is equal to

$$\langle(x_{k+1} - x_k)\rangle_L = \langle\mu\rangle_L\, l(t) = \sqrt{\pi}\, l(t) = 1/n(t).$$

Apparently, the same should be the Lagrangian average distance between 'zeros' as well. In fact, according to (5.8),

$$\langle\theta\rangle_L = \frac{1}{2}\int_0^\infty \Delta^2 \exp\left(-\frac{\Delta^2}{4}\right) d\Delta = \sqrt{\pi},$$

and hence

$$\langle\eta\rangle_L = \langle\theta\rangle_L\, l(t) = \sqrt{\pi}\, l(t).$$

2. Let us return to the discussion of BT evolution at finite Reynolds number. Substituting the external BT scale $l(t)$ according to (4.15) into (2.31), we have for the effective Reynolds number

$$R(t) = l(t)/\delta(t) = l^2(t)/4\mu t = R_0/\sqrt{\ln(t/t_n)}. \tag{5.14}$$

If $R(t) \gg 1$, which is fulfilled at $R_0 \gg 1$ and not too large t, the gradient field $q(x, t)$, as in the case of hydrodynamic turbulence, is extremely non-uniform over space and has an 'intermittent' pattern, i.e. is concentrated in chaotically scattered clusters with virtually no gradient and consequent dissipation in between.

From (5.1) and (5.8) one can write the following expression for the dimensionless moments of the gradient field:

$$M_n = \langle q^n\rangle/\langle q^2\rangle^{\frac{n}{2}} \approx (-1)^n\,(2n-1)!!\,[R(t)]^{(n-2)/2} A_n, \quad A_n \approx 1. \tag{5.15}$$

This suggests, for example, that the probability distribution of the field $q(x, t)$ is strongly non-Gaussian. As in hydrodynamic turbulence [18] the skewness factor $\gamma_3 < 0$, while the excess flatness factor $\gamma_4 = M_4 - 3 > 0$. The difference of the probability distribution q from Gaussian grows more pronounced with increase of $R(t)$.

Consider now the effect of Reynolds number finiteness on the BT spectrum behaviour at $k \gg 1/l(t)$. Obviously, the BT spectral density is related to the field $q(x, t)$ spectrum via the equality

$$G_v(k; t) = G_q(k; t)/k^2. \tag{5.16}$$

Therefore, we must find out first the spectrum of a pulse field $q(x, t)$ (5.1). In analogy with (3.2.4) and (3.2.3), at $k \gg 1/l(t)$, when one can neglect the inter-pulse interference, it is equal to

$$G_q(k; t) = \frac{n(t)}{2\pi}\langle|C_q(k, t)|^2\rangle \tag{5.17}$$

where

$$C_q(k, t) = \frac{\eta^2}{8\mu t^2} \int\limits_{-\infty}^{\infty} \frac{\cos ks \, ds}{\cosh^2 (s\eta/4\mu t)} = \frac{2\pi k\mu}{\sinh (2\pi k\mu t/\eta)}$$

is the Fourier transform of an individual pulse in (5.1). The angle brackets in (5.17) mean averaging over random values $\eta = \theta l(t)$ by means of the probability density (5.8). Substituting the latter average into (5.17) we have, owing to (5.16),

$$G_v(k; t) = 2\pi\mu^2 n(t) \int\limits_0^{\infty} \frac{W_\theta(\Delta) \, d\Delta}{\sinh^2 [2\pi\mu kt/\Delta l]}. \tag{5.18}$$

At $R(t) \gg 1$ the internal and external scales of BT differ drastically, and in the spectrum there exists an inertial range $1/\delta(t) \gg k \gg 1/l(t)$ within which viscosity does not affect the spectrum behaviour, allowing one to replace (5.18) with

$$G_v(k; t) = \frac{l(t)}{2\pi\sqrt{\pi}\,(kt)^2} \int\limits_0^{\infty} \Delta^2 \, W_\theta(\Delta) \, d\Delta = \frac{2l(t)}{\pi\sqrt{\pi}\,(kt)^2},$$

which coincides with the spectrum asymptotic behaviour (4.45) deduced earlier by another procedure. Within the viscous range $k > 1/\delta(t)$, (5.18) is transformed into

$$G_v(k; t) = 8\pi\mu^2 \, n(t) \int\limits_0^{\infty} W_\theta(\Delta) \exp(-4\pi\mu tk/\Delta l) \, d\Delta. \tag{5.19}$$

The exponential factor under the integral describes, in this case, the individual discontinuity spectrum decay, this being different at various shock front widths. Allowing for (5.8), (5.19) yields

$$\begin{aligned} G_v(k; t) &= \frac{16\sqrt{\pi}\,\mu^2}{l(t)} \int\limits_0^{\infty} z \exp(-z^2 - 2\pi\delta k/z) \, dz \\ &\approx \frac{16\pi\mu^2}{l(t)\sqrt{3}} (\pi k\delta)^{\frac{1}{3}} \exp[-3(\pi\delta k)^{\frac{2}{3}}], \quad k > 1/\delta, \end{aligned} \tag{5.20}$$

where

$$\delta(t) = \mu t/l(t) = l(t)/R(t) = l(t)\sqrt{\ln(t/t_n)}/R_0 .$$

is the internal scale of BT, coinciding with the average width of a shock front. Note that within the viscous interval $k \gg 1/\delta(t)$, the BT spectrum falls off slower than, for example, in the case of a periodic wave with spectrum damping rate proportional to k. This comparatively slow decrease of the BT spectrum is attributed to shock front width fluctuations or internal scale fluctuations, which are in fact the same.

3. With increase of t the relative thickness of the shock front $\delta(t)/l(t) = 1/R(t)$ enlarges (albeit logarithmically slowly), and at $R(t) \sim 1$ the internal scale can, at last, be compared with the external one. Here the role of nonlinear effects becomes unimportant, and BT enters the linear mode where its damping is mainly determined by linear dissipation. Let us discuss, qualitatively, at first, this final stage of the turbulence decay, i.e. the stage of linear damping. The time t_l to reach the linear stage can be estimated from $R(t_l) \sim 1$, which, allowing for (5.14), yields

$$t_l \sim t_n \exp(R_0^2). \tag{5.21}$$

we emphasise that for $R_0 \gg 1$ the time for BT to enter the linear stage is extraordinarily large, and far in excess of that for a periodic signal, in the case of which $t_l = t_n R_0$. This is related to multiple discontinuity coalescence, leading to energy transfer to the low spatial frequency range and to increase of the BT external scale.

At the final linear stage of decay, the processes of linear self-action and BT harmonic interaction seem to be frozen, and the BT spectrum evolution for $t > t_l$ can be roughly estimated as (see Section 6.1)

$$G_v(k; t) = G_v(k; t_l) \exp[-2\mu k^2(t - t_l)], \tag{5.22}$$

which describes the result of BT harmonic linear dissipation. Here $G_v(k; t)$ is the BT spectrum accounting for nonlinear effects accumulated by time t and is described asymptotically by the formulae of type (4.40), (4.45), (5.18), (5.20). Due to the absence of low-frequency harmonic parametric generation in the linear stage, the spectrum steepness in the low spatial frequency range $k \to 0$ does not vary when $t > t_l$, becoming almost invariant. The behaviour of the BT spectrum as $k \to 0$ and $t > t_l$ can be evaluated, substituting (4.45) into (5.22), as

$$G_v(k; t) \approx \frac{k^2\, l^5(t_l)}{t_l^2}\, \mathrm{e}^{-2\mu k^2 t} \approx \frac{l_0 k^2 v^2 \exp(R_0^2/2)}{\sqrt{R_0}}\, \mathrm{e}^{-2\mu k^2 t}. \tag{5.23}$$

Exponential growth of the spectrum steepness (at $k \to 0$) with R_0 is attributed, as was mentioned earlier, to multiple merging of the shock fronts and external scale growth at the nonlinear evolution stage $t < t_l = t_n \exp(R_0^2)$.

4. Rigorous analysis of BT behaviour at the final linear stage is convenient to be performed on the exact solution of BE. It is worth reminding the reader that the Hopf–Cole solution offers the sought field as

$$v(x, t) = \frac{\partial S(x, t)}{\partial x} = -2\mu \frac{\partial}{\partial x} \ln \varphi(x, t) \tag{5.24}$$

through the linear equation solution

$$\frac{\partial\varphi(x, t)}{\partial t} = \mu\frac{\partial^2\varphi(x, t)}{\partial x^2}. \tag{5.25}$$

If the initial field $v_0(x)$ proves statistically homogeneous and its spectral density as $k \to 0$ behaves as $G_0(k) \sim k^n$, $n > 1$, then the initial action $S_0(x) = \int^x v_0(y)\,dy$ is also statistically homogeneous, and has variance $\sigma_*^2 = \langle s_0^2\rangle$. The initial condition to equation (5.25) in this case can be represented as

$$\varphi_0(x) = \exp\,[-S_0(x)/2\mu] = \langle\varphi_0\rangle(1 + \psi_0(x)), \tag{5.26}$$

where $\psi_0(x)$ is a statistically homogeneous random field with zero average. Correspondingly, the solution to (5.25) is equal to

$$\varphi(x, t) = \langle\varphi\rangle\,[1 + \psi(x, t)], \tag{5.27}$$

where $\langle\varphi\rangle = \langle\varphi_0\rangle = \text{const}$ and $\langle\psi\rangle = 0$. As time goes on, the viscous dissipation and inhomogeneity smoothing causes the variance of the field $\psi(x, t)$ to become less, and at times when it amounts to $\sigma_\psi^2 = \langle\psi^2\rangle \ll 1$, the solution of (5.24) acquires the form

$$v(x, t) = -2\mu\frac{\partial}{\partial x}\ln\,[\langle\varphi\rangle(1 + \psi(x, t))] \approx -2\mu\frac{\partial}{\partial x}\,\psi(x, t). \tag{5.28}$$

As $\psi(x, t)$ satisfies the linear equation (5.25), then $v(x, t)$ also at these times fulfils the linear equation, which testifies precisely to the fact that BT has entered the linear stage. The accumulated nonlinear effects are described in this solution by a nonlinear inertial relation between the initial BT field $v_0(x)$ and the auxiliary field $\psi_0(x)$ of (5.26).

At times when (5.28) holds the BT spectrum is defined by

$$G_v(k; t) = 4\mu^2k^2G_\psi(k)e^{-2\mu k^2 t}, \tag{5.29}$$

where an exponential factor depicts the viscous short-wave dissipation. In accord with (5.26) the incorporated spectrum $G_\psi(k)$ of the field $\psi_0(x)$ is expressed by means of the two-dimensional characteristic function of the dimensionless field $S_0(x)$.

In the case of Gaussian statistics of $s_0(x)$ one obtains

$$G_\psi(k) = \frac{1}{2\pi}\int_{-\infty}^{\infty}[\exp\,(R_0^2R(z)) - 1]\,e^{ikz}\,dz \qquad (R_0 = \sigma_*^2/2\mu), \tag{5.30}$$

$$\sigma_\psi^2(t) = \int_{-\infty}^{\infty} G_\psi(k)\,e^{-2\mu k^2 t}\,dk, \tag{5.31}$$

where $R(z)$ is the correlation coefficient of $S_0(x)$. When $n \geq 2$, irrespective of the BT initial spectrum shape, the spectrum $v(x, t)$ has a universal form [138] a $t \to \infty$,

$$G_v(k;t) = 4\mu k^2 G_\psi(0) e^{-2\mu k^2 t},$$

and the field energy decays as

$$\sigma_v^2(t) = \sqrt{\frac{\pi\mu}{2}}\, G_\psi(0)\, t^{-\frac{3}{2}}.$$

Factor $G_\psi(0)$ depicts the spectrum steepness near $k = 0$, and is related nonlinearly to the initial action correlation function by

$$G_\psi(0) = \frac{1}{2\pi}\int_{-\infty}^{\infty} [\exp(R_0^2\, R(z)) - 1]\, \mathrm{d}z. \tag{5.32}$$

The connection of $G_\psi(0)$ with the initial spectrum can be represented by a power series in R_0^2. Cutting it off at the Nth term corresponds to the allowance for an $(N - 1)$-tuple nonlinear interaction of BT harmonics. For $R_0 \gg 1$, assuming $R(z) = 1 - z^2/2l_0^2$ and calculating (5.32) by the saddle-point technique, we have

$$G_\psi(0) = \frac{l_0}{\sqrt{2\pi}\, R_0}\, e^{R_0^2},$$

which proves consistent with the qualitative evaluation of (5.23). The condition of BT entering the linear regime $\sigma_\psi^2(t) \approx 1$ leads in this situation to the following expression for t_l:

$$t_l = t_n \exp(2R_0^2)/R_0.$$

from this, as from (5.21) , it is clear that the time for BT to enter the linear regime of damping is extremely large.

5.6 Statisitical properties of waves in a nondispersive medium with arbitrary nonlinearity

1. Analysis of BT by singling out a dominant particle, i.e. the one with the least action, can be extended to the BT examination in the presence of an exciting force [74] and to the study of random waves satisfying the modified BE [98, 139]:

$$\frac{\partial v}{\partial t} + c(v)\frac{\partial v}{\partial x} + Q(v,t) = \mu\frac{\partial^2 v}{\partial x^2}, \quad v(x,0) = v_0(x). \tag{6.1}$$

Equation (6.1) describes waves in a nondispersive medium with arbitrary dependence of velocity $c(v)$ on the field v, and function $Q(v, t)$ takes into account the medium inhomogeneities, low-frequency dissipation, or energy pumping into the wave. In what follows we confine ourselves to the case of $Q \equiv 0$, and discuss the field $v(x, t)$ behaviour as $\mu \to 0$ assuming that $c'(v) > 0$. In Section 2.2 an asymtotic solution to BT was obtained by a limiting process from an exact solution (2.1.13). One has to search for a solution to the modified BE by another method.

At $\mu = 0$ equation (6.1) is transformed into

$$\frac{\partial v}{\partial t} + c(v)\frac{\partial v}{\partial x} = 0, \quad c(v) = \frac{\partial \psi}{\partial v}, \tag{6.2}$$

the solutions of which become multi-valued in finite time. If, however, we interpret (6.2) as a corollary to the integral law of conservation

$$\frac{\mathrm{d}}{\mathrm{d}t}\int_{x_1}^{x_2} v(x, t)\,\mathrm{d}x + \psi(v(x_2, t)) - \psi(v(x_1, t)) = 0,$$

it seems possible to construct the generalised discontinuous solutions of (6.2). They are dealt with in detail in [4–6] where, in particular, it is shown that the generalised solution of (6.2) coincides with a solution to (6.1) as $\mu \to 0$. In the statistical analysis of such random waves, employing the following procedure [4] appears rather helpful. In order to fulfil the law of conservation let us express the generalised solution in terms of action $S(x, t)$:

$$v(x, t) = \partial S(x, t)/x,$$

taking the continuous branch of the function $S(x, t)$ satisfying the equation

$$\frac{\partial S}{\partial t} + \psi\left(\frac{\partial S}{\partial x}\right) = 0, \quad S_0(x) = \int^{x} v_0(y)\,\mathrm{d}y. \tag{6.3}$$

The set of characteristic equations below corresponds to equations (6.2) and (6.3):

$$\frac{\partial S}{\partial t} = 0, \quad \frac{\mathrm{d}X}{\mathrm{d}t} = c(V), \quad \frac{\mathrm{d}S}{\mathrm{d}t} = g(V),$$
$$g(v) = c(v)v - \psi(v) = v^2\frac{\mathrm{d}}{\mathrm{d}v}\left(\frac{y(v)}{v}\right), \quad g_v'(v) = c_v'v, \tag{6.4}$$

the solution of which,

$$V(y, t) = v_0(y),\ X(y, t) = y + c(v_0(y))t,$$
$$S(y, t) = S_0(y) + g(v_0(y))t, \tag{6.5}$$

describe the trajectories of 'particles' having the Lagrangian coordinate y, velocity $v_0(y)$ and action $S_0(y)$ at $t = 0$. When a point x is hit by several particles, the continuity of $S(x, t)$ can be provided by retaining the particle with minimal action. In this case (6.5) implies that the generalised solution of (6.2) can be written as

$$v(x, t) = c^{-1}\left(\frac{x - y(x, t)}{t}\right), \tag{6.6}$$

where $v = c^{-1}(\xi)$ is the function inverse to $\xi = c(v)$ and $y(x, t)$ is the absolute minimum coordinate of the function

$$\Phi(x, y, t) = S_0(y) + \tilde{g}\left(\frac{x-y}{t}\right)t,$$
$$\tilde{g}(\xi) = g(c^{-1}(\xi)),\ g''_{\xi} = 1/c'(c^{-1}(\xi)). \tag{6.7}$$

As in the case of BE, this solution permits one to perform an obvious graphical analysis of the field by searching for the first points where the initial action $S_0(y)$ is touched by the curve

$$\alpha(x, y, t) = -\tilde{g}\left(\frac{x-y}{t}\right)t + H,$$

H increasing from $-\infty$.

2. It seems instructive now to discuss the statistical properties of the random generalised solutions (6.6) of equation (6.2). At the stage prior to discontinuity formation, the one-dimensional probability density of the statistically homogeneous field is also conserved in a medium with arbitrary nonlinearity [97]. Nonlinearity changes the wave spectrum, while the one-point distribution remains the same. An analysis of the Fourier transform of a simple wave was carried out in [96], wherein it was demonstrated that the spectrum evolution laws are extensively determined by the form of nonlinear dependence of c on v. Nevertheless, the formation of discontinuities in a field must lead to the appearance of universal energy spectra asymptotic behaviour $\sim k^{-2}$ (see Section 3.2).

Later on in this section we shall consider a random field at the discontinuous stage, using the solution (6.6). Therefore, the problem of the field $v(x, t)$ statistical analysis is reduced to that of finding the statistics of the functional (6.7) absolute minimum coordinates. The generalised discontinuous solutions to (6.2) differ from the limit solution (2.2) of BE (2.1) by a function of form $C^{-1}((x - y)/t)$ in (6.6) and $\tilde{g}(\xi)$ in (6.7) alone. Utilising this circumstance we can immediately show the qualitative pattern of the field evolution at the stage of developed discontinuities.

At large times, when $\tilde{g}((x - y)/t)/t$ is a smooth function within the scale of the initial action $s_0(y)$, the coordinates of absolute minima of $\Phi(x, y, t)$ coincide, in essence, with certain local minima of $s_0(y) - y_k$, while $y(x, t) \approx y_k$ essentially does not change in the region between discontinuities $x \in (x_{k-1}, x_k)$, jumping to y_{k+1} on passing to an adjacent region $x > x_k$. In each of these regions the field has a universal structure

$$v(x, t) = c^{-1}\left(\frac{x - y_k}{t}\right), \quad x_{k-1} < x < x_k. \tag{6.8}$$

The discontinuity coordinates x_k are deduced from the equality

$$S_0(y_k) + \tilde{g}\left(\frac{x_k - y_k}{t}\right)t = S_0(y_{k-1}) + \tilde{g}\left(\frac{x_k - y_{k-1}}{t}\right)t.$$

Let us describe qualitatively, at first, the evolution of the field $v(x, t)$ given the initial noise field $v_0(x)$. In this instance the discontinuities have random velocities, resulting in their coalescence and an increase of the field scale $l(t)$ (equal to the characteristic inter-discontinuity distance). Over a rather lengthy period of time of multiple discontinuity coalescence, $l(t)$ grows very large, as compared with the initial correlation length l_0. A qualitative evaluation of the external scale $l(t)$ and the characteristic amplitude of the field $\sigma(t) = \sqrt{\langle v^2 \rangle}$ at this stage can be readily derived from (6.6) and (6.7) by noting that the minimum $y(x, t)$ coordinate may fall within the region $|x - y| \sim l$ on the boundaries of which the increment of the initial action $s_0(y)$ and the function $\tilde{g}((x - y)/t)$ are of the same order. Estimating an increment of $s_0(y)$ over the external scale l through a structure function $d_s(l) = \langle [s_0(x + l) - s_0(x)]^2 \rangle$, $s_0(x + l) - s_0(x) \sim \sqrt{d_s(l)}$, we obtain from (6.6) and (6.7) the following expressions for the external scale and the field amplitude

$$t\tilde{g}\left(\frac{l}{t}\right) \approx \sqrt{d_s(l)}, \; \sigma \approx c^{-1}\left(\frac{l}{t}\right). \tag{6.9}$$

Alternatively, allowing for the definition of $\tilde{g}(\xi)$ in (6.7), the possibility arises to directly write the equation for $\sigma(t)$

$$g(\sigma)t \approx \sqrt{d_s(t\, c(\sigma))}. \tag{6.9$'$}$$

These estimates only prove valid if $l(t) \gg l_0$. It follows from (6.9) that the asymptotic behaviour of the field is, along with the nonlinearity type $\psi(v)$, also determined by the structure function, $d_s(z)$, type as $z \to \infty$, i.e. by the initial spectrum behaviour within the large-scale range (see Section 5.3). So, if $d_s(z \gg l_0) = 2\sigma_*^2 = 2\langle s_0^2 \rangle$, then we have from (6.9) the equation

$$g(\sigma) \approx \sigma_*/t.$$

In particular, for a medium with quadratic nonlinearity ($c(v) = v$, $\tilde{g}(v) = v^2/2$), $\sigma^2 \sim \sigma_*/t$, which conforms to the result obtained for BT in Section 5.2.

3. In the following discussion we restrict our investigation to the case when $\langle s_0^2 \rangle = \sigma_*^2 < \infty$ and $S_0(y)$ is a statistically homogeneous Gaussian field. The procedure to construct an asymptotic solution here is similar to that concerning BT (Section 5.4). The only differences are in that the linear relation (2.1) between $v(x, t)$ and $y(x, t)$ is replaced by a nonlinear expression (6.6), while a parabolic term $(x - y)^2/2t$ in (2.2) transforms to a function $\tilde{g}((x - y)/t)t$. Thus, here we only consider the applicability conditions for the asymptotic approach regarding a medium with arbitrary nonlinearity, and offer a concluding expression for the probability density [98].

The asymptotic solution in Section 5.4 was valid at times when the parabola $(x - y)^2/2t$ became a smooth function on the initial action scale. As concerns the arbitrary nonlinearity, a similar requirement leads to the following condition (see (2.5.11)):

$$\tilde{g}_y = c^{-1}\left(\frac{x - y}{t}\right) \ll S_{0y} = v_0 \sim \sigma_0 = \sqrt{\langle v_0^2 \rangle}.$$

Allowing for (6.6) makes it equivalent to a condition: $\sigma(t) << \sigma_0$, $l(t) >> l_0$. Therefore, an asymptotic consideration holds when the field energy is much less, and the external scale much more, as compared with the initial energy and the initial correlation length l_0, respectively. Using a procedure analogous to that elaborated in Section 5.4, one can easily obtain, for the field $v(x, t)$ probability density,

$$W_v(v; t) = \frac{\psi''(v)}{f(\sigma_*/\xi t)} \exp\left[-v^2 \frac{\mathrm{d}}{\mathrm{d}v}\left(\frac{\psi(v)}{v}\right) \Big/ (\sigma_*/\xi t)\right], \tag{6.10}$$

where $\xi = \xi(t)$ is the solution of a transcendental equation

$$\frac{t}{2\pi l_0} f\left(\frac{\sigma_*}{\xi t}\right) \mathrm{e}^{-\xi^2/2} = 1, \quad \xi \approx \sqrt{\ln t}, \tag{6.11}$$

and the function $f(\eta)$ may be defined as

$$f(\eta) = \int_{-\infty}^{\infty} \exp\left[-\tilde{g}(y)/\eta\right] \mathrm{d}y = \int_{-\infty}^{\infty} \exp\left[-g(v)/\eta\right] c'(v)\, \mathrm{d}v. \tag{6.12}$$

Expression (6.10) describes the probability density of the field at a stage when its profile is formed as a result of multiple discontinuity coalescence. It is clear that the probability density at this stage is uniquely determined by the propagation velocity $c(v)$ dependence on the local value of v, i.e. eventually by the physical nature of the field nonlinearity $\psi(v)$ ($C(v) = \psi'(v)$). In the general case the distribution (6.10) proves non-Gaussian and non-symmetric. If, however, at low field amplitudes the nonlinearity degenerates into a quadratic one, $\psi(v) = a_2 v^2 + a_4 v^4 + \ldots$, then as $t \to \infty$ the distribution $W_v(v; t)$ will tend to a Gaussian one with variance $\sigma^2 = \sigma_*/a^2 \xi t$.

4. Let us undertake a more detailed discussion concerning the evolution of the realisations and the probability distribution of the field in a medium with a power-law nonlinearity

$$\psi(v) = a_m \frac{|v^m|}{m}, \quad c(v) = a_m |v|^{(m-1)} \operatorname{sgn} v,$$

$$v(x, t) = \left(\frac{x - y(x, t)}{t}\right)^{1/(m-1)} \tag{6.13}$$

$$\Phi(x, y, t) = S_0(y) + \frac{(m - 1)}{m}\left[\frac{(x - y)^m}{t}\right]^{1/(m-1)}.$$

In this case (6.10) implies that

$$W_v(v;t) = m\left|\frac{v}{v_*}\right|^{(m-2)} \exp\left(-\left|\frac{v}{v_*}\right|^m\right)\Bigg/ 2v_*\Gamma\left(\frac{m-1}{m}\right), \tag{6.14}$$

$$\langle v^2\rangle = v_*^2\,\frac{\Gamma((m+1)/m)}{\Gamma((m-1)/m)}, \quad v_*(t) = \frac{\sigma}{a_m t\xi}\left(\frac{m}{m-1}\right)^{1/m}, \tag{6.15}$$

$\Gamma(z)$ being the gamma function. It is obvious from (6.14) that at $m > 2$, due to a stronger nonlinearity at large field values, the distribution decrease is faster than that in a medium with quadratic nonlinearity. Simultaneously, in view of the pre-exponential factor the distribution $W_v(v;\,t)$ vanishes at $v \to 0$ and has its maximum at $v = \pm v_*((m-2)/m)^{1/m}$. The value of m being large enough, W_v is localised in the vicinity of $v = \pm v_*$ while the localisation zone width is equal to v_*/m. Provided $1 < m < 2$ the probability density (6.14) falls off as $v \to \infty$ at a slower rate as compared with a Gaussian distribution, and has a singularity as $v \to 0$. The latter phenomenon can be easily understood if one recalls the relation between the probability density and the random function realisation behaviour proposed in Section 3.2, and takes into account that the field in the intervals between discontinuities has a universal structure,

$$v(x,t) = \left(\frac{x - y_k}{a_m t}\right)^{1/(m-1)}. \tag{6.16}$$

Making use of (3.2.13), we have the above

$$W_v(v;\,t) = n(t)t\,a_m(m-1)\,|v|^{(m-2)}, \tag{6.17}$$

where $n(t) \sim 1/l(t)$ is the average number of discontinuities per unit length.

Let us now compare the energy damping of periodic and random fields. Along with the case of $\langle S_0^2\rangle = \sigma_*^2 < \infty$ we shall also consider the situation when $G_0(0) = G^0 \neq 0$ and the structure function of the initial action asymptotically tends to $d_s(z) = G^0|z|$ as $|z| \to \infty$. For the wave energy at the stage of developed discontinuities we have from (6.16), in all three cases,

$$\sigma^2(t) = \langle v^2(x,\,t)\rangle = (l(t)/a_m t)^{2/(m-1)}, \tag{6.18}$$

where $l(t)$ is the external field scale, constant and equal to l_0 for periodic fields, while increasing for random fields as

$$l(t) = \begin{cases} \sigma_*^{(m-1)/m}(a_m t)^{1/m}, & G^0 = 0, \\ G^{0\left(\frac{m-1}{m+1}\right)}(a_m t)^{2/(m+1)}, & G^0 \neq 0. \end{cases} \tag{6.19}$$

For the field energy decay law we have

$$\sigma^2(t) = At^{-\beta} \tag{6.20}$$

where, respectively,

$$\beta = \begin{cases} 2/(m-1) & \text{periodic signal,} \\ 2/m & G^0 = 0, \\ 2/(m+1) & G^0 \neq 0. \end{cases}$$

Obviously, the random field is always damped at a slower rate than the periodic one, this being associated with the increased external scale growth due to discontinuity merging. This difference between the energy decay laws of periodic and random fields is most striking for $1 < m < 2$ and diminishes gradually with m thereafter, since the greater is $m > 2$ the slower do discontinuities travel, this resulting eventually in a slower rate of their coalescence.

To illustrate the evolution of realisations of the field profile, we shall present the results of a numerical simulation performed on the basis of a generalised solution (6.6), (6.7). (This experiment was done in collaboration with I. Demin.) The initial field was assumed Gaussian, and the variance $\langle S_0^2 \rangle$ is bounded and taken equal to unity. Numerical simulation involved the following steps: generation of a statistically homogeneous function $s_0(y)$, searching for the function $\Phi(x, y, t)$ absolute minimum coordinates, and recovery of the field $v(x, t)$ profile by means of (6.6). In the process of numerical simulation we were dealing with times t far in excess of the characteristic nonlinearity time. This allowed us to model the initial action $s_0(y)$ in terms of a sequence of independent random values. When carrying out computations we considered media with power nonlinearity corresponding to values $m = 4/3$, $m = 2$, $m = 4$. The wave profiles and the appropriate histograms are demonstrated in Fig. 5.15 ($m = 4/3$), Fig. 5.16 ($m = 2$) and Fig. 5.17 ($m = 4$). The probability density averaging was performed over $N = 10^4$ realisations. Apart from this, such a simulation procedure permitted us to verify accurately enough the law of the field decay for $G^0 = 0$. The decay factor β obtained in this way matched the theoretical one, $2/m$, to within 5%.

As a qualitative measure describing the discrepancy between the probability density and the Gaussian one we can use the dimensionless cumulant coefficients γ_p [19]. So, for γ_4 the theory (γ_4^{th}) and experiment (γ_4^{e}) produce the following values

$$m = \tfrac{4}{3},\ \gamma_4^{\text{th}} = 6.942, \quad \gamma_4^{\text{e}} = 6.860,$$

$$m = 3,\ \gamma_4^{\text{th}} = -1.302, \quad \gamma_4^{\text{e}} = -1.322,$$

$$m = 2,\ \gamma_4^{\text{th}} = 0, \quad \gamma_4^{\text{e}} = 0.015.$$

If in a medium there exist odd powers along with even powers of nonlinearity, the probability density of a field will be non-symmetric. Consider a limiting case when the propagation velocity has different dependence for $v > 0$ and $v < 0$,

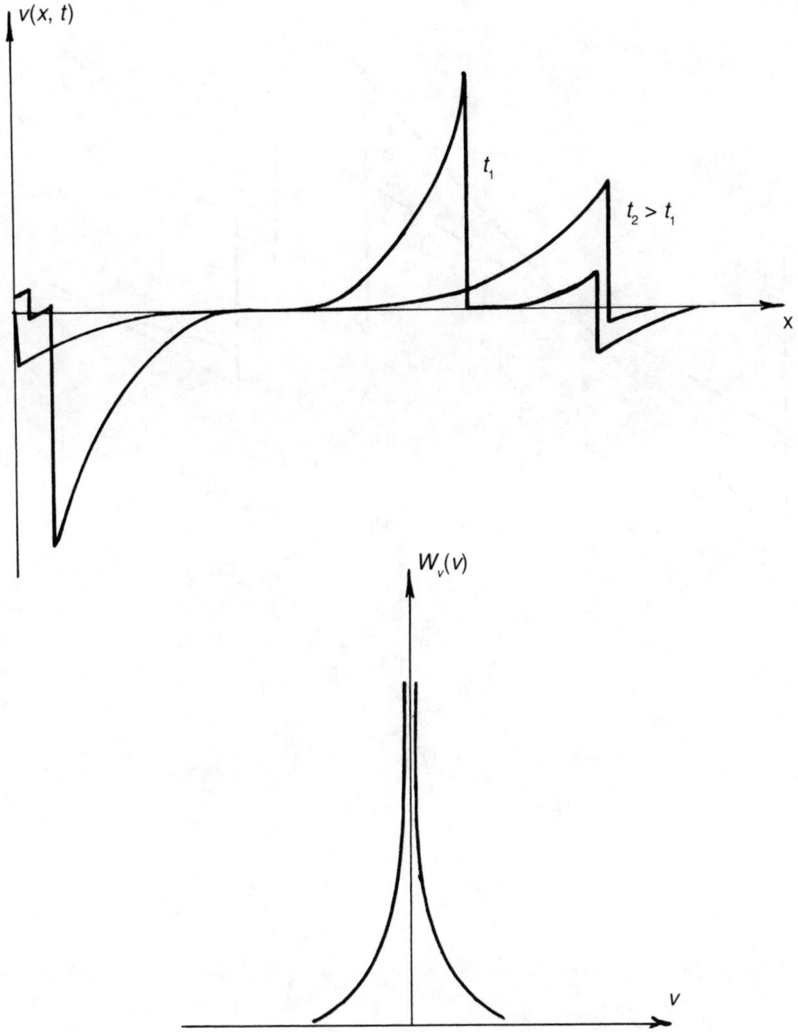

Fig. 5.15 Field $v(x, t)$ realisation and corresponding histogram at the stage of developed discontinuity interaction, for media with $m = 4/3$

$$C(v) = \begin{cases} a_+ v, & v > 0, \\ a_- v, & v < 0. \end{cases} \tag{6.21}$$

Then for the probability density we have from (6.10)

$$W_v(v; t) = \sqrt{\frac{2\xi t}{\pi\sigma_*}} \frac{1}{\left(\sqrt{a_+} + \sqrt{a_-}\right)} \begin{cases} a_+ \exp(-v^2 a_+ t\xi/\sigma_*^2), & v > 0, \\ a_- \exp(-v^2 a_- t\xi/\sigma_*^2), & v < 0. \end{cases} \tag{6.22}$$

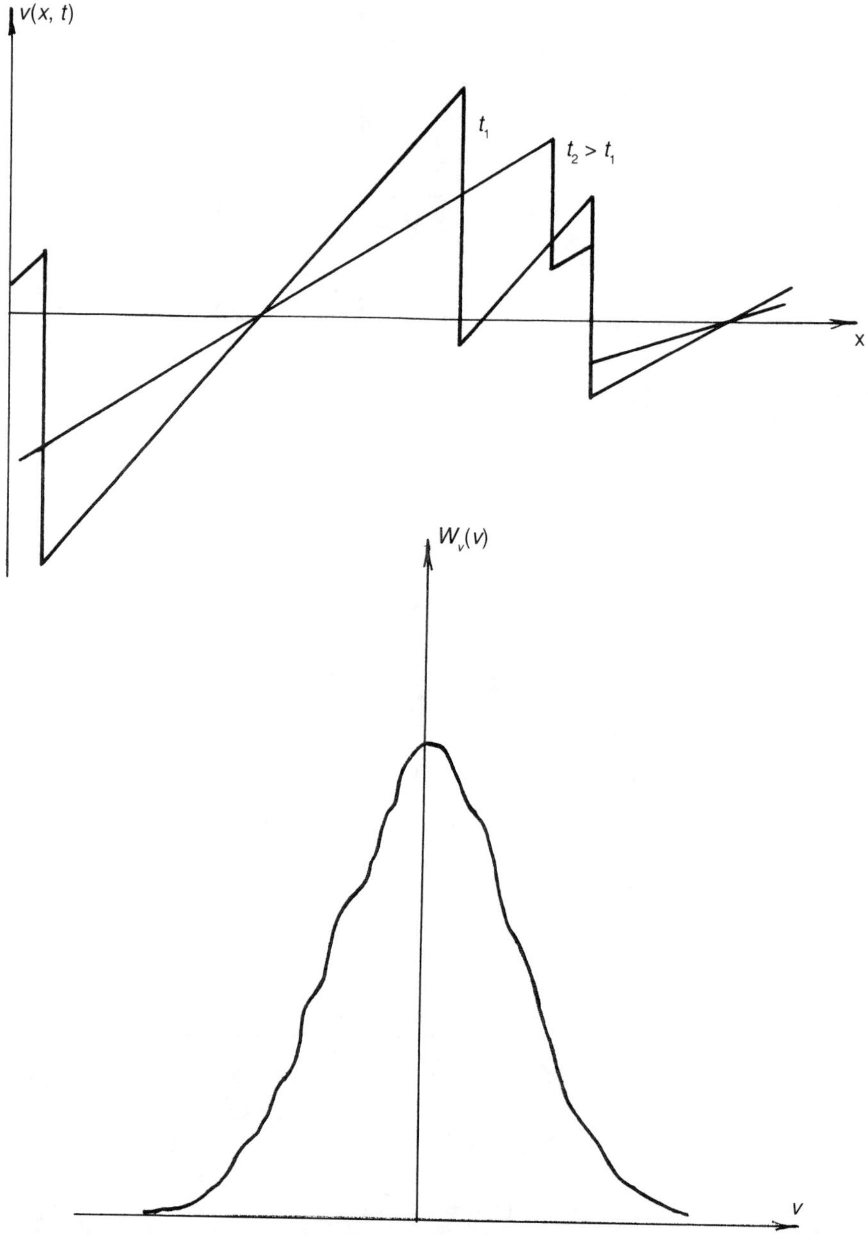

Fig. 5.16 Field $v(x, t)$ realisation and corresponding histogram at the stage of developed discontinuity interaction, for media with $m = 2$

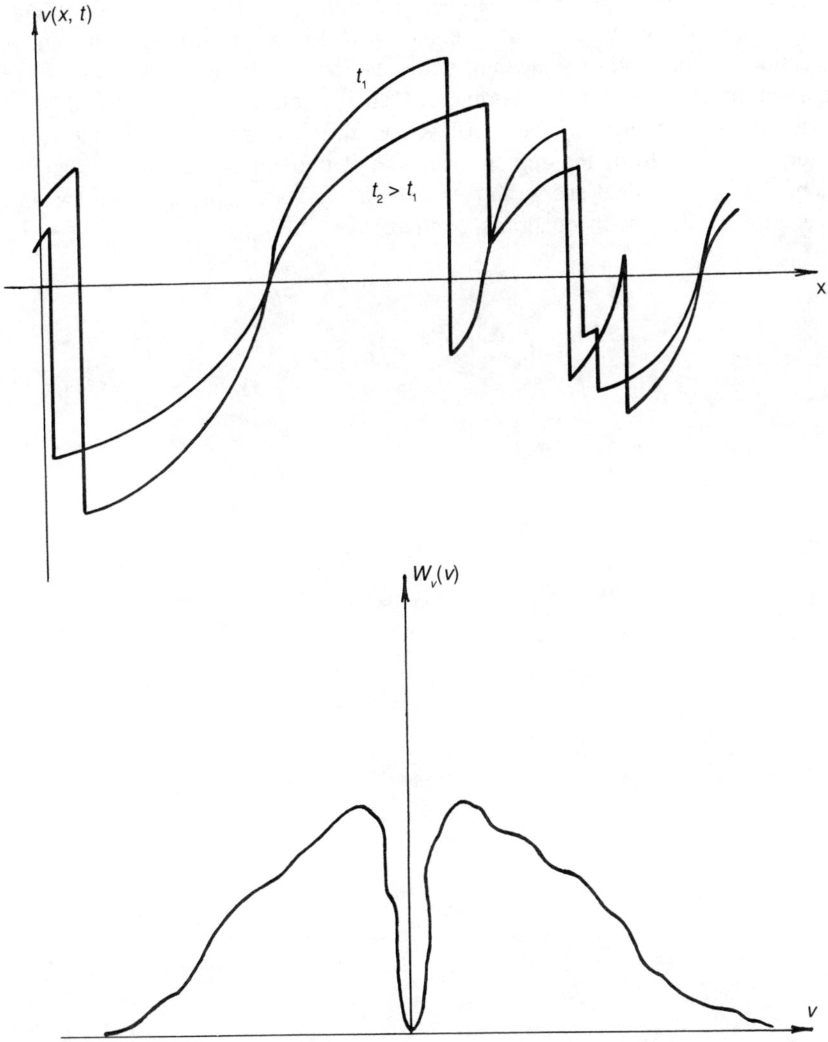

Fig. 5.17 Field $v(x, t)$ realisation and corresponding histogram at the stage of developed discontinuity interaction, for media with $m = 3$

i.e. it has Gaussian form both for $v > 0$ and for $v < 0$ as in a medium with a quadratic nonlinearity ($a_+ = a_-$). The width of these densities is different, however, and the distribution itself has a discontinuity at $v = 0$. For the field energy we have, from (6.22),

$$\langle v^2 \rangle = E_+ + E_- = \frac{\sigma_*}{\xi t}\left(\frac{1}{a_+ + \sqrt{a_+ a_-}} + \frac{1}{a_- + \sqrt{a_+ a_-}}\right). \tag{6.23}$$

The first term here describes the energy damping for $v > 0$ while the second one refers to $v < 0$. At $a_+ = a_-$ equation (6.23) is transformed into an expression for the case of a medium with quadratic nonlinearity. If $a_+ >> a_-$, i.e. the nonlinearity for $v > 0$ is stronger, then it is clear from (6.22) and (6.23) that the major energy of the wave is concentrated in $v < 0$, but due to the energy transfer from the region with $v < 0$ into the more quickly damped domain $v > 0$ the field energy $\langle v^2 \rangle \approx E_+ \approx \sigma_* / \xi t \sqrt{a_+ a_-}$ appears to be less than that in a medium with quadratic nonlinearity with $a = a_-$.

6 Three-dimensional potential turbulence; the large-scale structure of the Universe

This chapter is devoted to the study of statistical properties of a potential turbulence described by a three-dimensional Burgers equation (BE). We discuss the possibility of using the proposed model to qualitatively describe the development of the large-scale structure of the Universe at a nonlinear stage of gravitational instability. It is shown that in the framework of this model, increased density regions are formed which constitute a connected cellular structure, such that all the matter runs to the cell nodes, as time evolves. The growth laws of typical sizes of the structure, and their dependence on the spectral form of the small initial density fluctuations are estimated, and the probability properties of the chaotic velocity field are found. The results obtained on the basis of a model Burgers equation, simulating the nonlinear stage of gravitational instability, are compared with those of a straightforward numerical modelling of a gas of gravitationally interacting particles. An examination is also undertaken of the velocity and matter density fluctuations in a one-dimensional model gas, allowing for nonlinear effects, pressure forces and viscosity.

6.1 Cellular structure formation in three-dimensional potential turbulence

1. A natural generalisation of BE (2.1.1) is a three-dimensional vector BE [99]

$$\frac{\partial \mathbf{v}}{\partial \tau} + (\mathbf{v}.\nabla)\, \mathbf{v} = \mu \Delta \mathbf{v}, \quad \mathbf{v}(\mathbf{r}, 0) = \mathbf{v}_0(\mathbf{r}). \tag{1.1}$$

For potential fields, $\mathbf{v} = \nabla S$, it is reduced to the linear diffusion equation by means of a Hopf–Cole substitution $S = -2\mu \ln U$ and, consequently has the exact solution

$$\mathbf{v}(\mathbf{r}, \tau) = \frac{\int (\mathbf{r} - \mathbf{q}) \exp[-\Phi(\mathbf{r}, \mathbf{q}, \tau)/2\mu]\, d\mathbf{q}}{\tau \int \exp[-\Phi(\mathbf{r}, \mathbf{q}, \tau)/2\mu]\, d\mathbf{q}}, \tag{1.2}$$

$$\Phi(\mathbf{r}, \mathbf{q}, \tau) = S_0(\mathbf{q}) + \frac{(\mathbf{r} - \mathbf{q})^2}{2\tau}, \quad \mathbf{v}_0(\mathbf{q}) = \nabla \mathbf{S}_0. \tag{1.3}$$

At low ν the main contribution to (1.2) is that of small vicinities of the points where function Φ has an absolute minimum, and in the limit $\mu \to 0$ the solution (1.2) is transformed into

$$\mathbf{v}(\mathbf{r}, \tau) = \frac{\mathbf{r} - \mathbf{q}(\mathbf{r}, \tau)}{\tau}, \tag{1.4}$$

where $\mathbf{q}(\mathbf{r}, \tau)$ is the coordinate of the absolute minimum of Φ over $\mathbf{q}$.

The solution (1.4) can, as previously, be interpreted in terms of a hydrodynamic flow of noninteracting particles with initial velocities $\mathbf{v}_0(\mathbf{q})$ (see section 2.4). The coordinates of a particle situated at point $\mathbf{q}$ when $\tau = 0$, change according to

$$\mathbf{r} = \mathbf{q} + \mathbf{v}_0(\mathbf{q})\tau, \tag{1.5}$$

while the particle velocity remains invariant and equal to $\mathbf{v}_0(\mathbf{q})$. The velocity of a particle arriving at point $\mathbf{r}$ at time τ is given by the equality (1.4), where $\mathbf{q}(\mathbf{r}, \tau)$ is the Lagrangian coordinate of this particle serving as the root of (1.5) with respect to $\mathbf{q}$. Equation (1.5) coincides with the Φ minimum condition over $\mathbf{q}$: $\nabla_{\mathbf{q}}\Phi = 0$. As time τ goes on, (1.5) acquires several real roots, which correspond to a multi-stream motion of a flow of noninteracting particles. Here, a single-valued solution (1.4) of equation (1.1) coincides with the velocity of the particle whose action is absolutely minimal over $\mathbf{q}$. Before we begin investigating the behaviour of the three-dimensional BE solution and that of the continuity equation, let us discuss the possibility of using these equations to qualitatively model the large-scale structure of the Universe [109, 110, 114].

2. Given very large scales, the Universe is considered, nowadays, to be statistically homogeneous and isotropic, although if we view somewhat smaller scales, namely analyse the way galaxies are distributed, an extremely non-uniform distribution can be observed [103, 140]. Numerous measurements of some galaxy red shifts have recently allowed us to go over from the study of a visible galaxy distribution to the investigation of a real three-dimensional structure. The spatial distribution of galaxies implies the existence of large regions (up to 100 megaparsecs in diameter) where galaxies are not seen. Galaxy groups and clusters have a tendency to concentrate along relatively thin surfaces and lines or else to form superclusters. The problem of describing large-scale structure in the Universe is concerned with the explanation of the formation of the currently existing structure form initially small density fluctuations. As the current relative density fluctuations over the scales mentioned are large, they can be interpreted in the framework of the nonlinear theory of gravitational instability alone.

At present, there are several theories under consideration – scenarios of large-scale Universe structure formation containing different assumptions as to the initial density perturbation behaviour leading, meanwhile, to different routes for formation of large-scale structure: most familiar are fragmentation models [100, 101] and those of hierarchical clustering [102].

Under the fragmentation scenario the first to emerge are protosuperclusters ('pancakes') i.e. heavily flattened objects with a typical mass of $\sim 10^{15}$–$10^{16}\,M_{\odot}$ [100] where $M_{\odot}$ is the Sun's mass. At a later stage, under the effects of complex thermal, gasdynamical and gravitational processes, they fragment to form less massive objects: globular clusters and galaxies. The pancakes are considered as one of several types of structurally stable singularities of Lagrangian transforms which can be represented by a matter motion driving the particles from their initial positions to final ones. Today these singulazities are well-studied: their natural shapes have seen obtained, realisation conditions have seen outlined, and their geometrical shapes have seen elaborated [104, 105]. A few of these specific features were observed in particle distributions calculated during numerical tests simulating the gravitational instability process in a cold collisionless medium [103, 141]. Numerical experiments have also indicated that individual elements of a large-scale structure are not isolated from each other. They constitute a single connected cellular structure such that the increased density domains merge, leaving large dark domains in between with density significantly less than the average. Figure 6.1 shows the particle distribution resulting from the numerical simulation of a gas of cold gravitationally interacting particles [103].

Another, hierarchical, model which initially stemmed from a hypothesis of primary entropy-type perturbations suggests the opposite process of formation of objects of various scales. First, clouds with a mass of globular clusters are isolated which later on, affected by mutual attraction, unite to form galaxies and then galaxy clusters and superclusters [102].

Currently even more complicated models are set forth which combine both fragmentation scenario features and those of the consecutive clustering scenario. On the one hand, according to these models, in analogy with the hierarchical pattern, the first to appear are small-mass objects, i.e. globular clusters and galaxies, followed by galaxy clusters and superclusters, while on the other hand, a cellular structure emerges which is typical of the fragmentation version. Unfortunately, neither exact analytical study nor numerical simulation of this process seem feasible yet. Therefore, we are interested in the approximate theories and rather simple models of the density evolution at strongly nonlinear stages of the gravitational instability development.

3. Let us demonstrate that a three-dimensional BE (1.1) together with the equation of continuity presents a qualitatively adequate reflection of the increased density domain formation processes, their coalescence, and the emergence of a connected structure. The proposed model generalises the

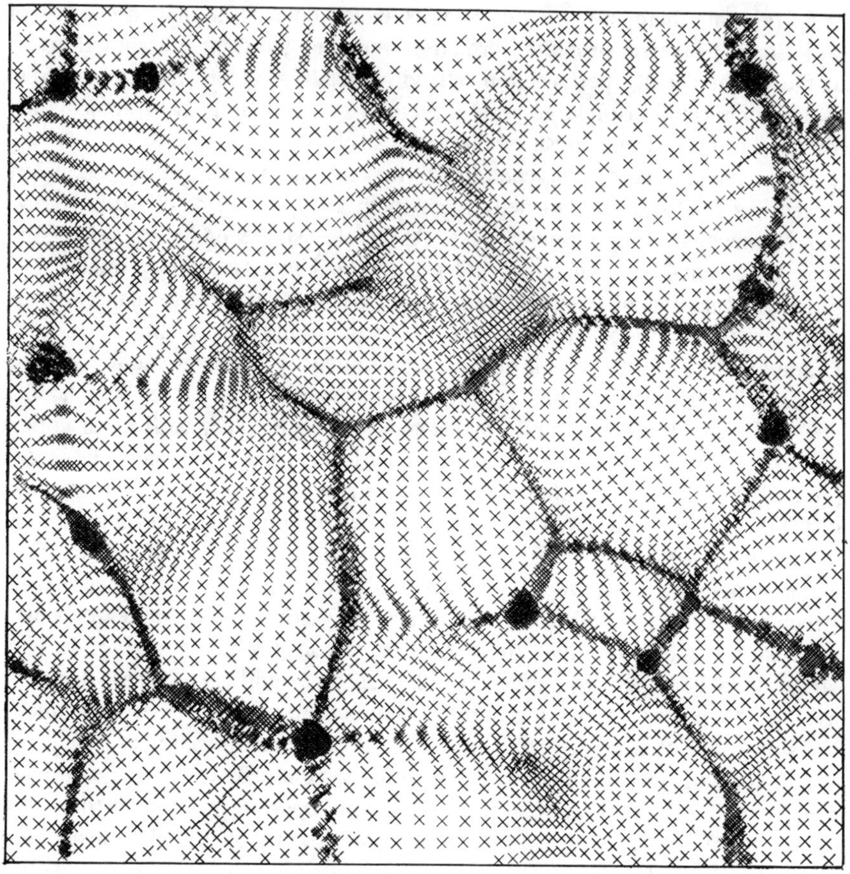

Fig. 6.1 Particle distribution as obtained from two-dimensional numerical simulation within the framework of the adiabatic theory of the origin of the structure of the Universe [103, 160]

well-known approximate Zel'dovich solution [103, 142] describing the nonlinear stage of gravitational instability with respect to later times.

We shall be concerned with the evolution of large-scale perturbations, with dimensions far in excess of the Jeans scale, for which pressure forces prove insignificant. It should be emphasised that, since such perturbations are enhanced due to gravitational instability to the same extent, the perturbation waves appear to be nondispersive and hence strongly nonlinear. As in the cases of hydrodynamic and acoustic turbulence, the formation of the density inhomogeneities out of the small random perturbations can be called 'gravitational turbulence' such that the latter, due to lack of dispersion, presents an example of strong turbulence.

In the expanding Universe, in order to describe the evolution of a dust-like (pressureless) medium it is found helpful to resort to a special velocity

$$V_i = u_i - \frac{\dot{a}}{a} R_i = a\frac{\mathrm{d}r_i}{\mathrm{d}t},\ r_i = \frac{R_i}{a}, \tag{1.6}$$

where u_i is a physical velocity, $a = a(t)$ is a scale factor describing the general Hubble expansion of the Universe, and the dot means a time derivative. In terms of these variables, the equations of motion, in the hydrodynamic approximation, have the form [102]:

$$\begin{aligned} &\frac{\partial V_i}{\partial t} + \frac{V_k}{a}\frac{\partial V_i}{\partial r_k} + \frac{\dot{a}}{a} V_i = -\frac{1}{a}\frac{\partial \varphi}{\partial r_i}, \\ &\frac{\partial^2 \varphi}{\partial r_i^2} = 4\pi G a^2 [\rho(\mathbf{r}, \tau) - \bar{\rho}(t)], \\ &\frac{\partial \rho}{\partial t} + 3\frac{\dot{a}}{a}\rho + \frac{1}{a}\frac{\partial}{\partial r_i}(\rho V_i) = 0, \end{aligned} \tag{1.7}$$

where φ is the gravitational potential perturbation, ρ is the density, $\bar{\rho}$ denotes the mean density related to the scale factor through $\ddot{a} = -4\pi G\bar{\rho}\, a/3$, and G is the gravitational constant. It should be remarked that in one dimension, assuming formally that $a = 1$ and $G < 0$, we pass from (1.6) to a set of equations (1.2.17) embodying nonlinear Langmuir oscillations of a cold plasma.

In the linear approximation, the solution of (1.6) has growing and decaying modes such that, due to the absence of dispersion, all the spatial scales of the initial perturbation grow at an equal rate. In the linear approximation one has, for the relative density fluctuations in a growing mode.

$$\delta = (\rho - \bar{\rho})/\bar{\rho} \sim b,$$

where $b(t)$ is the growing solution of equation $a\ddot{b} + 2\dot{a}\dot{b} + 3\ddot{a}b = 0$. In particular, in the Einstein–de Sitter cosmological model with $\Lambda = 0$ and $\Omega = 1$ (Λ is a cosmological term and Ω is the dimensionless average density of the Universe; for $\Omega < 1$ the Universe expands, being open, whereas if $\Omega > 1$ it is closed and the initial expansion is replaced by contraction),

$$a(t) \sim t^{\frac{2}{3}},\ b(t) \sim t^{\frac{2}{3}}. \tag{1.8}$$

The particle coordinates in a growing mode change as

$$r_i = q_i + b(t)S_i(\mathbf{q}),\quad S_i(\mathbf{q}) = \frac{\partial}{\partial q_i}\Phi_0(\mathbf{q}), \tag{1.9}$$

where r_i and q_i are the Eulerian and Lagrangian coordinates, respectively, and the potential field S is related to the density fluctuations at the initial stage. The density can be calculated provided the Lagrangian-to-Eulerian coordinate transformation Jacobian is known,

$$\eta = \rho a^3 = \eta_0/|D_{ij}|, \quad |D_{ij}| = |\partial r_i/\partial q_j|, \tag{1.10}$$

and in keeping with (1.9),

$$D_{ij} = \delta_{ij} + ba_{ij}, \quad a_{ij} = \frac{\partial^2 \Phi_0}{\partial q_i \partial q_j}. \tag{1.11}$$

Linearising (1.11) and making use of the second equation in (1.7), one can readily determine at the initial stage a connection between Φ_0, the density fluctuations δ, and the gravitational potential φ:

$$\delta = -b\frac{\partial^2 \Phi_0}{\partial q_i^2}, \quad \varphi = 3\ddot{a}ab\Phi_0. \tag{1.12}$$

Zel'dovich's approximation [142] suggests that the solution (1.9) be considered as valid also at a nonlinear stage, when $\delta > 1$. An elaborate study of this conjecture is given in a review [103]. Here we only not that for the one-dimensional perturbation the solution (1.9) remains exact until the particle trajectories intersect.

Approximation (1.9) permits one to calculate the acceleration of each particle,

$$\frac{\mathrm{d}}{\mathrm{d}t} V_i = \left(\frac{\dot{a}}{a} + \frac{\dot{b}}{b}\right) V_i. \tag{1.13}$$

Inasmuch as this solution is a self-consistent one, we may exclude the gravitational potential φ from (1.7) and, therefore, reduce the number of equations. Introducing new variables

$$\tau = b, \quad \eta = \rho a^3, \quad v_i = V_i/a\dot{b}, \tag{1.14}$$

instead of (1.7), we obtain

$$\frac{\partial v_i}{\partial \tau} + v_k \frac{\partial v_i}{\partial r_k} = 0, \tag{1.15}$$

$$\frac{\partial \eta}{\partial \tau} + \frac{\partial}{\partial r_i}(\eta v_i) = 0. \tag{1.16}$$

The first equation describes the 'inertial' motion of matter in terms of new variables. Its solution in Lagrangian coordinates is given by expressions (1.5) and (1.9), wherein $b(t) = \tau$, $\mathbf{s}(\mathbf{q}) = \mathbf{v}_0(\mathbf{q})$, $\Phi_0(\mathbf{q}) = S_0(\mathbf{q})$.

On account of the smoothness of the initial condition the mapping (1.9) is regular at small times and the density fluctuations are finite. As time evolves, however, at instants when some particles are overtaken by others, singularities arise in a mapping when $|D_{ij}| = 0$ and the matter density tends to infinity. In the works dealing with the nonlinear theory of gravitational instability (see review [103]) it is indicated that the first singularities to appear are those of a pancake type, i.e. extremely anisotropic clouds of increased density compressed in one (randomly oriented for each pancake) direction. This can be attributed to the fact that the probability of isotropic compression of the matter for random perturbations is low, and equals zero for Gaussian statistics of the initial perturbations [80].

An approximate solution (1.9) yields a quantitatively and qualitatively adequate description of the gravitational instability development only until the time of pancake formation. After the pancake appearance, the solution (1.9) becomes inadequate at least in respect of the pancake internal structure dependence on the constituent matter type. In the case of collisionless particles (stars, neutrino streams) inside a pancake, one can observe the initiation of initially three-stream and then of multi-stream motions. If, however, the pancakes emerge from gas, their thickening is prohibited by shock waves, although, even in the case of collisionless particles their type of motion is changed qualitatively after they fall into a pancake, such particles being retarded by the gravitational field of the pancake of increased density. Consequently, an oscillatory motion of the pancake-captured particles is set up due to gravitational interaction, and the pancake gains its thickness much more slowly than in the case of noninteracting particles. The numerical tests simulating the behaviour of gravitationally interacting particles inside a pancake demonstrated that, as a result of multiple particle oscillations, matter is localised in a finite anisotropic domain, slowly increasing in thickness [141].

In order to qualitatively simulate the density inhomogeneity development process within the framework of the nonlinear theory of gravitational instability, allowing for the deceleration of the particles that have fallen into a pancake, one can exploit the model equation of nonlinear diffusion (1.1) that is deduced by adding an artificial viscosity to the right-hand side of (1.15). As a matter of fact, as $\mu \to 0$ it determines the velocity field of a hydrodynamic beam of noninteracting particles.

The right-hand side of equation (1.1) gives a phenomenological description of particle retardation in the condensed regions. Certainly, the solution of (1.1) does not reveal the internal structure of the pancakes, while qualitatively embodying the effect stemming from the collisionless particle gravitational interaction inside the pancakes, i.e. the formation of localised domains with increased density. In this instance, as was observed for gravitationally interacting particles, particles falling into this domain lose their normal velocity component with smoothing of the tangential one.

Consequently, (1.1) along with the equation of continuity (1.16) presents a qualitatively correct averaged pattern of the matter motion and particle velocity field.

Apart from this, equation (1.1) is also an exceptional one, for it has an exact solution (1.2). This allows us to thoroughly investigate, in the framework of a model equation, the procedure of cellular structure formation and its statistical properties. From the analysis it follows that the global character of the structure for sufficiently low μ does not depend significantly on the value of the 'dissipation factor' μ which offers a phenomenological description of the particle gravitational and collisional interaction inside a high-density region.

4. Now let us draw our attention to the limiting solution of BE (1.1) and the basic features of the cellular structure formation. In analogy with the one-dimensional BE, Section 2.2, a convenient graphical procedure exists to seek the Eulerian velocity field $\mathbf{v}(\mathbf{r}, \tau)$ based on the solution (1.4). Evidently, the coordinate of the absolute minimum of $\Phi(\mathbf{r}, \mathbf{q}, \tau)$ (1.3) will simultaneously serve as the coordinate of the first point at while the paraboloid

$$\alpha = -\frac{(\mathbf{r} - \mathbf{q})^2}{2\tau} + H \tag{1.17}$$

touches the initial action hypersurface $S_0(\mathbf{q})$, i.e. touches the potential $\Phi_0(\mathbf{q})$, as H grows from $-\infty$. In other words, flattening the paraboloid (1.17) against the surface $S_0(\mathbf{q})$ one obtains at the point of touching the coordinates $\mathbf{q}(\mathbf{r}, \tau)$, i.e. the Lagrangian coordinates of a particle arriving at time τ to the point $\mathbf{r}$ which locates the axis of the paraboloid α. Putting the coordinates of the first targency point $\mathbf{q}(\mathbf{r}, \tau)$ into (1.4) one finds the solution to equation (1.1) as $\mu \to 0$. It is worth noting that if we go on 'lifting' the paraboloid (1.17) and reveal the other points of targency, on substituting their coordinates into (1.4) we get the velocities of all noninteracting particles reaching point $\mathbf{r}$ at time τ.

Let us trace back qualitatively the evolution stages of the velocity $\mathbf{v}(\mathbf{r}, \tau)$ and density $\eta(\mathbf{r}, \tau)$ fields. Let, at the initial time instant, $\eta(\mathbf{r}) = \eta_0 = \text{const}$ and the velocity field $\mathbf{v}_0(\mathbf{r})$ be random, statistically homogeneous and isotropic, while $\sigma = \sqrt{\langle \mathbf{v}_0^2 \rangle}$ and l_0 denote the initial field amplitude and scale, respectively. It is apparent that the kind of touching between the paraboloid $\alpha(\mathbf{r}, \mathbf{q}, \tau)$ (1.17) and the initial action $S_0(\mathbf{q})$ depends on their curvature ratio. The curvature of the paraboloid is $1/\tau$, while that of the initial action is defined by tensor $a_{ij} = \partial S_0/\partial q_i \partial q_j$. Being symmetrical, it can at each point be reduced to diagonal form by appropriately choosing a coordinate system, and the principal values β_i can be obtained, which we view ordered as follows,

$$\beta_1 \le \beta_2 \le \beta_3. \tag{1.18}$$

As a rough estimate of the initial action curvature one can take $|\beta| \sim \sigma_0/l_0$ which inverse quantity will coincide with a typical time of manifestation of nonlinear effects, $\tau_n = l_0/\sigma_0$.

To develop a picture of the matter distribution it is helpful to trace back the elementary volume evolution, i.e. the Jacobian of the Eulerian $\mathbf{r}(\mathbf{q}, \tau)$ to Lagrangian $\mathbf{q}$ coordinate transformation. The density in the neighbourhood of a fixed particle with Lagrangian coordinate $\mathbf{q}$ is defined by (1.10). With respect to noninteracting particles, i.e. concerning the particles which are described by equation (1.1) at $\mu = 0$ and have not yet experienced collisions, the Jacobian elements are equal (1,11), (1.14). After reducing tensor a_{ij} to the principal axes, one obtains for density

$$\eta(\mathbf{q}, \tau) = \eta_0/(1 + \beta_1(\mathbf{q})\tau)(1 + \beta_2(\mathbf{q})\tau)(1 + \beta_3(\mathbf{q})\tau). \qquad (1.19)$$

As long as the paraboloid curvature exceeds that of the initial action, i.e. for $\tau < 1/|\min \beta_1(\mathbf{q})|$, the paraboloid and the initial action have a single point of targency, i.e. the Lagrangian coordinate of a single particle arriving at time τ at point $\mathbf{r}$, the coordinate of the axis of the paraboloid α. Under change of $\mathbf{r}$ the point of targency $\mathbf{q}(\mathbf{r}, \tau)$ slides continuously along the initial action surface $S_0(\mathbf{q})$, the field $\mathbf{v}(\mathbf{r}, \tau)$ is continuous, and the density $\eta(\mathbf{r}, \tau)$ has no singularities. In fact, as long as perturbations are small ($|\beta_1|\tau$, $|\beta_2|\tau$, $|\beta_3|\tau \ll 1$) the expansion (1.19) leads to

$$\eta(\mathbf{q}, \tau) \approx \eta_0[1 - \tau(\beta_1(\mathbf{q}) + \beta_2(\mathbf{q}) + \beta_3(\mathbf{q}))] \qquad (1.20)$$

and the density fluctuations will be calculated within the framework of a linear theory.

With the growth of τ, however, the paraboloid curvature diminishes and at $\tau = \tau_* = 1/|\beta_1^*|$ where $\beta_1^*(\mathbf{q}_*)$ is a minimal value of the eigenvalue $\beta_1(\mathbf{q})$, the paraboloid curvature is comparable with the least one of the action principal curvature values at point $\mathbf{q}_*$. At this point, as is evident from (1.20), the matter density tends to infinity. Under Gaussian statistics of $\mathbf{v}_0(\mathbf{r})$ the probability of the principal value β_i coincidence equals zero [80], and, therefore, in the vicinity of a singularity the density tends to infinity due to one-dimensional compression along the principal axis corresponding to the least eigenvalue β_1. Thus, the solution (1.19) predicts that at the nonlinear stage the first to form are the heavily flattened clouds of condensed matter, i.e. pancakes.

As far as noninteracting particles are concerned, the time instant τ_* corresponds to mutual single-valuedness violation, when transforming the Lagrangian coordinates $\mathbf{q}$ into the Eulerian ones (1.5), as well as to appearance of singularities, i.e. to the emergence of regions with multi-stream regimes in the Eulerian velocity field, the so-called catastrophes [158]. The complete list of the generic singularities and their transformation of type (1.5) can be found in the works by Arnold [104, 105] and, as regards the problems of cosmology, in reference [143]. It is obvious that (1.5) also describes, in the small-angle approximation, the propagation of optical rays in vacuum, while (1.18) defines the wave intensity fluctuations. As to the problems of geometrical optics, the appearance of singularities corresponds to caustic formation, a process which is carefully classified in the review [145].

Statistical properties of the intensity fluctuations behind two- and one-dimensional phase screens have been clarified in sections 4.3–5 above.

Let us analyse the differences in singularity formation as to a medium of noninteracting particles and the field determined by equation (1.1) when overtaking is prohibited. As is obvious from (1.20), at $\tau \geq \tau_*$ in a Lagrangian space in the vicinity of the minima of $S_0(\mathbf{q})$ upon the surface $\beta_1(\mathbf{q}) = -1/\tau$, the matter density tends to infinity. In the Lagrangian space at the initial stage, the surface $\beta_1(\mathbf{q})$ = const make up ellipsoids with finite ratio of the axes. In the Eulerian space these particles constitute the surface of a pancake, an extremely anisotropic body inside which the motion is initially a three-stream one. After pancake birth ($\tau - \tau \ll \tau_*$) its thickness enlarges as $d_1 \sim (\tau - \tau_*)^{\frac{3}{2}}$ while the transverse dimensions scale as $d_2 \sim (\tau - \tau_*)^{\frac{1}{2}}$ and, consequently, a pancake is born thin [103].

In the framework of (1.1), when overtaking among particles is not allowed singularities prove to be of another type. In the case when the time τ slightly exceeds a critical value τ_*, the total domain of the Lagrangian space is divided into two subdomains. In the majority of points of the $\mathbf{q}$-space, paraboloid α and initial action S_0 have just one point of first tangency. In these subdomains, field $\mathbf{v}(\mathbf{r}, \tau)$ is continuous and the matter density is finite, since only one particle from these domains reaches the point $\mathbf{r}$. Meanwhile, there are regions that do not contain the points of targency of α and S_0. As the boundaries of these regions we consider the surfaces of the paraboloid α touching the initial action S_0 at two points simultaneously. In the Lagrangian space at τ (being slightly in excess of the critical time τ_* the corresponding surface of the double-touching points is closed, while in the Eulerian one it constitutes a surface which is a pancake of zero thickness. All the particles from the Lagrangian space bounded by this surface fall into the pancake. Figure 6.2 depicts a pattern of singularity formation with respect to velocity and density fields in one dimension for noninteracting particles and particles with overtaking forbidden. Figure 6.3 shows the singularity regions in Lagrangian and Eulerian coordinates in two dimensions.

The region sending particles to a pancake increases with τ, and so does the pancake itself in the Eulerian representation. This is followed by the initiation of yet more new pancakes that are chaotically oriented in space and gaining in size.

Another step of the field development is associated with the appearance of a triplet of points wherein the paraboloid α touches the initial action S_0. The triplet of points in the Lagrangian representation corresponds in the Eulerian representation to pancake intersection and formation of a one-dimensional structure, i.e. ribs of the initiating cells. Later on, the last type of singularity arises – points corresponding to simultaneous touching of the initial action surface by four points on the paraboloid. In the Eulerian space it is the intersection of ribs, at common vertices of the adjacent cells. Classification of singularity types is a typical problem of the theory of catastrophes [105].

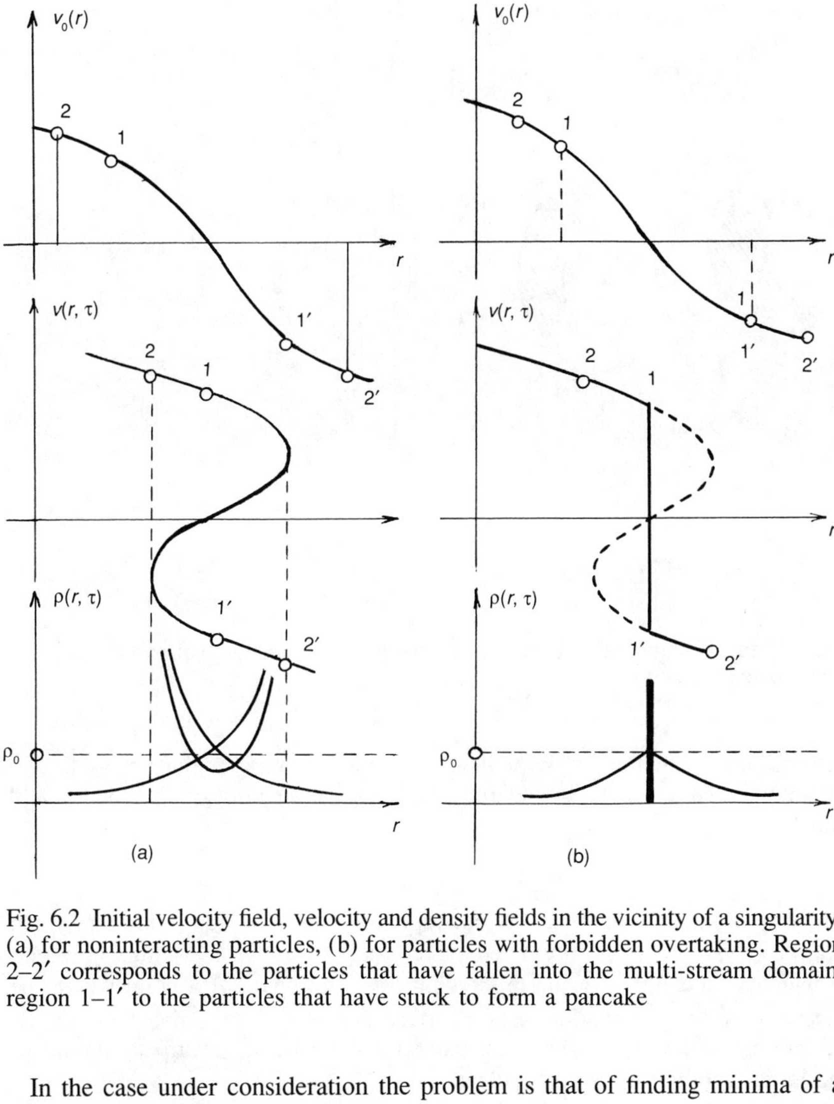

Fig. 6.2 Initial velocity field, velocity and density fields in the vicinity of a singularity: (a) for noninteracting particles, (b) for particles with forbidden overtaking. Region 2–2′ corresponds to the particles that have fallen into the multi-stream domain; region 1–1′ to the particles that have stuck to form a pancake

In the case under consideration the problem is that of finding minima of a three-parameter function $\Phi(\mathbf{q}, \mathbf{r}, \tau)$. We shall not engage ourselves in the process of structure formation, proceeding instead directly to the final stage, noting, meanwhile, that singularities of the three-dimensional BE with vanishing viscosity are classified elsewhere, in reference [159] and Supplement 2.

5. Consider now the case $\tau \gg \tau_n$. At such τ the paraboloid peak defines a smooth function as compared with the initial action $S_0(\mathbf{q})$, and, consequently, the absolute minimum of Φ coincides, virtually, with one of the local minima of $S_0(\mathbf{q})$. The absolute minimum coordinates $\mathbf{q}(\mathbf{r}, \tau)$ become in this instance a

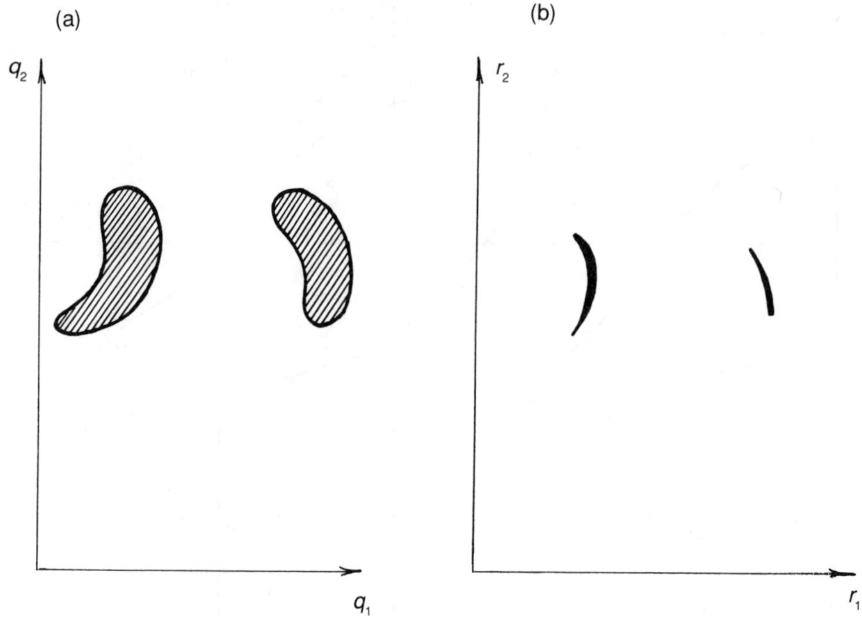

Fig. 6.3 Initial stage of the singularity appearance in (a) Lagrangian, and (b) Eulerian coordinates

discontinuous function of **r**, actually constant in some regions $\mathbf{r} \in \Sigma_i$ and jumping when crossing boundaries. Inside these regions the velocity field equals

$$\mathbf{v}(\mathbf{r}, \tau) = (\mathbf{r} - \mathbf{q}_i)/\tau, \ \mathbf{r} \in \Sigma_i \qquad (1.21)$$

where $\mathbf{q}_i$ is the coordinate of the corresponding initial action minimum which can be approximately considered as constant. Strictly speaking, however, provided we change **r** inside this region, the coordinate of a point where the paraboloid touches the initial action surface will also alter, although at $\tau \gg \tau_n$ deviations of the point where the paraboloid touches a 'sharp' minimum of the initial action will be insignificant.

Regions Σ_i with universal field structure inside will be called 'cells', characterised by the 'centres' $\mathbf{q}_i$ and actions $S_i = S_0(\mathbf{q}_i)$. In each one of the cells, particles fly away from a small region $\tilde{S}_i$ near the centre $\mathbf{q}_i$ such that for $\tau \gg \tau_n$, due to a quicker 'running away' of the faster particles, a universal cellular structure (1.21) is developed. It should be mentioned that in the Eulerian representation the cell centre can actually be outside the cell.

The cell walls (faces) are determined by the condition of simultaneous touching of the initial action at two points by paraboloid α of (1.17). At $\tau \gg \tau_n$ these points are close to the cell centres $\mathbf{q}_i$, and the boundary position of the ith and jth cells in contact is formulated via

$$\Phi(\mathbf{r}_{ij}, \mathbf{q}_i, \tau) = \Phi(\mathbf{r}_{ij}, \mathbf{q}_j, \tau),$$

leading to the following equation of the boundary surface $\mathbf{r}_{ij}$,

$$\left(\left(\mathbf{r}_{ij} - \frac{\mathbf{q}_i + \mathbf{q}_j}{2}\right) \cdot \Delta\mathbf{r}_{ij}\right) = \Delta S_{ij}\tau. \tag{1.22}$$

Here $\Delta\mathbf{r}_{ij} = \mathbf{r}_i - \mathbf{r}_j$ is a vector connecting the cell centres, $\Delta S_{ij} = S_i - S_j$ being the difference between the adjacent cell actions. Therefore, when $\tau \gg \tau_n$ the cell walls degenerate into planes orthogonal to vector $\Delta\mathbf{r}_{ij}$, their travel speeds being constant, parallel to vector $\Delta\mathbf{r}_{ij}$, proportional to the adjacent cell action difference $|\mathbf{v}_{ij}| = |\Delta S_{ij}/\Delta\mathbf{r}_{ij}|$ and directed away from the cell with the lower action. It is apparent that the adjacent cell walls intersect through a straight line, i.e. a rib. A set of linear equations governing the intersection of the three adjacent cell walls is, as comes out of (1.22), degenerate, and thus all three walls intersect through a straight common rib of three neighbouring cells. The ribs, in their turn, intersect in common nodes–vertices of the cells. The velocity field suffers a jump at the cell boundaries and for $\tau \gg \tau_n$ the jump surfaces form a connected structure.

For a vector field of velocity $\mathbf{v}$, longitudinal $v_L = (\mathbf{v}.\ \mathbf{s})$ and transverse $\mathbf{v}_N$ components are introduced, wherein $\mathbf{s}$ is a unit vector along a certain fixed axis x_1. Then it follows from (1.21) that these components inside the cells are equal to

$$v_L = (r_L - q_{iL})/\tau,\ \mathbf{v}_N = (\mathbf{r}_N - \mathbf{q}_{iN})/\tau. \tag{1.23}$$

Thus the longitudinal component of the velocity vector field appears as a sequence of sawtooth pulses, i.e. its structure coincides with the field of one-dimensional Burgers turbulence. The transverse component appears as a sequence of rectangular pulses with random amplitudes. In (1.23), q_{iL} and $\mathbf{q}_{iN}$ are the cell centre projections onto axis x_1 and a plane perpendicular to it, while the field discontinuities arise at the cell boundaries. A qualitative image of the velocity field cellular structure, and its longitudinal and transverse components in two dimensions, are depicted in Fig. 6.4.

In order to develop a picture of the matter distribution, it is helpful to employ relations (1.10) and (1.11) which describe the matter density variation in the vicinity of a fixed particle with Lagrangian coordinate $\mathbf{q}$. The simplest way is to estimate the matter density inside the cells, since the particles enclosed in them have not as yet experienced collisions and move according to the law (1.5), their density being determined by (1.10) and (1.18). Inasmuch as some cells swallow up the others, and hence only those with a sufficiently deep action minimum survive, one can, to estimate the density, use the asymptotic procedure outlined in subsection 6. It is clear, however, that, inasmuch as the matter inside the cells has been formed as a result of the particles flying away from a small isolated island in the vicinity of the minimum of $S_0(\mathbf{q}_i)$, then the principal values of tensor $\beta_i(\mathbf{q}_i)$ are positive, and for the matter density inside a cell we have from (1.21),

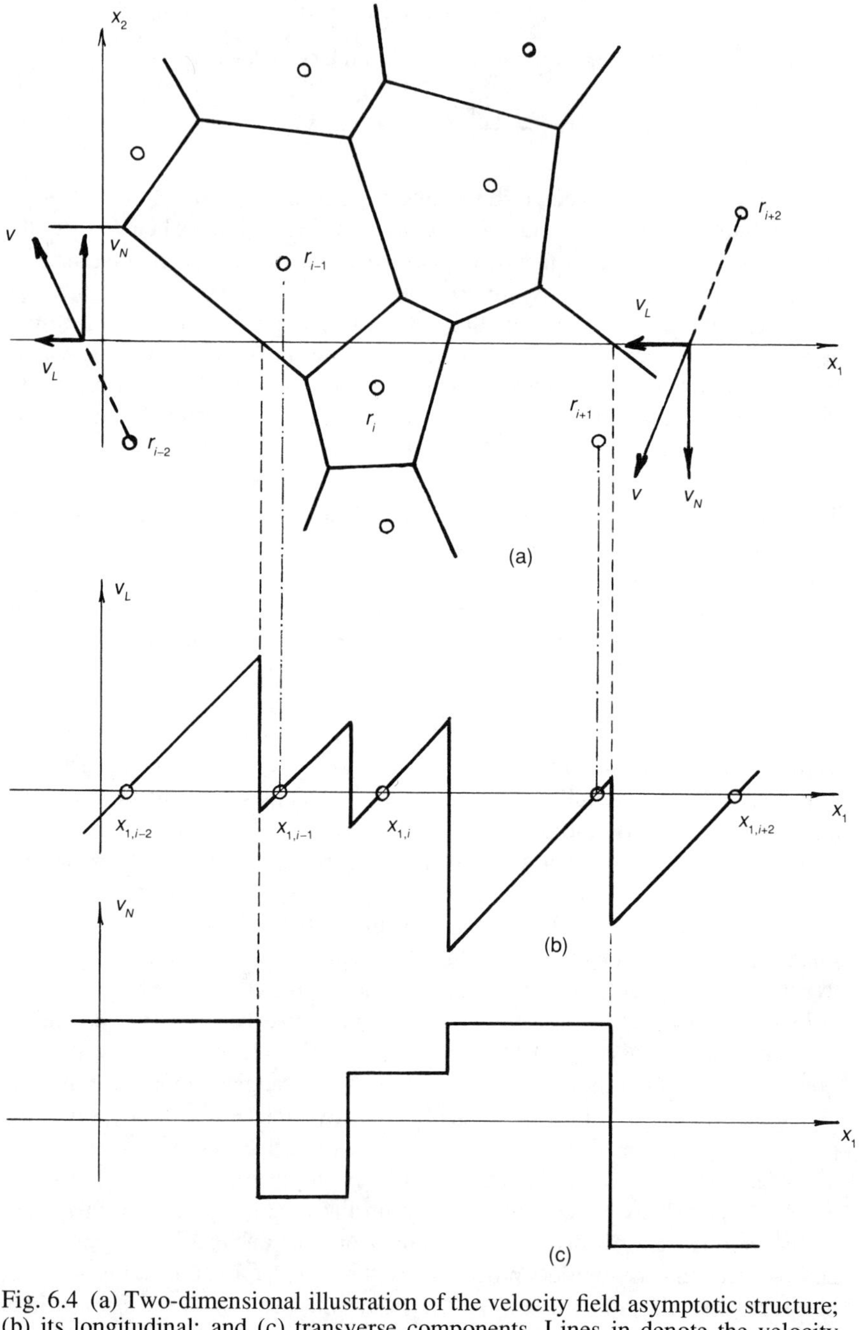

Fig. 6.4 (a) Two-dimensional illustration of the velocity field asymptotic structure; (b) its longitudinal; and (c) transverse components. Lines in denote the velocity field discontinuity domains

$$\eta(\mathbf{q}, \tau) = \eta_0/(\beta_1\beta_2\beta_3\tau^3) \approx \eta_0(\tau_n/\tau)^3. \tag{1.24}$$

As a consequence, the matter density inside the cells becomes much less, as compared with the average one η_0. In the Universe development scenarios, such regions are called 'dark' because, due to reduced density, they cannot give birth to stars.

It is a great deal more difficult to find the matter density in the walls, since the trajectories of the particles that arrive at the cells are very complicated. However, qualitative conclusions on the matter density distribution inside the walls can be inferred by studying the behaviour of certain particles which have fallen into the walls. To do this one needs to consider the fine structure of the walls. If $\mu \neq 0$ we have, in the vicinity of the walls, to consider the competition between the neighbouring cells, which leads from (1.2) to the following equation for the field in the vicinity of the wall:

$$\mathbf{v}(\mathbf{r}, \tau) = \frac{1}{\tau}\left(\mathbf{r} - \frac{\mathbf{q}_i + \mathbf{q}_j}{2}\right) - \frac{\Delta\mathbf{r}_{ij}}{2\tau}\tanh\left(\frac{\Delta\mathbf{r}_{ij}\cdot(\mathbf{r} - \mathbf{r}_{ij})}{4\mu\tau}\right), \tag{1.25}$$

$$\mathbf{r}_{ij} = \frac{\mathbf{q}_i + \mathbf{q}_j}{2} + \mathbf{V}_{ij}\tau, \quad \mathbf{V}_{ij} = \frac{\Delta S_{ij}\Delta\mathbf{r}_{ij}}{|\Delta\mathbf{r}_{ij}|^2}. \tag{1.26}$$

Therefore, at the stage $\tau \gg \tau_n$ the velocity field has two scales: the external one, a typical cell size $l(\tau)$, and the internal one $\delta(\tau)$ related to the former,

$$\delta \sim \mu\tau/l \tag{1.27}$$

and characterising the cell wall thickness.

Trajectories of individual particles $\mathbf{R}(\tau)$ satisfy the following equation:

$$\frac{d\mathbf{R}}{d\tau} = \mathbf{v}(\mathbf{R}, \tau), \tag{1.28}$$

where the field $\mathbf{v}(\mathbf{r}, \tau)$ is determined by the three-dimensional BE (1.1). Using the latter one can readily derive the expression for the force $\mathbf{F}$ acting upon the particle as follows:

$$\frac{d\mathbf{R}}{d\tau} = \mathbf{V}, \frac{d\mathbf{V}}{d\tau} = \mathbf{F}(\mathbf{R}, \tau), \quad \mathbf{F}(\mathbf{r}, \tau) = \mu\Delta\mathbf{v}(\mathbf{r}, \tau). \tag{1.29}$$

As can be seen from the field structure (1.25) and from equation (1.29), two stages in the individual particle motion can be distinguished. Inside the cells $\mathbf{F} = 0$, and the particle moves freely in accordance with (1.9) which, in physical variables, corresponds to the gravitational instability development. However, when it approaches the wall, at a distance of about δ the deceleration force directed normally from the inter-cell boundary comes into play, due to which the velocity normal component of the particle falling on the wall decreases and eventually becomes equal to that of the wall. The tangential component of the velocity is conserved, so as to make the particles which have fallen on the wall move from its centre to the periphery and concentrate at the cell ribs. The particle motion along the ribs favours a further concentration of matter at the points of rib intersection. Thus at $\tau \gg \tau_n$ the density field has a cellular structure where the matter concentration is mainly in the cell walls, with accumulation along the cell ribs attaining a maximum in the cell vertices produced by the rib intersections. The common trend is that in the course of time all matter tends to pass to the nodes, i.e. compact formations. Whereas in a continuous medium the walls and ribs can live an infinitely long life, the total mass enclosed in them vanishes with time. In particular, for discrete models the walls and ribs may contain no particles and, consequently, be invisible in the structure.

In the adiabatic scenario of the Universe evolution, the cell walls and ribs correspond to anisotropic superclusters of galaxies, while the nodes or rib intersections refer to compact galaxy condensations of the rich cluster type. As a result of the statistical character of the perturbations at not too large ratios τ/τ_n, different evolution stages can be observed in different space regions. In order to compare the results obtained from model (1.1) with those of the adiabatic scenario of the gravitational instability development, it seems useful to return to Figs 6.4 and 6.1 once again, where the particle distribution derived from the two-dimensional numerical simulation of a gas of gravitationally interacting particles is depicted. One can see in these figures all basic structural elements, i.e. cells, ribs and nodes predicted by the qualitative theory. A more careful study of the model (1.1) as against the results of a numerical calculation will be offered at the end of this section.

6. The wall motion results in a continuous change of cell shape. As the wall velocities are directed from the cell centres with a deeper minimum $S_0(\mathbf{q})$, these cells swallow up their neighbours, provoking the growth of the external scale $l(\tau)$. The growth law $l(\tau)$ is heavily influenced by the initial perturbation type. Because of this it seems helpful to start an elaborate discussion of the external scale growth rate dependence on the initial action spectrum $G_0(k)$ type. It will be assumed that it decreases rather rapidly, for example, by an exponential law, for $k > 1/l_0$. If applied to the adiabatic theory of the Universe large-scale structure formation, it is justified, since the small-scale modes are suppressed at early stages of evolution. The growth rate of $l(\tau)$ may differ qualitatively between the cases of bounded initial action variance $\langle S_0^2 \rangle = \sigma_s^2 < \infty$, and of unbounded variance, when

$$G_s(\mathbf{k}) \approx \alpha_\theta^2 k^{-\theta} b(\mathbf{k}), \quad 3 < \theta < 5, \tag{1.30}$$

here $0 < b_0(\mathbf{k}) < \infty$ and falls off rather quickly as $\mathbf{k} \to \infty$. As in one dimension (see section 6.2), a typical estimation of the external scale $l(\tau)$ will be performed from the condition of equality of increments of the paraboloid α (1.17) and the initial action in (1.3). This leads to the following equation:

$$l^2/\tau \approx \sqrt{d_s(l)},\ d_s(\mathbf{z}) = \langle [s_0(\mathbf{r}+\mathbf{z}) - \mathbf{s}_0(\mathbf{r})]^2 \rangle, \tag{1.31}$$

where $d_s(\mathbf{z})$ is the structure function of the initial action. In the first case, when $d_s(|\mathbf{z}| \gg l_0) = 2\sigma_s^2$, it follows from (1.31) that

$$l(\tau) \approx (\sigma_s \tau)^{\frac{1}{2}} \approx l_0 \left(\frac{\tau}{\tau_n}\right)^{\frac{1}{2}},\quad \tau_{\mathbf{n}} = \frac{l_0}{\sigma_0} = \frac{l_0^2}{\sigma_s}. \tag{1.32}$$

Here the growth law for the typical cell size is only defined by the integral properties of the initial perturbation spectrum, and does not depend on its fine structure.

In the second case, when $d_s(|\mathbf{z}| \gg l_0) \approx \alpha_\theta^2 \, |\mathbf{z}|^{\theta-3}$, (1.31) yields

$$l(\tau) = \alpha_\theta^{2/(7-\theta)} \tau^{2/(7-\theta)},\quad 3 < \theta < 5. \tag{1.33}$$

Physical distinction between the two initial perturbation cases considered lies in the absence for the former case, and in the presence for the latter one, of slowly decreasing correlations between the initial perturbation values at the far distant spatial points. On comparing (1.32) and (1.33), one can see that the existence of the long-distance correlations speeds up the external scale growth, i.e. the size of the cells.

Which case is realised in the Universe depends on the spectrum of the initial density fluctuations at the end of the recombination epoch [103]. Let their spectrum in the range of small wave numbers, i.e. the large scales, have the form

$$G_\rho(\mathbf{k}) \sim k^{-n}\tilde{b}(\mathbf{k}),\quad 0 < \tilde{b}(\mathbf{k}) < \infty, \tag{1.34}$$

where $\tilde{b}(\mathbf{k})$ is rapidly decreasing for $|\mathbf{k}| > l_0^{-1}$. All the theories of the density fluctuation development offer, by the end of the recombination epoch, a power spectrum in the large-scale region, but with different indices n. Gravitational instability of the density perturbations leads to density fluctuations. The action spectrum calculation, in the linear approximation (1.12), gives ($S_0(\mathbf{q}) \equiv \Phi_0(\mathbf{q})$):

$$G_s(\mathbf{k}) \sim G_\rho(\mathbf{k})\,|\mathbf{k}|^{-4} \sim |\mathbf{k}|^{n-4}. \tag{1.35}$$

This indicates that $\theta = n - 4$, and if $n \geq 1$ then the first case is realised: $\sigma_s^2 < \infty$.

On passing to the real coordinates and time (1.14), the formulae (1.32) and (1.34) allow us to evaluate the laws of matter mass growth in an individual structure element, $m_l = \rho_0 l_0^3$. In the two situations under consideration it yields

$$m_l(t) \approx \begin{cases} b^{\frac{3}{2}}(t),\ n > 1,\ \theta \leq 3, \\ b^{6/(3+n)}(t),\ -1 < n < 1,\ 3 < \theta < 5. \end{cases} \tag{1.36}$$

With respect to the flat Universe, when $b(t) \sim t^{\frac{2}{3}}$ we obtain, respectively,

$$m_l(t) \sim l^3(t) \sim \begin{cases} t, \ n > 1, \\ t^{4/(3+n)}, \ -1 < n < 1. \end{cases} \qquad (1.37)$$

Compare the derived formulae for the typical scale and mass of the cells with the results of the quasi-linear theory of gravitational instability [103]. Neglecting pressure, the gas of gravitationally interacting particles will appear as a nondispersive medium and all the scales will grow to the same extent. In the expanding flat Universe, when $a(t) \sim b(t) \sim t^{\frac{2}{3}}$, we have from (1.12), for the density spectrum in linear approximation

$$G_\rho(\mathbf{k}; t) \sim t^{\frac{4}{3}}\, G_\rho(\mathbf{k}). \qquad (1.38)$$

Within the quasi-linear theory, the typical scale of a structure element $l \sim 1/|\mathbf{k}_*|$ is deduced from the condition claiming that the relative mass fluctuations in this element are of order unity,

$$\frac{\Delta M}{M} \sim t^{\frac{4}{3}} \int_0^{k_*} G_\rho(k) k^2 \mathrm{d}k \sim 1. \qquad (1.39)$$

For a power spectrum $G_\rho(\mathbf{k})$ of type (1.34) this leads to

$$l(t) \sim t^{4/[3(3+n)]} \qquad (1.40)$$

which coincides with (1.38), if $m_l \sim l^3$ is accounted for. This law holds for $n < 4$ only; for initial spectra with $n < 4$ a universal spectrum $G_\rho(k;\, t) \sim A(t)k^4$ develops, due to nonlinear effects, in the low-frequency range. The quasi-linear theory suggests that in (1.39), for the spectra with $n < 4$, one should set $G_\rho(k) \sim Ak^4$, where $A = \text{const}$ [147, 148] which leads to the law $l \sim t^{\frac{4}{21}}$. Thus in the quasi-linear theory we have for the external scale

$$l(t) \sim t^\gamma,\ \gamma = \begin{cases} \dfrac{4}{21},\ n \geq 4 \\ \dfrac{4}{3(3+n)},\ -1 < n \leq 4. \end{cases} \qquad (1.41)$$

It follows from (1.32) and (1.33), however, that the transition from the value $\gamma = \frac{4}{3}(3+n)$ to the universal law $\gamma = \frac{1}{3}$ occurs at $n = 1$. The more rapid growth $l \sim t^{\frac{1}{3}}$, of the external scale in the nonlinear theory relates to the fact that in the quasi-linear theory, where $l \sim t^{\frac{4}{21}}$ the systematic nonlinear generation of low-frequency components is not considered. Within the interval $1 < n < 3$ the density spectrum in the region $k \to 0$ is really conserved $G_\rho(k;\, t) \sim k^n$, but this law is quite quickly replaced by $G_\rho(k;\, t) \sim k^4$ which exactly governs the scale $l(t)$ growth law. It is worth noting that the transition region with respect to n also exists for the velocity spectrum in one-dimensional BT (see section 5.3). Figure 6.5 depicts the plots for the

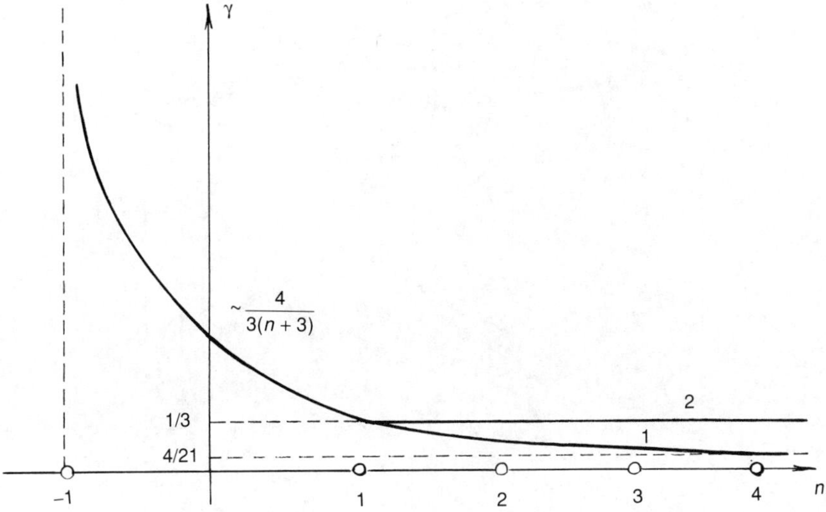

Fig. 6.5 Plots of the external scale $l(t)$ growth rate index against power index n of the initial density fluctuation spectrum: quasilinear theory 1, nonlinear theory 2

external scale $l(t)$ index dependence on the index n of the density fluctuation initial spectrum form (1.34), in the framework of the quasi-linear theory (1.41) and the nonlinear theory described by relations (1.37).

7. To conclude, we demonstrate the results of the two- and three-dimensional problem simulation for the gravitational instability evolution based on BE (1.1). In references [111, 112] an asymptotic solution (1.4) was employed which reduced the solution of the BE to a procedure of seeking the absolute minimum coordinates with respect to the function $\Phi(\mathbf{r}, \mathbf{q}, \tau)$ (1.3). As was shown above, this solution in its turn is reduced to a procedure of searching for the first point in which paraboloid $\alpha = -(\mathbf{r} - \mathbf{q})^2/2\tau + H$ of (1.17) touches the initial potential $S_0(\mathbf{q})$. The two-dimensional problem involves three situations, as follows. Given a single contact point, the particle travels according to (1.5) with no collisions until the current time τ. The paraboloid touches the initial action at two points which, in two dimensions, corresponds to the formation of pancake that later on grow to form the interfaces between the cells (these points are depicted by squares in Figs 6.6). And finally, three contact points are possible, equivalent to an intersection of ribs (crosses in the figures). During a numerical experiment the initial field $S_0(\mathbf{q})$ was supposed to be Gaussian and to have a power spectrum

$$G_s(k) \sim \begin{cases} k^{-2}, \ k \in [k_1, k_2], \\ 0, \ k \notin [k_1, k_2]. \end{cases}$$

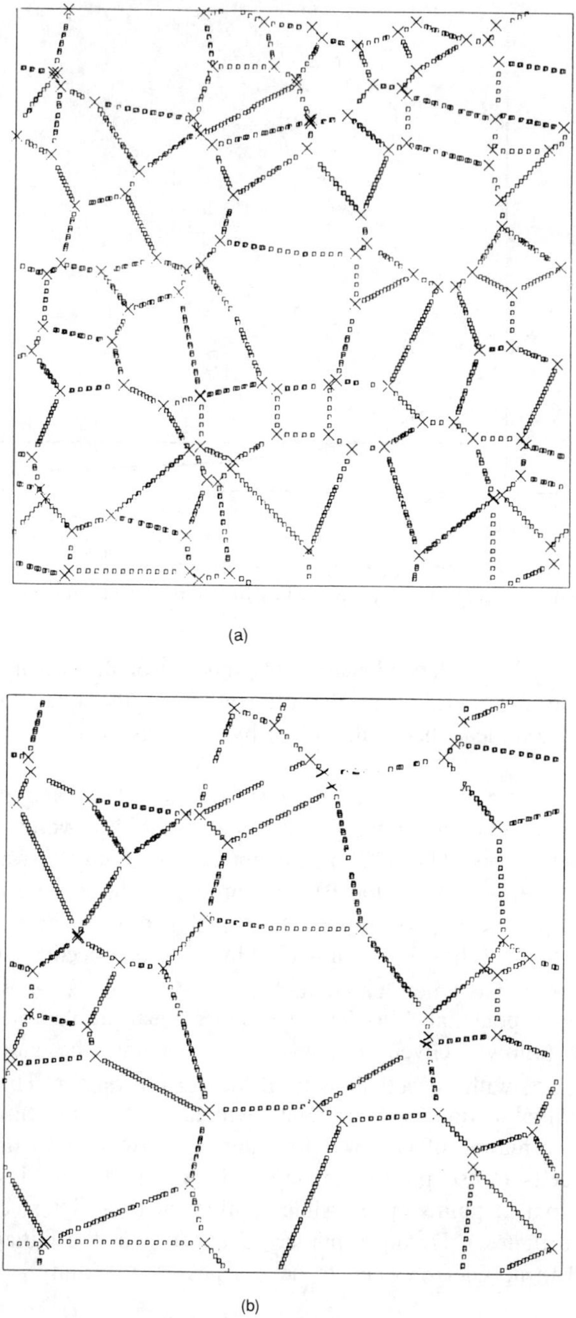

Fig. 6.6 Eulerian structure of the density field (b) at $\tau/\tau_n = 40$; (b) at $\tau/\tau_n = 80$

It follows from the linear theory that the relative density fluctuations grow in conformity with the linear law (1.20),

$$\delta\rho/\rho \approx \tau/\tau_n,\ 1/\tau_n \approx \sqrt{\langle |\Delta \mathbf{s}_0^2| \rangle}, \tag{1.42}$$

where τ_n is the characteristic time of the singularity formation. The numerical experiment results correspond to the case when, formally, $\delta\rho/\rho \gg 1$, and a sufficiently developed structure has been already formed. Numerical tests showed that at large times, all matter is indeed concentrated in singularities, while the cell interfaces appear as straight lines, the cell number diminishing with time. Figure 6.6 depicts the density field Eulerian structure corresponding to extremely large times, $\tau/\tau_n = 40, 80$.

However, of much more interest is another set of numerical tests, proposed by the present authors, wherein, along with a BE-based simulation, a straightforward numerical calculation of the self-consistent motion of gravitationally interacting particles was carried out. Further, the initial perturbations of the gravitational potential φ and potential $\Phi_0 \equiv S_0$ in (1.3) were given as related through (1.12). Figure 6.7 demonstrates the particle and singularity positions in the straightforward and model problems. It is evident from these figures that the model problems. It is evident from these figures that the model gives a strikingly adequate description of the locations of the two singularity types – ribs and nodes. In order to emphasise the fact that the 'dark' regions really lack particles (luminous galaxies) the points are given by white spots against black background. In section 6.3 we shall compare qualitative properties of the model and straightforward problems considered, in one dimension.

In reference [112] there was also more detailed comparison of the BE-based simulation (this model is also called the 'adhesion' model) with the direct simulation of the motion of gravitationally interacting particles at rather late stages. In Figs 6.8, 6.9 and 6.10 are represented the results of this comparison at the epochs when the relative linear density fluctuations (1.42) are $\delta\rho/\rho \approx 4.2$, 6.0 and 12, respectively. Each figure consists of three plots: a, b and c. The set of figures (a) shows the distribution of the particles calculated from the direct simulation. The figures (b) show the skeleton of the structure obtained in the framework of the adhesion simulation, constructed with the geometrical procedure. The points show the positions of the paraboloid when it touches the action surface at two points and they show the positions of 'filaments' in two dimensions. Circles show here the position of the paraboloid contact with the action at three points; here are the positions of clumps. In order to make the comparison between the two descriptions easier and to stress the differences more clearly, the authors of [112] combined these two pictures into the series (c). The disagreement between the model description and the direct simulation begins to be significant at a rather later stage of evolution, but the authors of [112] say

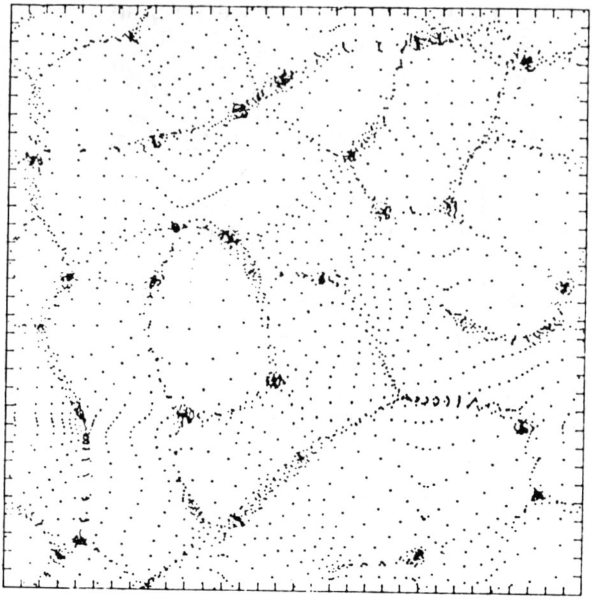

(a)

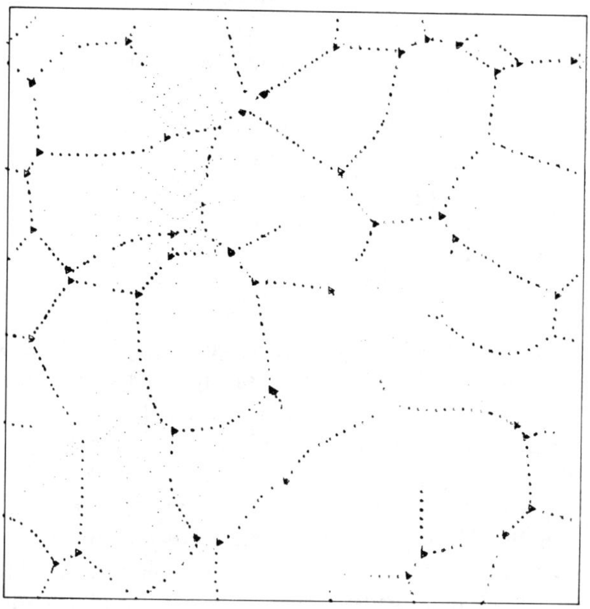

(b)

Fig. 6.7 Results of the numerical calculation of the gravitationally interacting particle motion. Locations of particles in a self-consistent problem, a; locations of particles and singularities in a model based on the Burgers equation

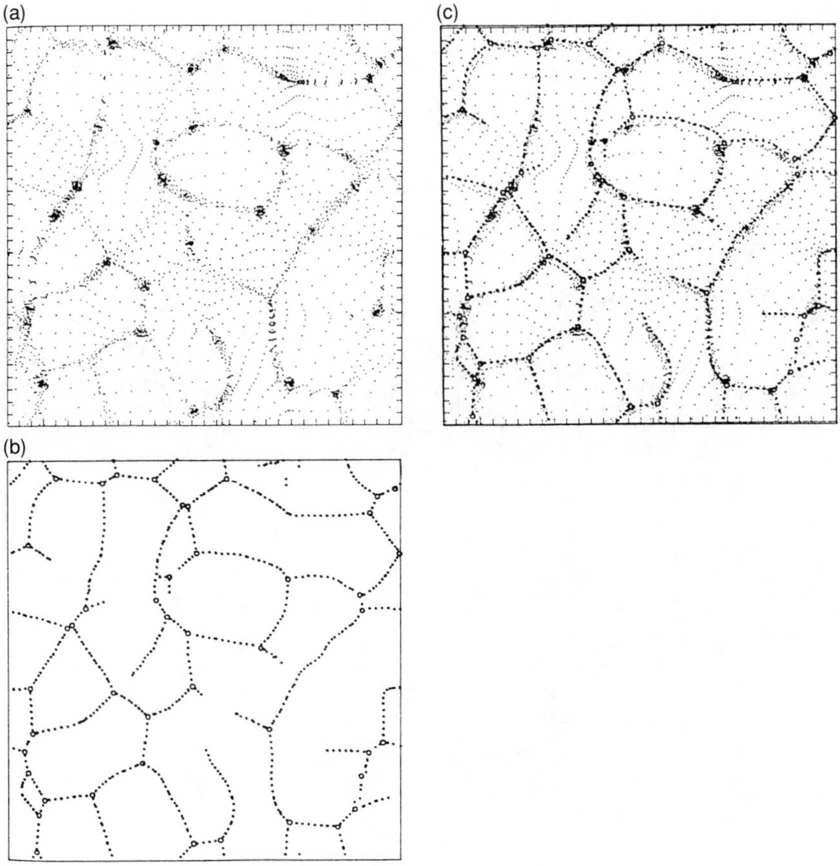

Fig. 6.8 The comparison of (a) direct simulation with (b) adhesion simulation, and (c) the composite picture. The epoch is such that $(\delta\rho/\rho)_{\text{lin. theory}} = 4.2$

that this may be connected to the growth of errors in the numerical simulations.

The comparison between the model and direct descriptions of gravitationally interacting particles was carried out in [111, 112], in the framework of the asymptotic solution of the Burgers equation. But it is also possible to use in numerical simulation the Hopf–Cole transformation, which reduces the three-dimensional Burgers equation (1.1) to the linear diffusion equation, and thus yields the general solution of BE (1.2). The positions of the particles may then be found by integration of the equation (1.28). This method was used in the recent work [167, 168], in which the three-dimensional velocity field was calculated by using the relation (1.2),

$$\mathbf{v}(\mathbf{r}, t) = -2\mu\nabla \ln U, \tag{1.43}$$

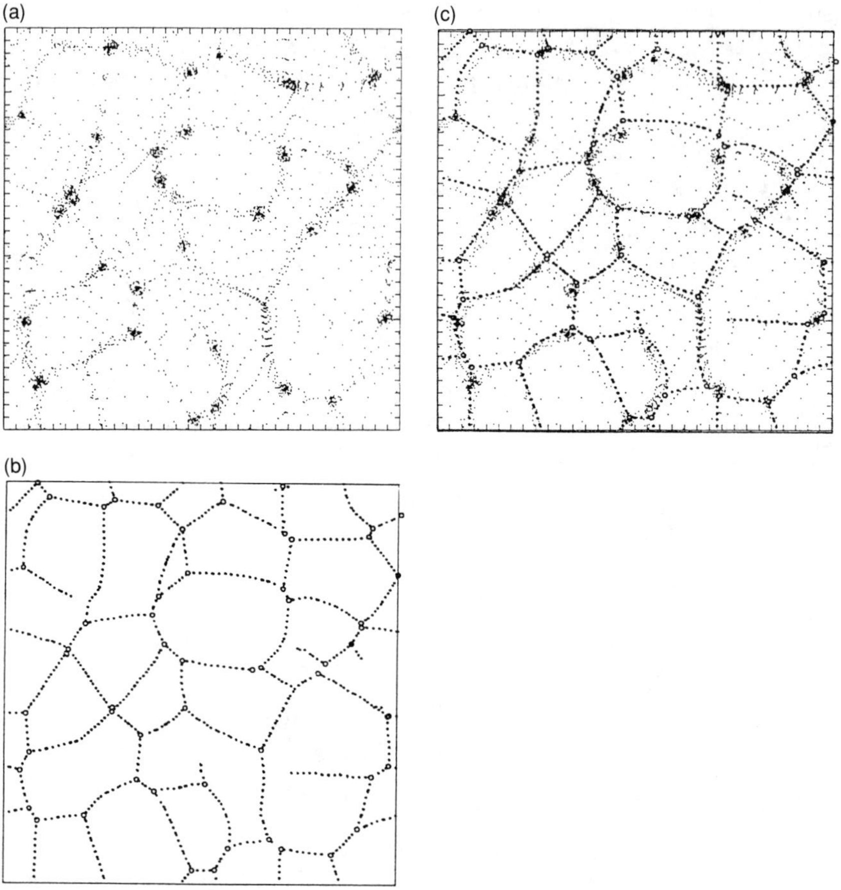

Fig. 6.9 As in Fig. 6.8, but at the epoch when $(\delta\rho/\rho)_{\text{lin. theory}} \approx 6.0$

and the solution of the linear equation for U was obtained by numerical convolution of the initial field

$$U(\mathbf{r}, 0) = \exp\{-S_0(\mathbf{r})/2\mu\}$$

with the kernel function which is Gaussian. The adhesion simulation was also compared with results of direct three-dimensional simulations of gravitationally interacting particles. In all the experiments it was assumed that the initial conditions were the same in both simulations. Reference [167] also gave some estimates as to the cosmological scales on which one may use the adhesion simulations. Here we introduce some results from that reference which illustrate the structures of density and velocity fields in three dimensions. The comparison was done for different kinds of initial perturba-

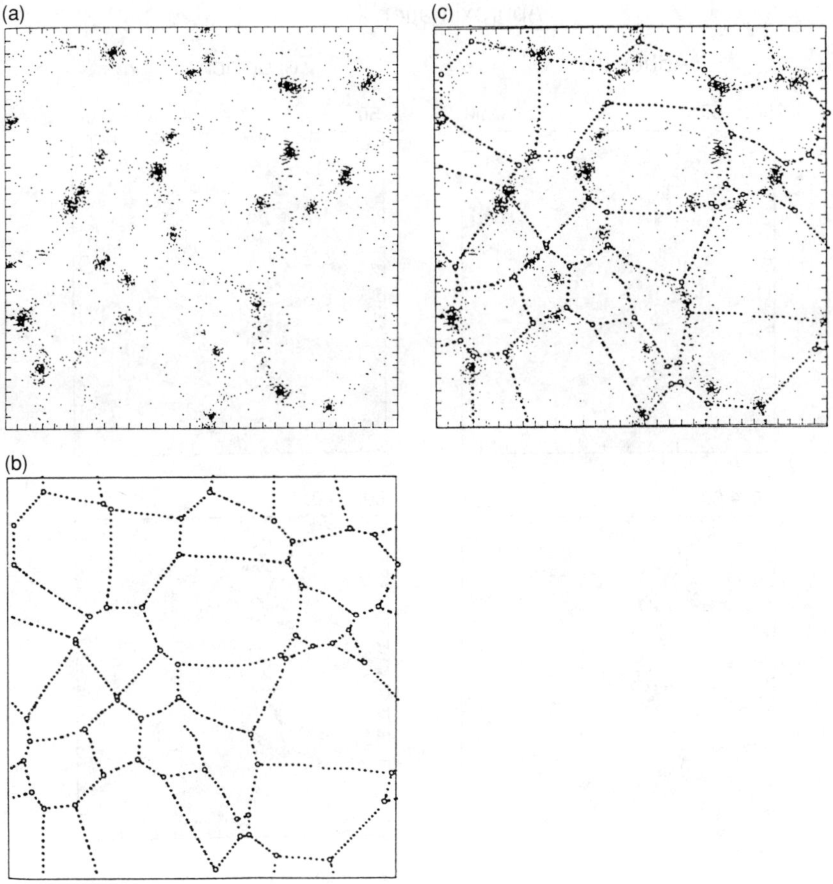

Fig. 6.10 As in Fig. 6.8, but at the epoch when $(\delta\rho/\rho)_{\text{lin. theory}} \approx 12$

tion, known in cosmology as cold dark matter (CDM), hot dark matter (HDM), and biased CDM, and which differ from each other in the cut-off of the initial spectra at small scales.

In Fig. 6.11 one can see the comparison between the adhesion model (left panels) and the Zel'dovich approximation (right panels), with the same initial conditions. From the top to the bottom of the picture the effective time increases by a factor of four. At the early times the two simulations agree, but after pancakes collapse the adhesion simulation keeps them thin and coherent, while in the Zel'dovich approximation, particles pass through pancakes unperturbed.

In Figs 6.12, 6.13 and 6.14 one can see the comparison between the adhesion model (panels on the left) and the direct simulation (panels on the

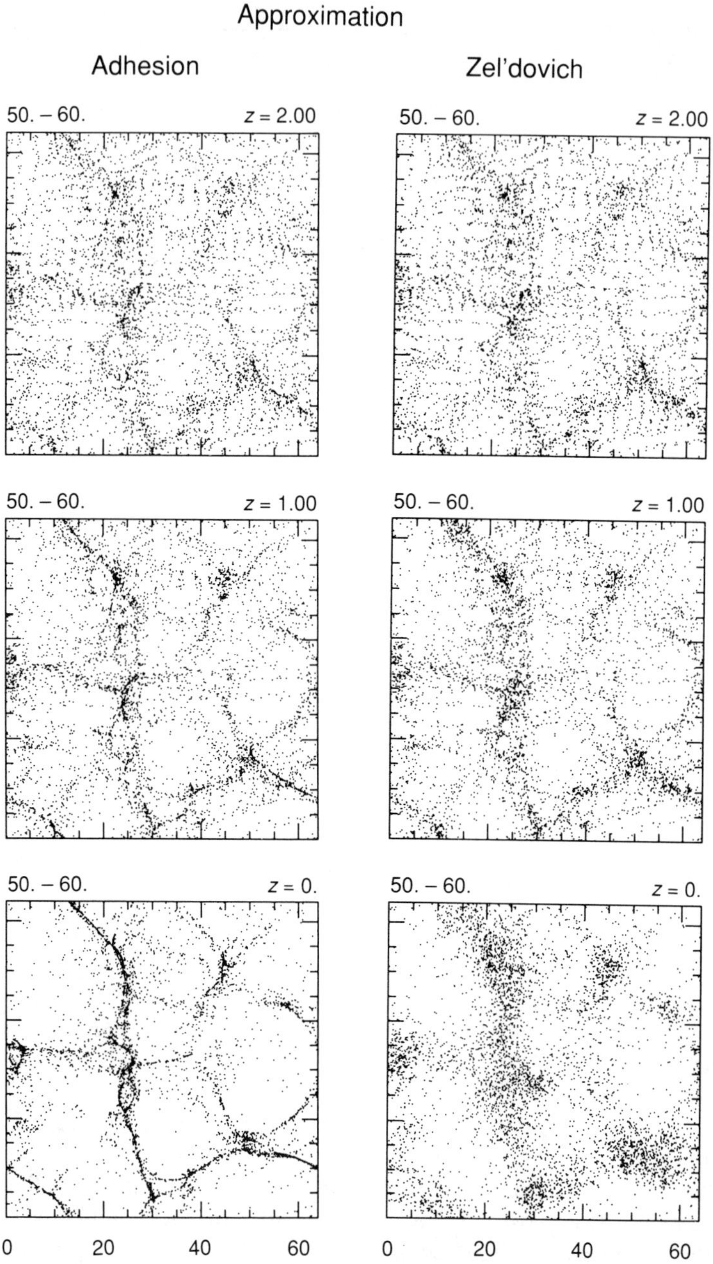

Fig. 6.11 Comparison of the adhesion simulation (left panel) with the results of the Zel'dovich approximation (right panel). Here z is the cosmological redshift and is connected with 'time' t in adhesion simulation by the relation $C = B_0/(1 + z)$. The redshift is z = 2 (top), z = 1 (middle), and z = 0 (bottom)

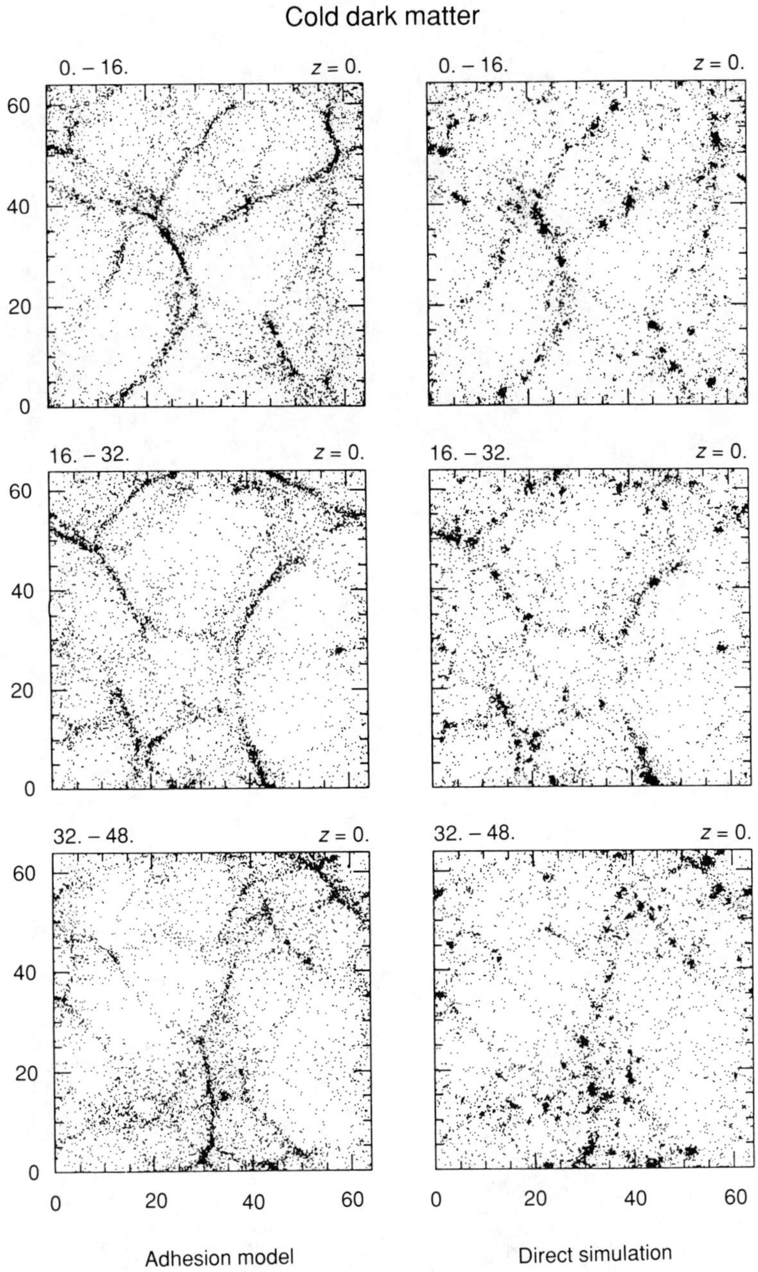

Fig. 6.12 Comparison between direct simulation (right panels) and adhesion simulation (left panels) with identical initial conditions. The redshift is $z = 0$
Cold Dark Matter initial conditions, with power density spectrum with slope exponent $n = 1$ on very large scales and slope exponent $n = -3$ on very small scales.

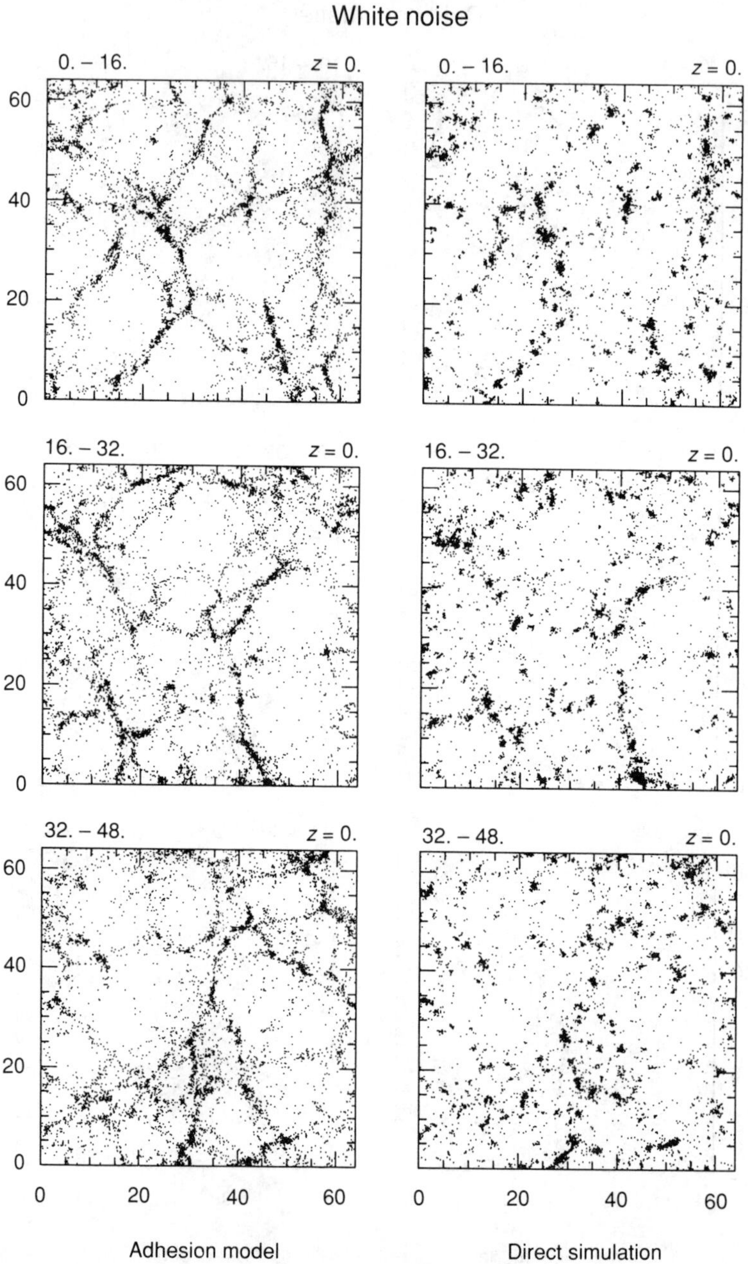

Fig. 6.13 As in Fig. 6.12, but with white noise initial conditions

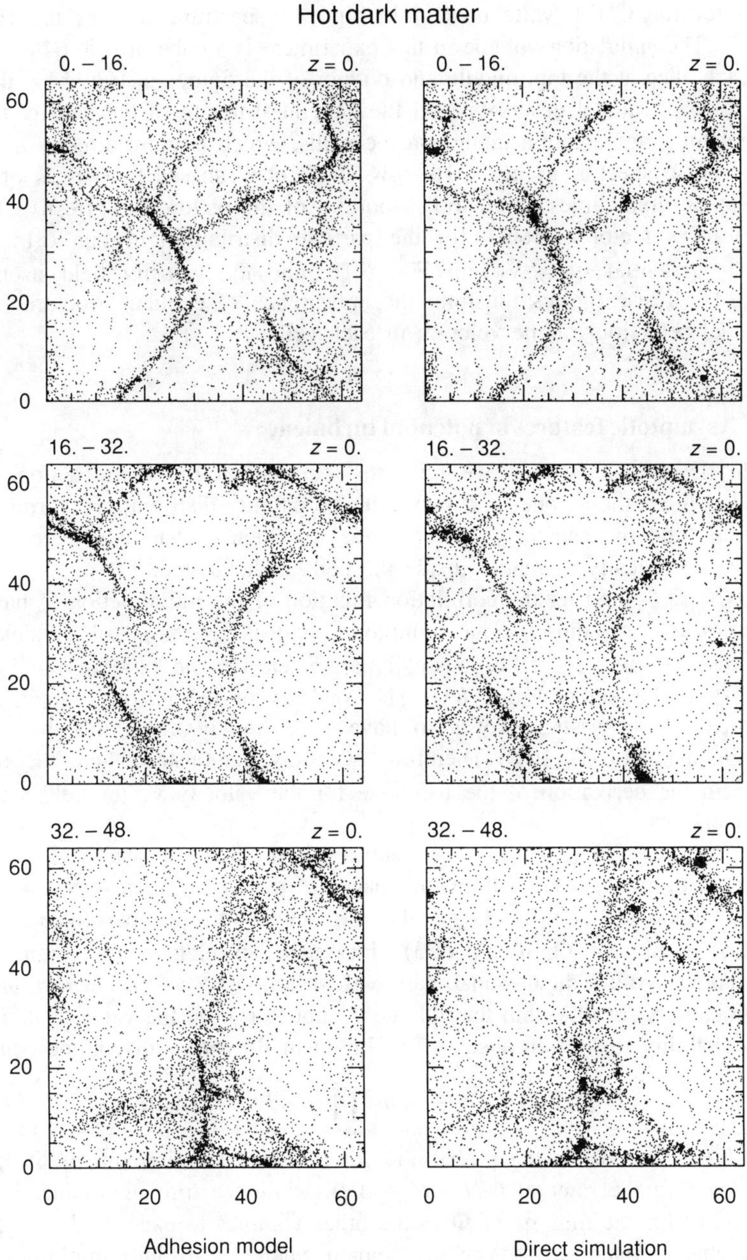

Fig. 6.14 As in Fig. 6.12, but with the hot dark matter initial conditions, where there is no initial power on small scales

right) for the CDM white noise initial power spectrum and for the HDM models. The simulation volume in this experiment is a cube of side $64h^{-1}$ Mpc, and each slice at the top, middle and bottom of the figures is $16h^{-1}$ Mpc thick. The adhesion model describes well the general structure of the density field, but not the fragmentation matter on the walls, which is more directly related to the noise spectra. In figs 6.15, 6.16, 6.17 one can see two slices of the cube at different times (from top to bottom of the picture the effective time grows by a factor of four), for the particle distribution in Fig. 6.15, the velocity potential ($\mathbf{v} = \nabla\phi$) in Fig. 6.16, and the velocity field itself in Fig. 6.17. These figures illustrate the appearance of cellular structure, with universal behaviour of the velocity in each cell.

6.2 Asymptotic features of potential turbulence

1. In the present section we find, for the case of the initial action with bounded fluctuations, one- and two-point probability distributions, correlation functions and the spectra of the potential turbulence chaotic velocity field. An average density of matter inside the cells is calculated.

Let $\sigma_s^2 < \infty$, and let the correlation function of the initial action S_0 have a finite correlation radius l_0. The asymptotic analysis here is possible thanks to the presence, at $\tau \gg \tau_n$, of a small parameter, i.e. the ratio of the initial scale l_0 to the external one $l(\tau)$. The analysis of the chaotic vector field statistical characteristics proves to have a lot in common with that of a one-dimensional BT [109]. Therefore, we only demonstrate here the main points in the derivation of the formulae for the velocity vector field $\mathbf{v}(\mathbf{r}, \tau)$ characteristics.

For the sake of generality we handle the problem of searching for the statistical properties of $\mathbf{v}$ within the spaces of different dimensions $n = 1, 2, 3$. We utilise the solution (1.4) where $\mathbf{q}$ stands for the coordinate of the absolute minimum of Φ in (1.3). Introduce an elementary volume V_m represented by an n-dimensional cube with side $\Delta \approx l_0 \ll l(\tau)$. Let $F(h, m)$ be the probability for Φ within the volume V_m not to exceed the value of h. Then the probability to find an absolute minimum of Φ in an mth volume equals

$$P(\mathbf{q} \in V_m) = \int W(-H, m) \prod_{j \neq m} [1 - F(-H, j)]\, \mathrm{d}H. \tag{2.1}$$

Here $W(-H, m)\,\mathrm{d}H$ is the probability for an absolute minimum of Φ in V_m to fall within the interval $(-H, -H - \mathrm{d}H)$, while the infinite product is the probability for the minima of Φ in the other volumes to exceed $-H$. At $\tau \gg \tau_n$ the characteristic size $l(\tau)$ of the domain where an absolute minimum can be observed is far in excess of l_0, (1.32), and a large number of the local minima of S_0 compete for the right to become an absolute minimum. Consequently, $F(-H, j) \ll 1$ and the following asymptotic formula holds true:

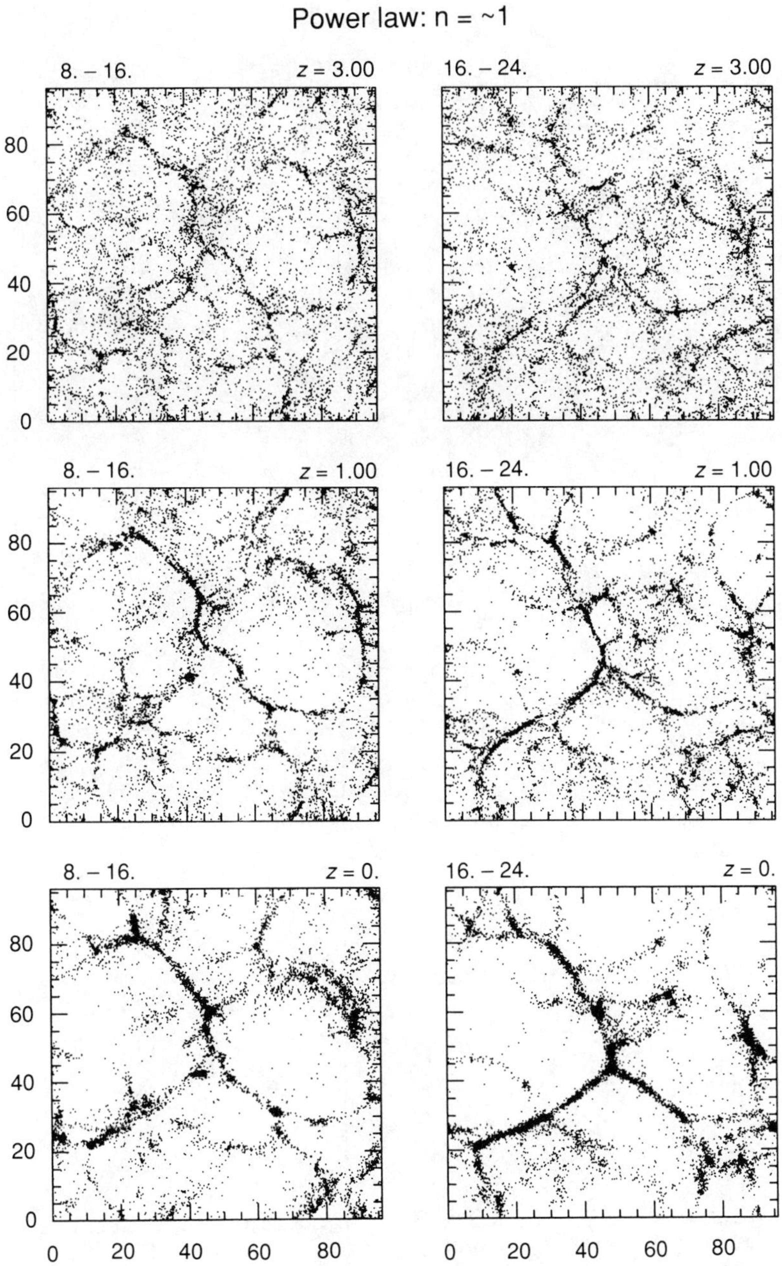

Fig. 6.15 Two slices of the particle distribution from an $n = -1$ power-law simulation at redshifts $z = 3$ (top), $z = 1$ (middle), and $z = 0$ (bottom). The characteristic scale of the structure grows with increasing $\tau = t_0/(1 + z)$

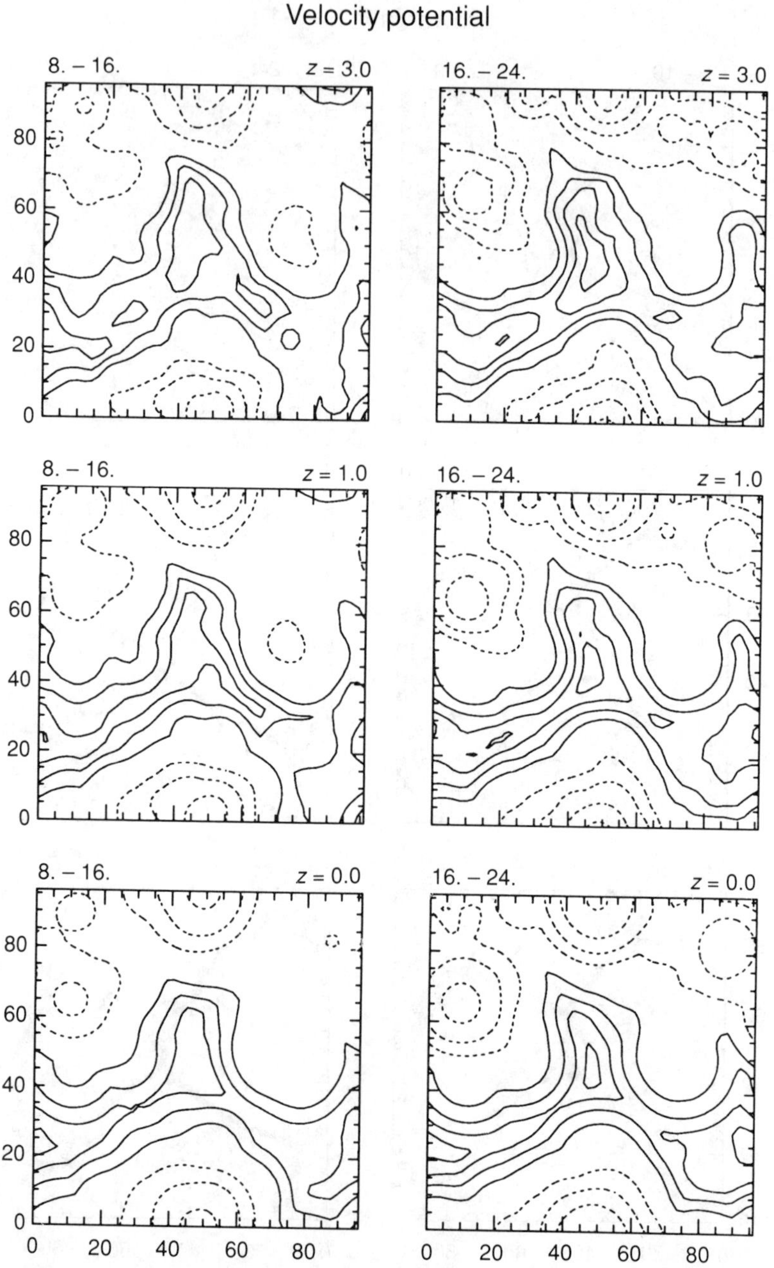

Fig. 6.16 Evolution of the Burgers velocity potential S for the simulation shown in Fig. 6.15, averaged vertically over the same slices

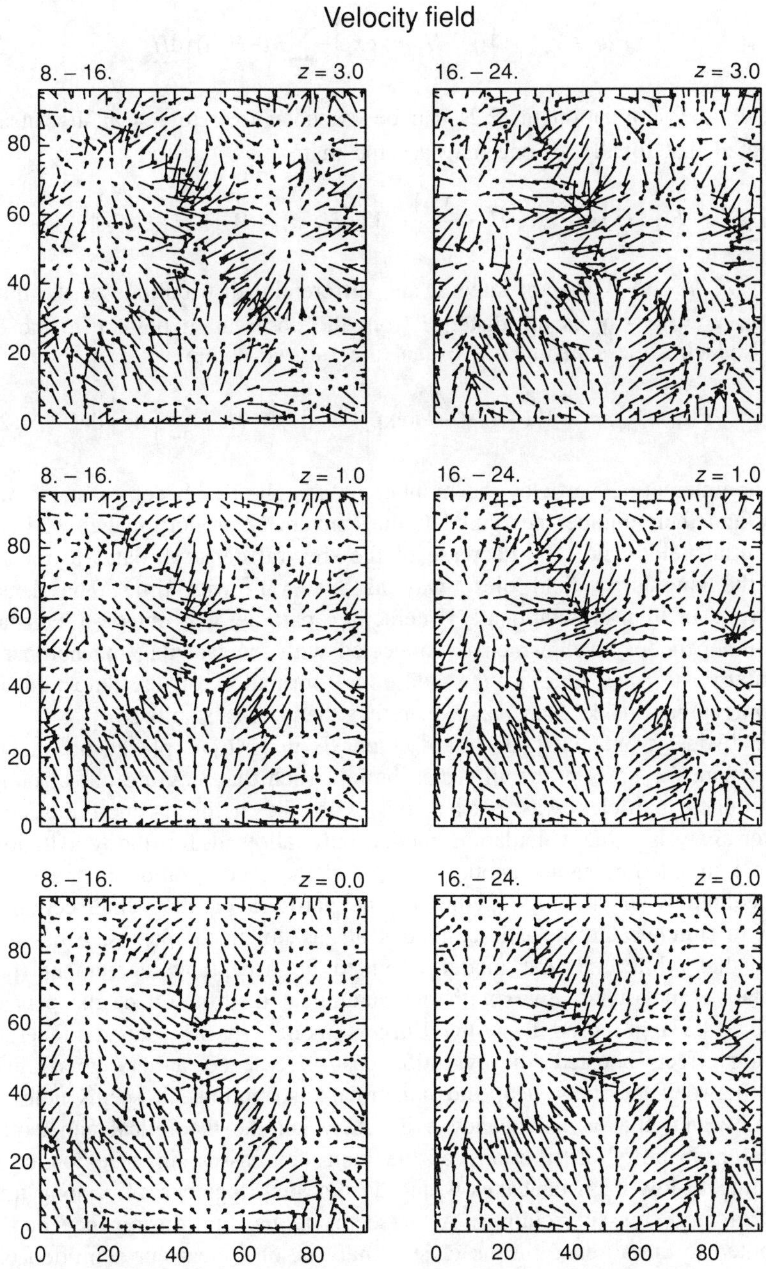

Fig. 6.17 Evolution of the Burgers velocity field $\mathbf{v} = \nabla S$ for the simulation of Fig. 6.16, averaged vertically over the same slices and projected onto the (x, y)-plane. Dots mark the heads of the velocity vectors. Vector magnitudes are scaled to show the distance a particle would move in an interval $\Delta t = 0.2$. As the simulation evolves, the velocity field settles into a pattern of regular outflow from low-density regions, with sharp discontinuities at high-density caustics

$$P(q \in V_m) = \int W(-H, m) \exp\left[-\sum_j F(-H, j)\right] \mathrm{d}H. \tag{2.2}$$

Further, the summation in (2.2) can be extended over to $j = m$. It can also be noted that for $\Delta \approx l_0$ one may assume that

$$F(-H, j) = F_0\left(-H - \frac{y_j^2}{2\tau}\right), \quad W(-H, m) = W_0\left(-H - \frac{y_m^2}{2\tau}\right),$$

where $F_0(h)$ and $W_0(h)$ are integral and differential distributions of the initial action, $\mathbf{y}_m = \mathbf{r} - \mathbf{q}_m$ is the distance from the observation point $\mathbf{r}$ to the cell $\mathbf{q}_m$. Replacing the sum in (2.2) by an integral, we obtain

$$P(\mathbf{q} \in V_m) = \int W_0(-H - \frac{\mathbf{y}_m^2}{2\tau}) \exp\left[-\frac{1}{\Delta^n}\int F_0\left(-H - \frac{\mathbf{y}^2}{2\tau}\right) \mathrm{d}\mathbf{y}\right] \mathrm{d}H. \tag{2.3}$$

For noninteracting particles the number of singularities becomes stable with time, and their number per unit length is inversely proportional to l_0, i.e. to the initial field scale. The decrease of the spectral density relates to the fact that the singularity field dies away as $\rho \sim (xt)^{-\frac{1}{2}}$ with time. The density spectrum of the gravitationally interacting particle gas enters a constant, time-independent level, which is connected with the fact that the number of singularities *grows*, due to the additional formation of oscillatory many-stream regions inside each pancake, rather than being stabilised.

As distinct from [115], the main emphasis in [146] is placed on analysis of the pancake mass probability distribution, such that singling out of clouds of grouped particles was carried out with the help of one-dimensional cluster-analysis. One calculation model only allowed for the gravitational particle interaction, while another suggested the incorporation in the evolution equations of sticking-together of the particles on their intersection, in order to simulate the inelastic collisions of gas-stellar complexes. It has been noted that while the first model provided an acceptable agreement (both qualitative and quantitative) with the theoretical distribution of the pancake masses (3.23) proposed from the Burgers model, in the second model the sticking effect 'boosted' the evolution ensuring, hence, a very good agreement between the theoretical and numerical distributions. It seems noteworthy, however, that a discrepancy exists between the theoretical and experimental probability distributions of the pancake masses, in a region corresponding to low-mass pancakes which do not stick together, for a long time, to form more massive clumps. As it seems to us, this discrepancy can be attributed to the absence, in the cluster-analysis, of a clear-cut definition of a massive pancake (usually having a complicated internal structure).

It will be conjectured that $S_0(\mathbf{r})$ has a Gaussian distribution with variance $\sigma_s^2 = \sigma_0^2 l_0^2$. As in one dimension, the integration domain in (2.3) is restricted to a narrow interval $\Delta H \ll \sigma_s$ in the vicinity of H_0, where H_0 is a characteristic value of the absolute minimum of Φ equal to a solution of the transcendental equation

$$\left(\frac{\sigma_s\tau}{\Delta^2}\right)^{\frac{n}{2}}(2\pi)^{(n-1)/2}\left(\frac{\sigma_s}{H_0}\right)^{(n+2)/2}\exp\left(-\frac{H_0^2}{2\sigma_s^2}\right)=1, \qquad (2.4)$$

$$H_0 \approx \sigma_s\sqrt{n\ln(\tau/\tau_n)},\quad \tau_n = l_0/\sigma_0 = \Delta^2/\sigma_s.$$

Proceeding to the absolute minimum coordinate probability density $W_q(\mathbf{q}; \mathbf{r}, \tau) = P(\mathbf{q} \in V_m)/\Delta^n$, we finally have from (2.3)

$$W_q(\mathbf{q}; \mathbf{r}, \tau) = \frac{1}{(2\pi l^2(\tau))^{n/2}}\exp\left[-\frac{(\mathbf{r}-\mathbf{q})^2}{2l^2(\tau)}\right] \qquad (2.5)$$

$l(\tau)$ being the turbulence external scale given by

$$l^2(\tau) = \frac{\sigma_s^2\tau}{H_0} \approx \frac{\sigma_0 l_0\tau}{\sqrt{n\ln(\tau/\tau_n)}} = l_0^2\frac{\tau}{\tau_n}\left(n\ln\frac{\tau}{\tau_n}\right)^{-\frac{1}{2}}. \qquad (2.6)$$

One can see from (1.8) and (2.6) that the one-point probability distribution of the vector field v,

$$W_v(\mathbf{v}; \tau) = \frac{1}{(2\pi b^2)^{n/2}}\exp\left(-\frac{\mathbf{v}^2}{2b^2}\right),\quad b(\tau) = \frac{l(\tau)}{\tau}, \qquad (2.7)$$

in a space of any dimension is self-preserving and Gaussian, and the field 'energy'

$$\langle\mathbf{v}^2\rangle = n\frac{l^2}{\tau^2} = \frac{\sigma_0 l_0}{\tau}\sqrt{\frac{n}{\ln(\tau/\tau_n)}} \qquad (2.8)$$

falls off, essentially according to a power law τ^{-1}. Observe that the decay law (2.8) differs from that derived in (1.32) from the qualitative considerations by the slow logarithmic correction alone.

2. Let us consider the two-point probability distribution of the field **v**, giving information on its spectral–correlation properties and allowing us to clarify finer details of the cellular structure. Consider the field **v** and the absolute minimum of Φ coordinates **q** at points $-\mathbf{z}$ and **z**,

$$\mathbf{v}_1 = \mathbf{v}(-\mathbf{z}, \tau),\ \mathbf{v}_2 = \mathbf{v}(\mathbf{z}, \tau),\ \mathbf{q}_1 = \mathbf{q}(-\mathbf{z}, \tau),\ \mathbf{q}_2 = \mathbf{q}(\mathbf{z}, \tau). \qquad (2.9)$$

Above all it is worth noting that, as in the one-dimensional case (see Section 6.4), they satisfy the inequalities

$$(\mathbf{z}\cdot(\mathbf{q}_1 - \mathbf{q}_2)) \le 0,\ (\mathbf{z}\cdot(\mathbf{v}_1 - \mathbf{v}_2)) < 2z^2/\tau,$$

and the two-point probability distribution of the field at points $-\mathbf{z}$ and **z**, respectively, satisfies the condition

$$W_2 \equiv 0 \text{ at } (\mathbf{z}\cdot(\mathbf{v}_1 - \mathbf{v}_2)) < 2z^2/\tau. \qquad (2.10)$$

This common property of the field two-point probability distribution holds at any τ. At the stage of developed cells it can be written as

$$W_2 = PW^{\mathrm{d}} + (1 - P)W^{\mathrm{c}}, \tag{2.11}$$

where W^{d} is the two-point probability distribution of the field provided that the observation points $-\mathbf{z}$ and $\mathbf{z}$ fall within one cell, W^{c} refers to the case when they belong to different cells while P is the probability for the points $-\mathbf{z}$ and $\mathbf{z}$ to fall within the same cell.

A further inference of an asymptotic expression for W_2 is similar to that described above concerning the two-point probability distribution in the one-dimensional case. Introduce the dimensionless fields and coordinates

$$\mathbf{u}_1 = \frac{\tau}{l(\tau)}\mathbf{v}(-\mathbf{z}, \tau), \quad \mathbf{u}_2 = \frac{\tau}{l(\tau)}\mathbf{v}(\mathbf{z}, \tau), \quad \mathbf{s} = \frac{\mathbf{z}}{l(\tau)}, \tag{2.12}$$

to be followed immediately by a final formula for the dimensionless two-point probability density $\tilde{W}_2(\mathbf{u}_1, \mathbf{u}_2; \mathbf{s})$ [109]:

$$\tilde{W}_2(\mathbf{u}_1, \mathbf{u}_2; \mathbf{s}) = \frac{1}{(2\pi)^{(n-1)/2}} \frac{\delta(\mathbf{u}_1 - \mathbf{u}_2 - 2\mathbf{s})}{g(-(\mathbf{u}_1 \cdot \mathbf{N}))\exp(\mathbf{u}_1^2/2) + g((\mathbf{u}_2 \cdot \mathbf{N}))\exp(\mathbf{u}_2^2/2)}$$

$$+ \frac{1}{(2\pi)^{(n+1)/2}}\exp(-\mathbf{u}_1^2/2 - \mathbf{u}_2^2/2)\int_{-s-(u_1\cdot N)}^{s-(u_2\cdot N)} \frac{2s\,dz}{[g(z+s)\exp(sz) + g(s-z)\exp(-sz)]^2},$$

$$g(z) = \int_{-\infty}^{z}\exp(-y^2/2)\,dy. \tag{2.13}$$

Here $s = |\mathbf{s}|$, $\mathbf{N} = \mathbf{s}/|\mathbf{s}|$ designate the length and direction of a vector connecting the observation points. The probability $P(2s)$ for two points to fall within one and the same cell, involved in (2.11), is identical to the analogous property in one dimension,

$$P(2s) = \int_{-\infty}^{\infty} \frac{dz}{g(s+z)\exp((s+z)^2/2) + g(s-z)\exp((s-z)^2/2)}. \tag{2.14}$$

Therefore for $\tau \gg \tau_{\mathrm{n}}$ the probability distribution W and, hence, the spectral–correlation properties of the potential field all turn out to be self-preserving. The self-preserving regime arises following the formation of a cellular structure and multiple absorption of some cells by others, which results in the fact that the external scale $l(\tau)$ becomes much than the initial correlation radius l_0. The self-preserving stage of the field evolution is analogous to the establishment of a local thermodynamic equilibrium in a system with slight losses due to multiple-particle interaction (in the case under consideration due to multiple cell absorption). In analogy with the quasi-equilibrium state, the memory of the initial perturbation fine structure is erased in the cellular structure. It is the external scale $l(\tau)$ alone that retains memory of the scale l_0 and amplitude σ_0 of the initial velocity field

$\mathbf{v}_0(\mathbf{r})$. Moreover, should the initial velocity field be anisotropic, it would become isotropic owing to the nonlinear effects. Indeed, the procedure to derive the probability distributions showed that for $\tau \gg \tau_n$ the probability distribution pattern is only defined by the asymptotic behaviour of the initial action one-point probability distribution being independent of its scales.

Let us introduce the longitudinal $U_L = (\mathbf{U}.\ \mathbf{N})$ and transverse U_N components of a vector field $\mathbf{U}$. Integrating (2.13) with respect to U_{1N}, U_{2N} we reveal the two-point probability distribution of the longitudinal component. Its shape is the same in the spaces of different dimensions, and coincides with the analogous probability distribution of the one-dimensional BT (5.4.34). After integration with respect to U_{1L}, U_{2L}, (2.13) yields, for the two-point probability distribution of the transverse components, the following:

$$\begin{aligned}\tilde{W}_2(u_{1N}, u_{2N}; 2s) = \delta(u_{2N} - u_{1N})\, P(2s)\left(\frac{1}{2\pi}\right)^{(n-1)/2} \exp\left(-\frac{1}{2}u_{1N}^2\right) \\ + (1 - P(2s))\left(\frac{1}{2\pi}\right)^{(n-1)} \exp\left(-\frac{1}{2}u_{1N}^2 - \frac{1}{2}u_{2N}^2\right).\end{aligned} \tag{2.15}$$

Therefore, the transverse components (unlike the longitudinal ones) are statistically independent in different cells and have Gaussian probability distribution inside them. Here, $P(2s)$ is the probability (2.14) for two points to fall within the same cell.

The two-point probability distributions describe all the two-point moment functions of the field. As to the normalised longitudinal and transverse correlation functions of the field, we have from (2.13)

$$\begin{aligned}R_{LL}(2s) = \langle u_{1L} u_{2L}\rangle = \frac{d}{ds}(sP(2s)), \\ R_{NN}(2s) = \frac{1}{2}\langle u_{1N} u_{2N}\rangle = P(2s).\end{aligned} \tag{2.16}$$

One can easily observe that they are connected through the potentiality condition [18]

$$R_{LL}(s) = \frac{d}{ds}(sR_{NN}(s)). \tag{2.17}$$

With respect to the arbitrary higher two-point moment functions of the transverse component, (2.15) yields

$$\mu_{m,k} = \langle \tilde{u}_{1N}^m \tilde{u}_{2N}^k\rangle = P(2s)\,(\langle\xi^{m+k}\rangle - \langle\xi^m\rangle\langle\xi^k\rangle) + \langle\xi^m\rangle\langle\xi^k\rangle. \tag{2.18}$$

Here $\tilde{u}_N$ is the field projection on the same transverse vector, while $\langle\xi^m\rangle$ represents the moment of the Gaussian probability distribution, normalised to unit variance: $\langle\xi^m\rangle = (m-1)!!$ with even m; $\langle\xi^m\rangle = 0$ with odd m. In particular, for the fourth cumulant function, we have from (2.18)

$$\kappa_{2,2} = \mu_{2,2} - 2\mu_{1,1}^2 - 1 = 2P(2s)\,(1 - P(2s))\,. \tag{2.19}$$

It equals zero at $s = 0$ and as $s \to \infty$, and attains its maximum at $2s \approx 0.5$ (Fig. 6.18).

As far as the potential isotropic fields are concerned, the normalised energy spectrum $e(\kappa)$ is formulated via a one-dimensional spectrum $G_{NN}(\kappa)$ of the transverse component [18]:

$$e(\kappa) = \kappa^3 \frac{\mathrm{d}}{\mathrm{d}\kappa}\left(\frac{1}{\kappa}\frac{\mathrm{d}}{\mathrm{d}\kappa} G_{NN}(\kappa)\right), \tag{2.20}$$

$$G_{NN}(\kappa) = \frac{1}{2\pi}\int_{-\infty}^{\infty} R_{NN}(s) \exp(\mathrm{i}\kappa s)\,\mathrm{d}s = \frac{1}{2\pi}\int_{-\infty}^{\infty} P(s) \exp(\mathrm{i}\kappa s)\,\mathrm{d}s. \tag{2.21}$$

The properties of $P(2s)$ imply from (5.4.41) that as $s \to 0$ the correlation function is non-analytic, on account of the field discontinuities as $\mu \to 0$. The discontinuity initiation leads to the power asymptotic behaviour $e(\kappa) \sim \kappa^{-2}$. It should be emphasised that the appearance of the κ^{-2} asymptotic behaviour does not mean that the field $\mathbf{v}(\mathbf{r}, \tau)$ has a hierarchy of different spatial scales, but it is related to the presence, in a separate-discontinuity spectrum, of a large number of spatial harmonics such that their phase matching results in discontinuities of the field $\mathbf{v}(\mathbf{r}, \tau)$. As for the small wave number range, the spectrum is also characterised by a universal behaviour $e(\kappa) \sim \kappa^4$, which has to do with the nonlinear generation of low-frequency components.

The probability $P(2s)$ also conveys information on the fine features of the cellular structure. Consider an arbitrary straight line, for example, the x-axis. Denote the coordinates of the points where this straight line is intersected by the cell walls as x_k, and introduce $W_y(y)$, being the probability distribution of the normalised interval lengths $y_k = (x_k - x_{k-1})/l$. The theory of renewal processes [94] claims that $W_y(y)$ is connected with $P(2s)$ through an equation

$$\langle y\rangle\, P(s) = \int_s^{\infty} (y - s)\, W_y(y)\,\mathrm{d}y, \quad \langle y\rangle = \int_0^{\infty} y W_y(y)\,\mathrm{d}y,$$

which gives

$$W_y(y) = \sqrt{\pi}\; P''(y)\,. \tag{2.22}$$

It seems instructive to compare $W_y(y)$ with the probability distribution $W_0(y) = \exp(-y)$ that fully corresponds to random splitting into intervals of a straight line. The comparison of (2.22) with $W_0(y)$ demonstrates that $W(y)$ vanishes for $y \gg 1$ at a higher rate than $W_0(y)$. This confirms the presence of a certain order, i.e. the inner coherence of the chaotic field cellular structure.

3. At low, albeit finite, μ, the viscosity has only to be allowed for at the cell walls having thickness $\delta = \mu t/l$. Due to finite thickness of the wall the

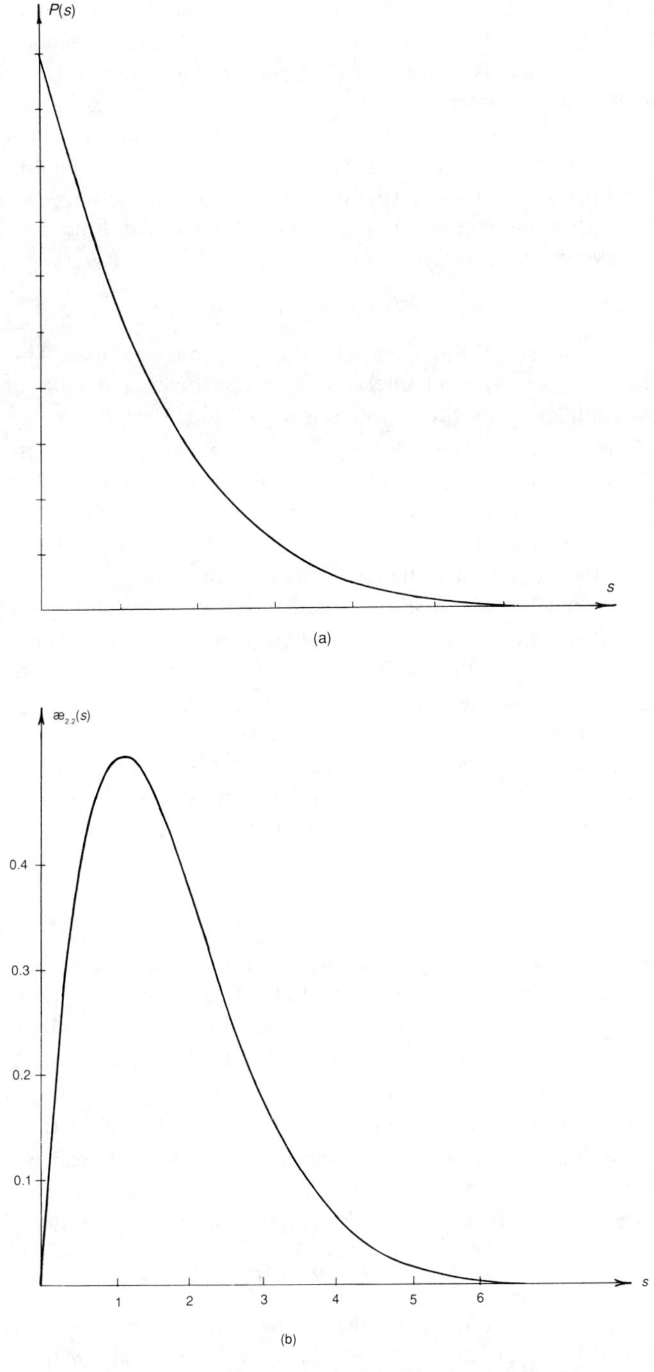

Fig. 6.18 (a) Transverse correlation function; and (b) fourth cumulant function; of the potential turbulence velocity field at the self-preserving stage

field correlation function becomes analytic within the viscous interval $|\mathbf{z}| < \delta$, but for $|\mathbf{z}| > \delta$ is again described by (2.16). The above-mentioned spectrum power-law decay is only retained within the inertial interval $1/l < |\mathbf{k}| < 1/\delta$, while for the viscous interval $|\mathbf{k}| > 1/\delta$ it is replaced by an exponential decay. Under this condition the viscous structure itself can only exist provided $\delta(\tau) \ll l(\tau)$. For $\tau \gg \tau_l$, where τ_l is determined from the condition $\delta(\tau_l) \approx l(\tau_l)$, the cellular structure is destroyed and a linear decay stage is set up. As was the case for one dimension, (2.6) implies that the time for the linear regime to develop is extremely long:

$$\tau_l \approx \tau_{\mathrm{n}} \exp(R_0^2),$$

where $R_0 = \sigma_0 l_0/\mu$ is the initial perturbation Reynolds number. This is attributed to the fact that the wall thickness gain due to viscosity is compensated for almost completely by the characteristic cell size growth.

At the linear decay stage $\tau \gg \tau_l$ (which exists if $\langle S_0^2 \rangle = \sigma^2 < \infty$), an analysis of the field spectral-correlation properties is convenient to be held, as in one dimension, on the basis of the Hopf–Cole solution [109]. This analysis, in particular, leads to the field $\mathbf{v}(\mathbf{r}, \tau)$ becoming isotropic independently of the degree of initial perturbation anisotropy.

Finally we note that the qualitative and even quantitative behaviour of the chaotic potential field $\mathbf{v}(\mathbf{r}, \tau)$, as distinct from the eddy turbulence, does not depend on the space dimension $n = 1, 2, 3$. The stage of shock front formation, $\tau \leq \tau_{\mathrm{n}}$, and the cellular structure initiation see the energy flow to be directed towards the small-scale spectrum components. At the self-preserving stage of the cellular structure, $\tau_{\mathrm{n}} < \tau < \tau_l$, the external scale increase leads, irrespective of the space dimension, to partial energy redistribution into the domain of the slowly damped large-scale spectrum components. For $\tau > \tau_l$ the field spectrum continues to narrow, but this time due to the linear dissipation.

4. Let us now assess the matter density inside the cells, using the expression (2.18) for the matter density in the vicinity of a Lagrangian particle which left the $\mathbf{q}_m$ cell centre neighbourhood. Inasmuch as there were no particle collisions inside the cell, the Jacobian $|D_{ij}|$ in (1.18) can be formulated using tensor $a_{ij} = \partial^2 S_0/\partial q_i \partial q_j$ of the initial action (1.19). At $\tau \gg \tau_{\mathrm{n}}$, only cells with a sufficiently deep initial action minimum $-H \approx H_0 = \sigma_s\sqrt{3\ln(\tau/\tau_{\mathrm{n}})}$ survive, such that the initial action minimum of these cells can be shown as lying within a narrow band $\Delta H \approx \sigma_s/\sqrt{3\ln(\tau/\tau_{\mathrm{n}})}$. Therefore the averaging of (1.18) should be effected under the conditional probability density

$$W_0(a_{ij}|-H_0) = W_{as}(a_{ij}, -H_0)/W_0(-H_0). \tag{2.23}$$

Here $W_{as}(a_{ij}, H)$ is the joint probability density of tensor $a_{ij} = \partial^2 s_0/\partial q_i \partial q_j$ and the initial action $S_0(\mathbf{q})$. Let us write the correlation function of the initial Gaussian action as

$$\langle S_0(\mathbf{q})\, S_0(\mathbf{q}+\mathbf{z})\rangle = \sigma_s^2(1 - \mathbf{z}^2/(2l_1^2) + \mathbf{z}^4/(4!l_2^4) - \ldots),$$

where $\sigma_s^2 = \sigma_0^2 l_0^2$, $l_1 \sim l_2 \sim l_0$ being the initial correlation radius. Then for the averages corresponding to the probability distribution (2.23), the following equalities hold:

$$\langle a_{ij} S_0\rangle = -\delta_{ij}\sigma_s^2/l_1^2,$$
$$\langle a_{ij} a_{mk}\rangle = (\delta_{mk}\delta_{ij} + \delta_{kj}\delta_{mi} + \delta_{jm}\delta_{ki})\sigma_s^2/3l_2^2,$$

where δ_{ij} is the Kronecker symbol. Consequently, for $\tau \gg \tau_n$ the conventional average of the tensor diagonal elements $\langle a_{ij}\rangle_H = \delta_{ij} H_0/l_1^2$ is greatly in excess of its root-mean-square deviation $\sqrt{\langle a_{ij}^2\rangle} \approx \sigma_s/l_2$. Thus from (1.10) and (1.18) it follows that the average density in the cells drops as

$$\langle\rho\rangle = \rho_0\langle 1/D(a_{ij})\rangle \approx \rho_0/D(\langle a_{ij}\rangle)$$
$$\approx \rho_0/\tau^3\langle a_{11}\rangle_{H_0}\langle a_{22}\rangle_{H_0}\langle a_{33}\rangle_{H_0} \approx \rho_0\left(\tau_n/\tau\sqrt{\ln(\tau/\tau_n)}\right)^3, \tag{2.24}$$

while the volume occupied by the 'dark' domains enlarges with time due to the absorption of the weaker neighbours. Nevertheless the mass of matter in them decreases as

$$m \approx \rho_0 l_0^3(\tau_n/\tau)^{\frac{3}{2}},$$

since still more matter is concentrated in 'bright' domains, i.e. in the cell walls, ribs and nodes.

5. Now it can be shown that provided the action variance is not bounded, and its spectrum is described by (1.30) with $3 < \theta < 5$, the field becomes statistically self-preserving. The initial action structure function then acquires the form

$$d_s(\mathbf{z}) = \alpha_\theta^2\, a(\mathbf{z})|\mathbf{z}|^{\theta-3},$$

where $a(|\mathbf{z}| \gg l_0) = \text{const}$. One can always assume $a(\infty) = 1$, when choosing the normalisation α_θ^2 in (1.30). Just as in the one-dimensional case (Section 5.3), we introduce new variables using the external scale of the turbulence,

$$\mathbf{r} = \boldsymbol{\xi} l,\ \mathbf{q} = \boldsymbol{\eta} l,\ \mathbf{v} = \mathbf{u} l/\tau.$$

Then for the dimensionless field $\mathbf{u}$ the solution of (1.6) appears as

$$\mathbf{u} = \frac{\int(\boldsymbol{\xi} - \boldsymbol{\eta})\exp[-R(\tau)\,\tilde{\Phi}(\boldsymbol{\xi}, \boldsymbol{\eta}, \tau,)]\,d\boldsymbol{\eta}}{\int\exp[-R(\tau)\,\tilde{\Phi}(\boldsymbol{\xi}, \boldsymbol{\eta}, \tau,)]\,d\boldsymbol{\eta}}, \tag{2.25}$$

where $R(\tau)$ is the Reynolds number defined as

$$R(\tau) = l^2(\tau)/4\tau,$$
$$\tilde{\Phi}(\boldsymbol{\xi}, \boldsymbol{\eta}, \tau) = \frac{(\boldsymbol{\eta} - \boldsymbol{\xi})^2}{2} + \tilde{S}_0(\boldsymbol{\eta}, \tau),\ \tilde{S}_0(\boldsymbol{\eta} - \tau) = \frac{\tau}{l^2}\, s_0(\boldsymbol{\eta} l). \tag{2.26}$$

The function $\tilde{S}_0(\boldsymbol{\eta}, \tau)$ incorporated in (2.26) possesses the following structure function:

$$d_{\tilde{s}}(\mathbf{z}, \tau) = \langle [\tilde{S}_0(\eta + \mathbf{z}, \tau) - \tilde{S}_0(\eta, \tau)]^2 \rangle = \frac{\tau^2}{l^4} d_s(\mathbf{z}l) = (\tau^2 \alpha_\theta^2 l^{\theta-7}) a(l\mathbf{z}) \, |\mathbf{z}|^{\theta-3}. \quad (2.27)$$

Assuming that the turbulence external scale $l(\tau)$ is determined from the condition $\tau^2 \alpha_\theta^2 l^{\theta-7} = 1$, one can see it to grow as (1.30)

$$l(\tau) = \alpha_\theta^{2/(7-\theta)} \tau^{2/(7-\theta)}, \quad 3 < \theta < 5, \quad (2.28)$$

and the effective Reynolds number (2.26),

$$R(\tau) = \alpha_\theta^{4/(7-\theta)} \tau^{(\theta-3)/(7-\theta)} / \mu \quad (2.29)$$

is also observed for $3 < \theta < 5$ to increase with time. At times when $R(\tau) \gg 1$ the integral (2.34) is calculated by the saddle-point technique, the contribution only being that of a point where $\tilde{\Phi}$ attains an absolute minimum. Consequently, the solution (2.25) has the form

$$\mathbf{u}(\boldsymbol{\xi}, \tau) = \boldsymbol{\xi} - \boldsymbol{\eta}(\boldsymbol{\xi}, \tau), \quad (2.30)$$

where $\boldsymbol{\eta}(\boldsymbol{\xi}, \tau)$ is the coordinate of an absolute minimum of $\tilde{\Phi}(\boldsymbol{\xi}, \boldsymbol{\eta}, \tau)$. At $l(\tau) \gg l_0$ the structure function s_0 has the universal form, over virtually the whole domain $|\mathbf{z}| > l_0/l \ll 1$:

$$d_{\tilde{s}}(\mathbf{z}, \tau) = |\mathbf{z}|^{\theta-3},$$

i.e. has no time-dependence and no spatial scales. This means that statistical properties of $\mathbf{u}(\boldsymbol{\xi}, \tau)$ do not depend on time τ and, thus, the field $\mathbf{v}(\mathbf{r}, \tau)$ becomes statistically self-preserving.

6. Let us speculate as to an alternative approach to the matter density calculation and concerning the construction of a model equation allowing us to qualitatively describe the nonlinear stage of the gravitational instability. As $\mu \to 0$ the BE (1.1) solution has the form (1.8)

$$\mathbf{v}(\mathbf{r}, \tau) = \frac{\mathbf{r} - \mathbf{q}(\mathbf{r}, \tau)}{\tau}. \quad (2.31)$$

In (2.1), $\mathbf{q}$ is interpreted as a Lagrangian coordinate of a particle falling onto point $\mathbf{r}$ at time τ. The matter density is determined then via the Eulerian-to-Lagrangian coordinate transformation Jacobian (1.10),

$$\eta(\mathbf{q}, \tau) = \eta_0 / |\partial \mathbf{r} / \partial \mathbf{q}|. \quad (2.32)$$

With the aid of (2.1) we formally introduce function $\mathbf{q}(\mathbf{r}, \tau)$, and at $\nu \neq 0$:

$$\mathbf{q} = \mathbf{r} - \mathbf{v}(\mathbf{r}, \tau)\tau. \quad (2.33)$$

Attaching to it the sense of a Lagrangian coordinate in the Eulerian space, we shall handle it appropriately assuming, namely, that density is defined by relation (2.2), or its equivalent

$$\eta(\mathbf{r}, \tau) = \eta_0\left|\frac{\partial \mathbf{q}(\mathbf{r}, \tau)}{\partial \mathbf{r}}\right|. \tag{2.34}$$

It is an easy matter to show that q satisfies the equation

$$\frac{\partial \mathbf{q}}{\partial \tau} + (\mathbf{v}.\nabla)\ \mathbf{q} = \mu\Delta\mathbf{q}. \tag{2.35}$$

It proves quite natural that the equation for density will no longer coincide with that of continuity. Consider first a one-dimensional case, expressing q in terms of y and r in terms of x. In this case we have the following with respect to the Eulerian density,

$$\eta(x, \tau) = \eta_0\frac{\partial y(x, \tau)}{\partial x} = \eta_0\left(1 - \tau\frac{\partial v(x, \tau)}{\partial x}\right), \tag{2.36}$$

and the density equation itself reads

$$\frac{\partial \eta}{\partial \tau} + \frac{\partial}{\partial x}(\eta v) = \mu\frac{\partial^2 \eta}{\partial x^2}. \tag{2.37}$$

The emergence of an additional term in the equation for density lends itself easily to interpretation. Indeed, with respect to the gravitationally interacting particles the redistribution of matter is not only due to the average velocity v described in equation (1.1), but to the oscillatory velocity components as well, which are simulated by a diffusion term in (2.37). It follows from (2.36) that this solution is not only responsible for the conservation of the velocity integral $\int v\, \mathrm{d}x$ (resulting from the BE divergence form), but also provides the conservation of momentum,

$$\int[\eta(x, \tau) - v(x, \tau)]\, \mathrm{d}x = \eta_0\int\left[v(x, \tau) - \tau v(x, \tau)\frac{\partial v(x, \tau)}{\partial x}\right] \mathrm{d}x,$$

$$\eta_0\int v(x, \tau)\, \mathrm{d}x = \eta_0\int v_0(x)\, \mathrm{d}x.$$

Let us outline the path to density calculation in three dimensions and estimate the density distribution at the stage when the velocity field has a developed cellular structure. It is evident from (1.6) and (2.33) that the Lagrangian coordinate $\mathbf{q}$ is expressed in terms of the initial action S_0 as follows:

$$\mathbf{q}(\mathbf{r}, \tau) = \frac{\int \mathbf{q}\exp[-\Phi(\mathbf{r}, \mathbf{q}, \tau)/2\mu]\, \mathrm{d}\mathbf{q}}{\int \exp[-\Phi(\mathbf{r}, \mathbf{q}, \tau)/2\mu]\, \mathrm{d}\mathbf{q}},$$

$$\Phi(\mathbf{r}, \mathbf{q}, \tau) = S_0(\mathbf{q}) + \frac{(\mathbf{r} - \mathbf{q})^2}{2\tau}. \tag{2.38}$$

The substitution of (2.38) into (2.34) gives us a formula providing an efficient numerical analysis of density at all times, since in this case, to calculate density in Eulerian space, it suffices to calculate an integral of type (2.38) instead of solving the partial differential equations. First of all it should be noted that at times $t \gg t_n$ and under the self-preserving velocity field formation condition the density field proves self-preserving as well.

We confine ourselves to discussing the stage $\mu \to 0$, when the field $\mathbf{v}(\mathbf{r}, \tau)$ has a developed structure and is divided to form cells Σ_i such that in each cell the field is universal, (1.21), and is characterised by a cell centre $\mathbf{q}_i$, i.e. the point which 'sends away' particles from its neighbourhood. The cell walls and ribs degenerate at this stage to form planes and straight lines. In order to assess the matter density inside the cells as well as in the walls, ribs and vertices, one needs to define the Lagrangian coordinate field $\mathbf{q}(\mathbf{r}, \tau)$ behaviour in their vicinities. Inside the cell, when an absolute minimum of Φ is known to dominate over the other minima, the Lagrangian coordinate $\mathbf{q}(\mathbf{r}, \tau) = \mathbf{q}_i$, where $\mathbf{q}_i$ is the local minimum coordinate of $S_0(\mathbf{q})$. The physical meaning of this equality is very simple, meaning that all the volume inside a cell is occupied by particles which initially belonged to its centre point. As in the given approximation $\mathbf{q}$ does not depend on $\mathbf{r}$ the matter density equals zero. Indeed, the cell interior is occupied by particles initially belonging to an extremely small vicinity of the centre. Because of this the matter density inside a cell, albeit very low, does not quite equal zero. The density estimates inside the cells are given above in (2.24).

When approaching the cell walls one has to allow for the competition between the two centres of the adjacent ith and jth cells, i,e, the two minima of Φ close in value, As was the case for the field $\mathbf{v}(\mathbf{r}, \tau)$ of (1.25), we have the following asymptotic expression for the Lagrangian coordinate field in the vicinity of the walls

$$\mathbf{q}(\mathbf{r}, \tau) = \frac{\mathbf{q}_i + \mathbf{q}_j}{2} - \frac{\Delta\mathbf{r}_{ij}}{a}\tanh\left[\frac{\Delta\mathbf{r}_{ij}\cdot(\mathbf{r} - \mathbf{r}_{ij})}{4\mu\tau}\right]. \tag{2.39}$$

Here, as in (1.25), $\Delta\mathbf{r}_{ij} = \mathbf{r}_i - \mathbf{r}_j$ is a vector connecting the centres of the adjacent cells, and $\mathbf{r}_{ij}$ is defined by (1.26).

First, let us calculate the wall density in a one-dimensional version of (2.6), when concerning the Lagrangian coordinate $y(x, t)$ and density $\eta(x, \tau)$ in the discontinuity neighbourhood we have, respectively,

$$y(x, \tau) = \frac{y_i + y_j}{2} - \frac{\Delta_{ij}}{2}\tanh\left[\frac{\Delta_{ij}(x - x_{ij})}{4\mu\tau}\right], \tag{2.40}$$

$$\eta(x, \tau) = \eta_0\,\Delta_{ij}^2\Big/\left\{8\mu\tau\cosh^2\left[\frac{\Delta_{ij}(x - x_{ij})}{4\mu\tau}\right]\right\}. \tag{2.41}$$

Here $\Delta_{ij} = y_i - y_j$ is equal in order of magnitude to the characteristic distance between adjacent cells, or to the external scale of the BT, Consequently, the maximum matter density is attained in the wall centre, and is of order

$$\eta_{max} = \eta_0 \Delta_{ij}^2 / 8\mu\tau \approx \eta_0 l/\delta.$$

where $\delta = \mu\tau/l$ is the wall thickness. If we neglect the fine structure of matter distribution inside the wall, then we have from (2.4)) as to the matter mass inside the wall,

$$M_i = \int \eta(x, \tau)\, dx = \eta_0 \Delta_{ij} = \eta_0 |y_j - y_i|. \tag{2.42}$$

Formula (2.42), however, can be deduced much more easily if one takes into account that at $\tau \gg \tau_n$ all the matter initially enclosed between the adjacent cell centres y_j and y_i is now concentrated within this wall.

Let us reveal now the wall density in the three-dimensional case. To this end it can be remarked that $\mathbf{q}(\mathbf{r}, \tau)$ of (2.39) under variation of $\mathbf{r}$ only 'scans' across the coordinate values of the points lying on a straight line segment connecting the adjacent cell centres. In other words, the wall separating two cells attracts matter initially belonging to the straight line connecting the two centres, the wall density being equal to zero. The same result can be easily obtained from (2.34) by calculating the Jacobian of (2.39) and allowing for the fact that the transformation (2.39) is singular, i.e. when $\mathbf{r}$ changes, variation of $\mathbf{q}$ is observed but only along a single direction. It is natural that a more elaborate examination might show 'zero' wall density to be many time greater than 'zero' density inside the cells.

Analogous reasoning demonstrates that in any rib, matter is concentrated that initially belonged inside a plane triangle with vertices formed by the cell centres, this rib being common to the cells considered. In this case the transformation from the Eulerian coordinates to Lagrangian ones proves also singular, and the matter density in ribs is zero as well.

As a final remark it can be said that the cell vertices contain matter initially enclosed within a tetrahedron with vertices formed by the four centres of the appropriate cells. Let $\Delta\mathbf{r}_{12}$, $\Delta\mathbf{r}_{13}$, $\Delta\ \mathbf{r}_{14}$ be vectors coming out of the centre 1 and connecting it with the three centres of the adjacent cells. Then the tetrahedron volume will be expressed in terms of the triple scalar product of these vectors, and the matter mass concentrated within this node (Fig. 6.19) is equal to this tetrahedron volume multiplied by the initial density,

$$m_1 = \frac{\eta_0}{6}(\Delta\mathbf{r}_{12} \cdot \Delta\mathbf{r}_{13} \wedge \Delta\mathbf{r}_{14}). \tag{2.43}$$

The node radius is of the order of the internal scale δ, and its characteristic mass, defined via an external scale of turbulence is $m \sim \eta_0 l^3$.

To sum up, one can say that two cellular structures exist: the velocity field

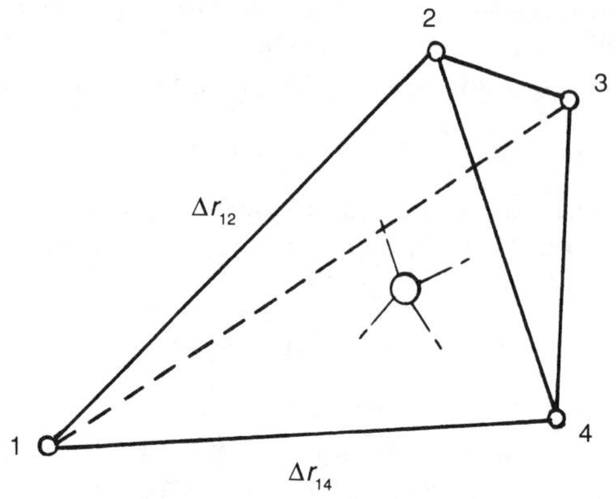

Fig. 6.19 Tetrahedron in a Lagrangian space, defining the mass of matter concentrated in a node formed by rib intersection

cellular structure and that of the matter density. Moreover, while the invisible cellular structure of the velocity field grows more perfect with time, the visible cellular structure of matter density tends to be destroyed, forming compact superclusters. Figuratively speaking, the cellular structure of the matter is melting down before one's eyes as it is formed. When the pancakes just begin to stick together, the matter starts to concentrate in ribs so as to form the separate fragments of the cellular structure frame scattered in space, i.e. the matter forms chains. Later on these chains disappear too, as they unite to form clumps, i.e. rib nodes. These two structures, however, are rigidly connected, such that the movement of clumps situated in the cell vertices is along the trajectories defined from the velocity field cell vertex motion.

One can also speak about another structure, suggesting that the Lagrangian space is split, at $\tau \gg \tau_n$, into tetrahedra which compactly cover the whole space. These tetrahedron vertices are defined from the condition of the paraboloid α of (1.17) simultaneously touching the initial action $S_0(\mathbf{q})$ vertices. This paraboloid centre $\mathbf{r}$ corresponds in this case to the Eulerian coordinate of a node, i.e. to a clump, whereas the tetrahedron volume amounts to the clump mass. At a time of clump collision, the tetrahedron system is being transformed in the Lagrangian representation, which corresponds to the matter mass exchange in colliding clumps. Figure 6.20 depicts the Lagrangian and Eulerian spaces being divided into cells in the two-dimensional case when the Lagrangian plane is split to form triangles, and the matter clump lies at the three-rib intersection, its mass being proportional to the corresponding triangle area.

Figure. 6.21 shows an elementary collision of two clumps and the triangle system transformation process in the Lagrangian representation. On collision,

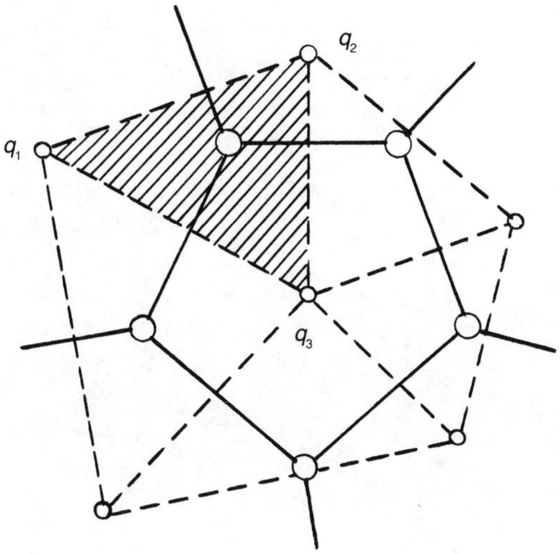

Fig. 6.20 Cellular structures: velocity field discontinuity structure in the Eulerian representation – solid lines, structure of triangles in Lagrangian representation whose matter goes into the nodes of Eulerian structure

mass exchange occurs, and it seems quite easy to prove that the collision results in the conservation of both the clump total mass and the total momentum. However, the calculations demonstrated that the kinetic energy of the two-clump translational motion increases after the collision, and the more elongated are the appropriate Lagrangian triangles the higher is the growth. This phenomenon was discovered by V. Vazin and V. Savin (personal communication), who observed it as a result of a numerical test. As it seems to us, the translational motion kinetic energy growth of the clumps after collision can be interpreted under the assumption that a clump possesses, along with the translational motion energy, a certain amount of internal energy of rotation as well. In two dimensions an intense dissipation of the translational motion energy only occurs as a consequence of triple collisions when three clumps are stuck together to from a single one.

In reference [146], analytical calculations are suggested concerning the clump distribution over velocities and masses at sufficiently large times when all matter has been concentrated in the clumps, i.e. structure nodes. It is shown, in particular, that independently of the space dimension the clump velocity has Gaussian distribution, while distribution over masses depends on the space dimension. A common feature for both is that statistical properties of the clumps at large times are self-preserving. As regards the two-dimensional case, we also calculated the probability distribution of the moments of inertia of the triangles which are the Lagrangian structure

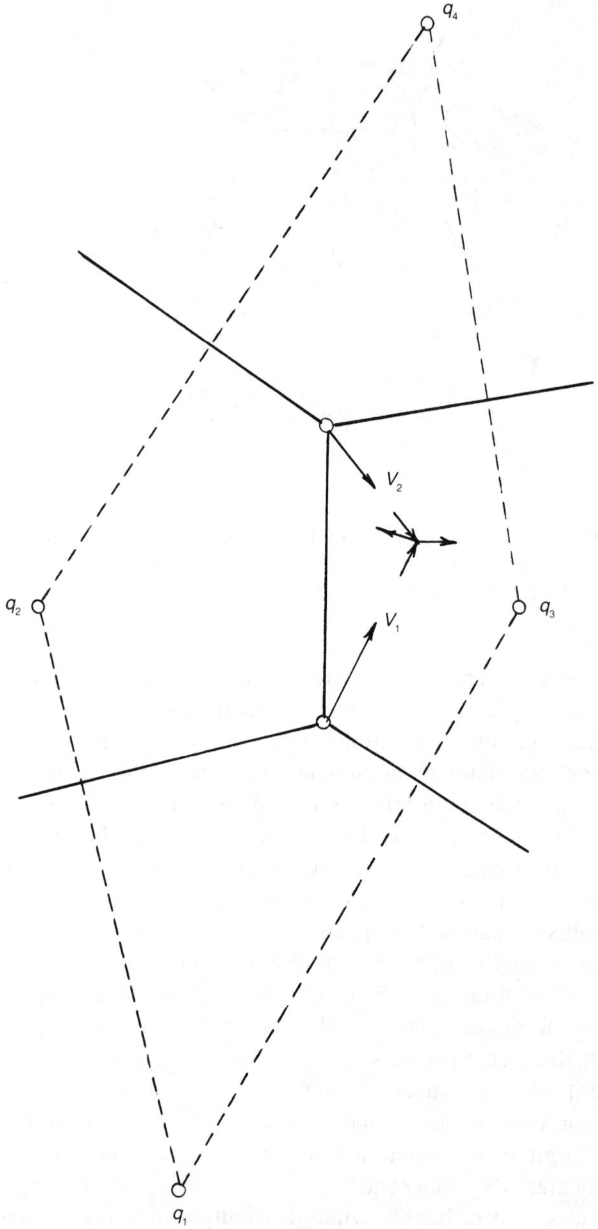

Fig. 6.21 Elementary event of two-clump collision accompanied by mass exchange. Clump coordinates and velocities (a) before collision; (b) after collision. Cross with arrows denotes a point of clump collision

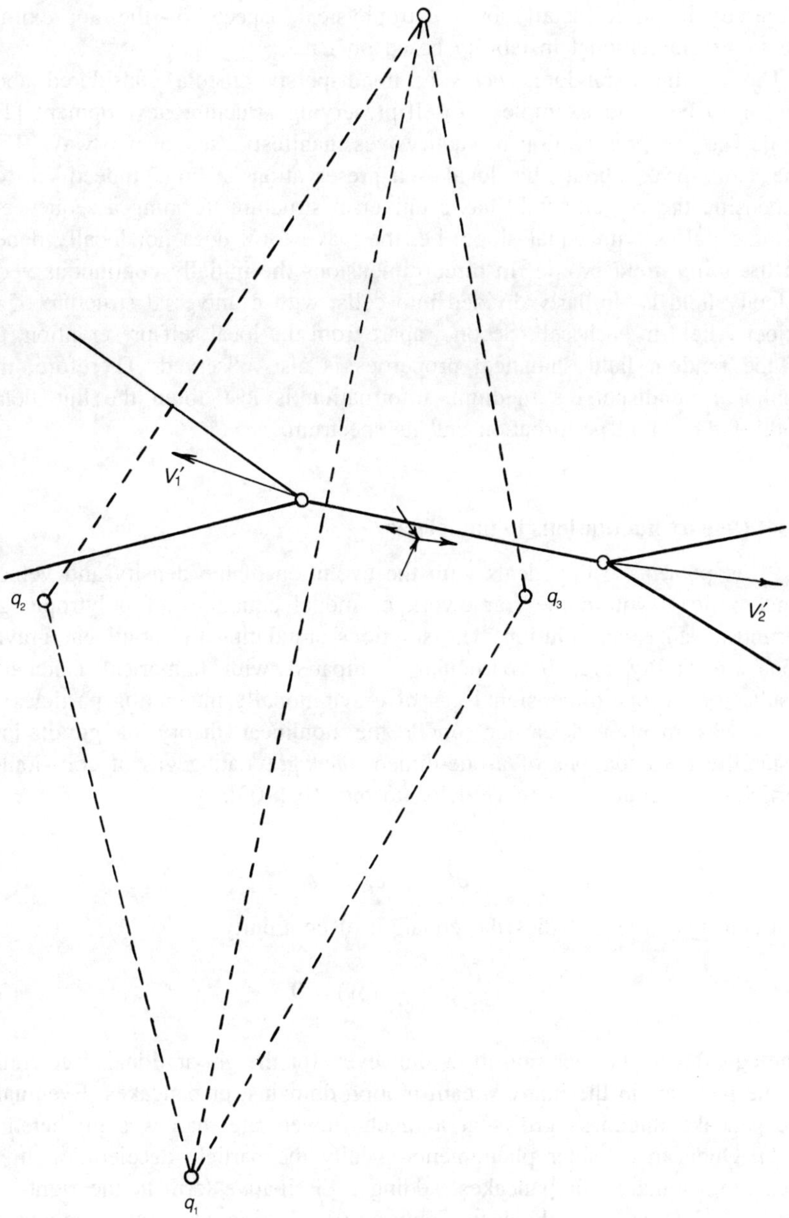

q4
V1'
q2
q3
V2'
q1

elements, and showed that non-equilateral triangles dominate. The same work addresses in more detail some astrophysical aspects of the approximate theory of gravitational instability based on BE.

The nonlinear random waves in nondispersive media considered above appear to be vivid examples of self-preserving structure development [149, 150]. The self-preservation of such waves manifests itself in two ways. First, one can speak about the local self-preservation setting. Indeed, in one dimension the velocity field has a universal structure, forming a sequence of triangle pulses with equal slope, i.e. the wave form does not locally depend on the initial field profile. In three dimensions the initially continuous vector velocity field is similarly divided into cells, with a universal structure of the velocity field in each cell. Second, apart from the local self-preservation, that of the random field statistical properties is also observed. Therefore, in a nonlinear nondispersive medium, information is lost about the fine details both of the initial perturbation and its spectrum.

6.3 Density fluctuations in model gas

1. The present section deals with the evolution of the density and velocity random fields within the framework of model equations of polytropic gas, permitting an exact solution. The solutions simulating the nonlinear gravitational instability are, in particular, compared with numerical calculation results for the one-dimensional gas of gravitationally interacting particles.

As was mentioned earlier, under the nonlinear theory of gravitational instability the equations of a one-dimensional gravitating gas of collisionless particles are reduced, as to velocity, to the RE [103]:

$$\frac{\partial}{\partial t}v + v\frac{\partial v}{\partial x} = 0, \tag{3.1}$$

and density $\rho(x, t)$ satisfies the equation of continuity

$$\frac{\partial}{\partial t}\rho + \frac{\partial}{\partial x}(\rho v) = 0. \tag{3.2}$$

The equation (3.1) does not allow, however, for the gravitational deceleration of the particles in the many-stream motion domains, or pancakes. Eventually, the pancake thickness grows at a much slower rate than was predicted by (3.1). One can consider phenomenologically the particle deceleration in the clustering domains, or pancakes, adding a dissipative term to the right-hand side of (3.1) as was done in Section 6.1. Taking also into account the pressure forces, we obtain for the velocity field

$$\frac{\partial v}{\partial t} + v\frac{\partial v}{\partial x} + \frac{1}{\rho}\frac{\partial P}{\partial x} = \mu\frac{\partial^2 v}{\partial x^2}. \tag{3.3}$$

Let us analyse a polytropic gas with an adiabatic index $\gamma = 3$, when $P = \kappa^2\rho^3$ and equation (3.3) is transformed into

$$\frac{\partial v}{\partial t} + v\frac{\partial v}{\partial x} + \kappa^2\rho\frac{\partial \rho}{\partial x} = \mu\frac{\partial^2 v}{\partial x^2}. \tag{3.4}$$

To render it more convenient, introduce the local sound velocity $c = \kappa\rho$, evolution of which is described by an equation analogous to (3.2). Complicate it by adding a dissipative term, the same as in (3.4), to the right-hand side. After this the set of equations for v and c acquires the form

$$\frac{\partial v}{\partial t} + v\frac{\partial v}{\partial x} + c\frac{\partial c}{\partial x} = \mu\frac{\partial^2 v}{\partial x^2}, \tag{3.5}$$

$$\frac{\partial c}{\partial t} + v\frac{\partial c}{\partial x} + c\frac{\partial v}{\partial x} = \mu\frac{\partial^2 c}{\partial x^2}. \tag{3.6}$$

As was noted in section 6.2, the appearance of a dissipative term in the equation of continuity (3.6) describes the matter density redistribution due to oscillatory components of the particle motion velocity in the clustering domain (2.37). In this case equation (3.5) embodies a certain average velocity of the flow. The choice of the dissipation factor μ in (3.6) to be equal to the factor μ in (3.5) is justified by the fact that at $c(x, t = 0) = c_0$, this system provides the conditions to fulfil the integral law of conservation of momentum which is, no doubt, valid for a gas of gravitating particles as well.

It can be easily proved from (3.5) and (3.6) that the fields

$$u_+ = v + c, \quad u_- = v - c$$

satisfy the BE

$$\frac{\partial u_\pm}{\partial t} + u_\pm\frac{\partial u_\pm}{\partial x} = \mu\frac{\partial^2 u_\pm}{\partial x^2}. \tag{3.7}$$

The sought fields of velocity and density are given through the solution of equations (3.7) as follows:

$$v(x, t) = \frac{u_+(x, t) + u_-(x, t)}{2}, \tag{3.8}$$

$$\rho(x, t) = \frac{u_+(x, t) - u_-(x, t)}{2\kappa}. \tag{3.9}$$

Let us thoroughly investigate the case when the initial density is constant and equals ρ_0, and the initial velocity is $v(x, t = 0) = v_0(x)$. Here the initial conditions in (3.7) appear as

$$u_+(x, 0) = v_0(x) + c_0, \quad u_-(x, 0) = v_0(x) - c_0,$$

$c_0 = \kappa\rho_0$ being the linear sound velocity, and the solutions of (3.7) will take the form

$$u_+(x, t) = u(x - c_0 t, t) + c_0, \quad u_-(x, t) = u(x + c_0 t, t) - c_0. \qquad (3.10)$$

Correspondingly, the velocity and density fields (3.8) and (3.9) are written as follows:

$$v(x, t) = \frac{u(x + c_0 t, t) + u(x - c_0 t, t)}{2}, \qquad (3.11)$$

$$\rho(x, t) = \rho_0\left[1 - \frac{u(x + c_0 t, t) - u(x - c_0 t, t)}{2c_0}\right]. \qquad (3.12)$$

It has to be emphasised that these equation give not only the law of integral velocity conservation $\int v \, dx = \text{const}$, which is typical of BE, but also ensure the conservation of integral momentum,

$$\int \rho(x, t) v(x, t) \, dx = \text{const}.$$

Consider now the limiting cases of (3.11) and (3.12). Given that the pressure forces tend to zero, which is formally similar to $c_0 \to 0$, we have

$$v(x, t) = u(x, t), \; \rho(x, t) = \rho_0\left[1 - t\frac{\partial v(x, t)}{\partial x}\right]. \qquad (3.13)$$

Prior to collision times these expressions describe the velocity and density fields with respect to noninteracting particles, whereas after collisions they refer to the particle sticking and deceleration in clustering domains.

In the limiting case of infinitesimal viscosity, $\nu \to 0$, the solution of (3.7) has the form

$$u(x, t) = \frac{x - y(x, t)}{t}, \qquad (3.14)$$

where $y(x, t)$ is the coordinate of an absolute minimum of the function

$$\Phi(x, y, t) = S_0(y) + \frac{(x - y)^2}{2t}, \quad S_0(y) = \int^y v_0(x) \, dx. \qquad (3.15)$$

Substituting (3.14) into (3.11), expression (3.12) yields

$$v(x, t) = \frac{1}{t}\left[x - \frac{y(x + c_0 t, t) + y(x - c_0 t, t)}{2}\right], \qquad (3.16)$$

$$\rho(x, t) = \rho_0\left[\frac{y(x + c_0 t, t) - y(x - c_0 t, t)}{2c_0 t}\right]. \qquad (3.17)$$

Provided the pressure forces tend to zero ($c_0 \to 0$) we obtain

$$v(x, t) = \frac{x - y(x, t)}{t}, \quad \rho(x, t) = \rho_0\frac{\partial y(x, t)}{\partial x}. \qquad (3.18)$$

In terms of the noninteracting particle flows, $y(x, t)$ is an initial (Lagrangian) coordinate of the particle with minimal action among those noninteracting ones that might have arrived at point x at time t. The rest of the particles stick together, during overtaking, according to the law of absolutely inelastic collisions. Taking into account that $\partial y/\partial x$ is the Jacobian of the Lagrangian-to-Eulerian coordinate transformation, we come to (2.36) again.

2. Let us proceed now to an analysis of the model gas density fluctuations under different limiting situations. The initial action $S_0(x)$ and the velocity field $v_0(x)$ will be treated as Gaussian, statistically homogeneous and possessing correlation functions which allow one to perform the following expansions:

$$\begin{aligned} B_s(z) &= \langle S_0(x)\, S_0(x+z)\rangle = \sigma_s^2(1 - z^2/(2l_s^2) + z^4/(24 l_s^2 l_v^2) - \dots), \\ \sigma_v^2 R_0(z) &= \langle v_0(x)\, v_0(x+z)\rangle = -B_s''(z) = \sigma_v^2(1 - z^2/(2l_v^2) + \dots). \end{aligned} \tag{3.19}$$

Here l_v and $l_s(l_v \le l_s)$ are characteristic scales of the initial fields, $\sigma_v^2 = \sigma_s^2/l_s^2$ being the variance of the initial velocity field, and $R_0(z)$ stands for its correlation factor. Everywhere in the ensuing discussion the initial density will be assumed constant and equal to ρ_0.

Consider first the density fluctuation evolution in gas without pressure that simulates the gravitational instability development. Let initially $\mu \to 0$. A characteristic time spent on the wave steepening and massive pancake formation amounts to $t_n = l_v/\sigma_v$.

At a stage before discontinuity formation, $t < t_n$, the density fluctuations are investigated in sections 4.3 and 4.4, where the energy spectra (4.3.4) and (4.3.6) are the density probability distributions (4.4.2) and (4.4.5) are revealed. Due to the density fluctuation growth, the probability distribution is broadened, such that the probability distribution maximum shift towards $\rho < \rho_0$ reflects the initiation of 'dark' domains, i.e. domains of reduced density because of the gas particle 'running away'. On the other hand, the power-law drop of the density, $W_\rho(\rho;\, t) \sim \rho^{-3}$, (4.4.6), is associated with the emergence of singularities of type $\rho(x, t) \sim x^{-\frac{1}{2}}$. These singularities can also lead to the appearance of the density spectrum power asymptotics $G_\rho(k;\, t) \sim k^{-1}$ (4.3.7).

Let us consider the properties of macroparticles formed owing to the light-weight particles sticking together. The masses of these particle clumps might come from (3.13), taking into account the fact that the field $u(x, t)$ has, in sticking points, discontinuities such that in their vicinity $u_x = -\Delta v_k \delta(x - x_k)$, where Δv_k is the discontinuity amplitude and x_k its coordinate. The appropriate macroparticle, as follows from (3.13), has a mass $m_k = \rho_0 \Delta v_k t$. According to (3.18) the velocity and density fields are determined by function $y(x, t)$, being the Lagrangian coordinate of the particle reaching a point with Eulerian coordinate x at time t. In the neighbourhood of a macroparticle with coordinate x_k, the magnitude of $y(x, t)$ jumps from y_k^+ to y_{k+1}^-. As a result of this, all the gas particles falling within the interval (y_k^+, y_{k+1}^-) at the initial time $t = 0$ stick together to form a macroparticle with coordinate x_k (Fig. 3.1). In this case the macroparticle mass is equal to

$$m_k = \rho_0\eta_k, \quad \eta_k = y^-_{k+1} - y^+_k, \tag{3.20}$$

and its velocity coincides with that of the discontinuity, At $t \gg t_n = \sigma_v/l_v$ the field $v(x, t)$ reveals itself as a sequence of sawtooth pulses with discontinuities at $x = x_k$ and 'zeros' at $x = y_k \approx y^+_k \approx y^-_k$. Between the discontinuities the field has a universal structure (Fig. 6.22)

$$v(x, t) = \frac{x - y_k}{t}, \quad x_k < x < x_{k+1}, \tag{3.21}$$

where y_k are the coordinates of the corresponding local minima of the initial action $S_0(y)$. For the velocity divergence we have

$$\frac{\partial}{\partial x} v(x, t) = \frac{1}{t}\left[1 - \sum_k \eta_k \delta(x - x_k)\right], \quad \eta_k = y_{k+1} - y_k,$$

and, thus, the density field acquires the form

$$\rho(x, t) = \sum_k m_k \, \delta(x - x_k), \quad m_k = \rho_0\eta_k.$$

Because of the chaotic collisions of macroparticles, their number diminishes as the characteristic mass rises. In order to assess the statistical properties and the macroparticle mass growth rate, one needs to know the interval length η_k statistics. In Section 5.5 we found an asymptotic ($t \gg t_n$) probability distribution of η_k, (5.5.8),

$$W_\eta(\eta; t) = \frac{\eta}{2l^2} \exp\left(-\frac{\eta^2}{4l^2}\right),$$
$$l = l(t) = l_v\sqrt{t/t_n\sqrt{\ln(t/t_n)}}\,. \tag{3.22}$$

Hence we see that the macroparticle masses $m_k = \rho_0\eta_k$ are distributed according to the Rayleigh law

$$W_m(m; t) = \frac{m}{2\rho_0^2 l^2} \exp\left(-\frac{m^2}{2\rho_0^2 l^2}\right). \tag{3.23}$$

Thus, the average mass of a macroparticle increases as

$$\langle m \rangle = \sqrt{\pi}\, \rho_0 l(t) \approx \rho_0 l_v \sqrt{t/t_n},$$

and the average mass per unit length $n(t)\langle m \rangle = \rho_0$ equals the initial gas density. Here $n(t) = 1/\sqrt{\pi}\, l(t)$ denotes the average particle number per unit length.

Analysing the density spectrum, it seems convenient to proceed from the relation (3.13). Provided $G_v(k; t)$ is the velocity fluctuation spectrum, the density fluctuation spectrum according to (3.13) has the form

$$G_\rho(k; t) = \rho_0^2[\delta(k) + t^2k^2G_v(k; t)]. \tag{3.24}$$

At a stage before discontinuity formation, the field $v(x, t)$ spectrum coincides with that of a Riemann wave and is described by (4.2.2) and (4.2.3). The

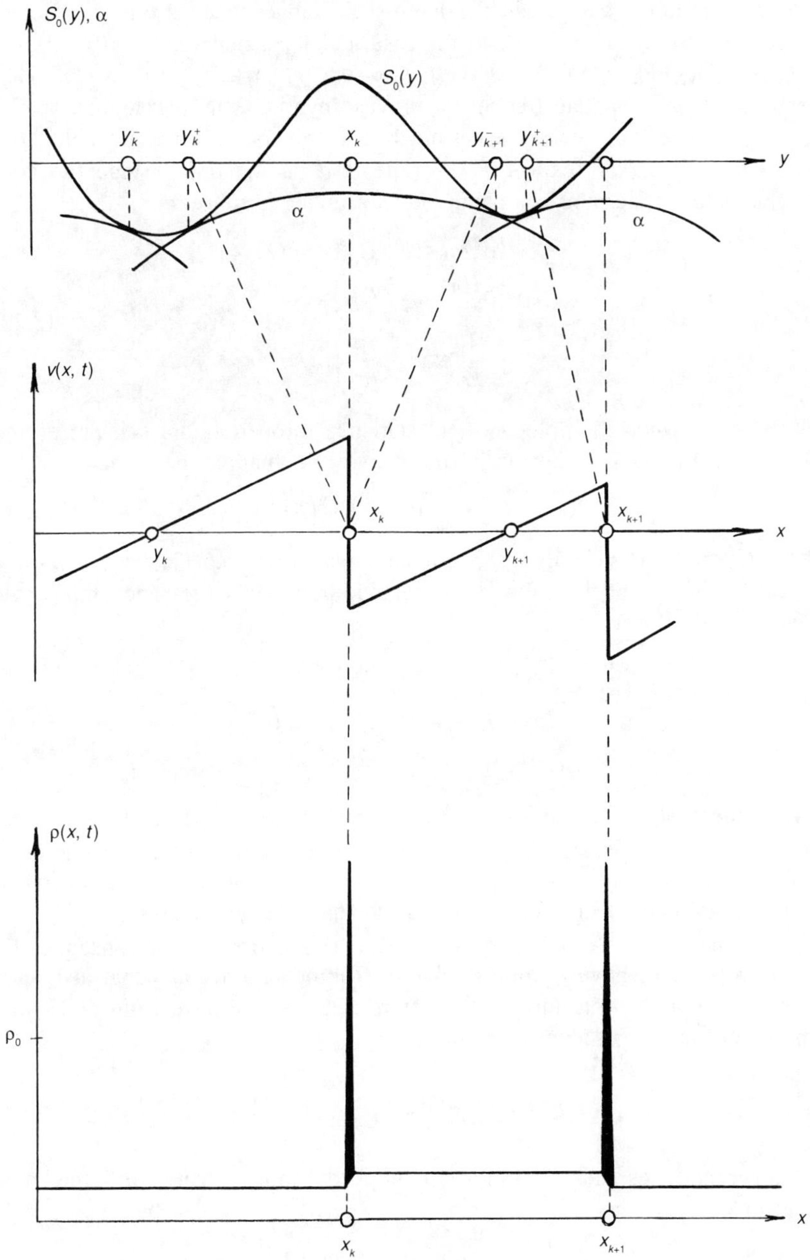

Fig. 6.22 A set of critical parabolas, velocity and density fields for $t \gg t_n$ and $\mu \to 0$

density spectrum behaviour in the large-scale range depends on the velocity initial spectrum value at $t = 0$. In the case under consideration $G_0(0) = 0$ and (4.2.6) implies that $G_v(k; t) \sim k^2t^2$. If, however, $G^0(0) = G^0 \neq 0$, the velocity spectrum at zero spatial frequency proves invariant. In the higher spatial frequency range the velocity spectrum has a universal asymptotic behaviour (4.2.9), $G_v(k; t) \sim (kt)^{-3} \exp(-t_n^2/2t^2)$. Therefore at the initial stage the density fluctuation spectrum has asymptotic formulae as follows:

$$G_\rho(k; t) \sim \begin{cases} \rho_0^2 t^2 k^2 G^0, & G_0(0) = G^0 \neq 0, \\ \rho_0^2 t^4 k^4, & G^0 = 0, \end{cases}$$
$$G_\rho(k; t) \sim \frac{1}{kt} \exp\left(-\frac{t_n^2}{2t^2}\right), \quad k \to \infty. \tag{3.25}$$

At $t \gg t_n$, when the velocity field statistical properties are self-preserving (Section 6.4) the same happens to the density fluctuation spectrum,

$$G_\rho(k; t) = \rho_0^2[\delta(k) + l^3k^2\tilde{G}(kl)], \tag{3.26}$$

where $\tilde{G}(\kappa)$ is the velocity spectrum universal form (5.3.17), (5.4.40). At a self-preserving stage the asymptotic behaviour in the large- and small-scale ranges has the form

$$G_\rho(k; t) = \begin{cases} \rho_0^2 k^2 l^3 G^0 \sim k^2 t^{\frac{3}{2}}, & G^0 \neq 0, \\ \rho_0^2 k^4 l^5 \sim k^4 t^{\frac{5}{2}}, & G^0 \neq 0, \end{cases} \quad kl \ll 1, \tag{3.25'}$$
$$G_\rho(k; t) \sim \rho_0^2 l \sim t^{\frac{1}{2}}, \quad kl \gg 1.$$

The emergence of a flat spectrum (no dependence on k) at $kl \gg 1$ is associated with the appearance of macroparticles having zero thickness at $\nu \to 0$. The spectrum universal law arising in the large-scale domain is attributed to large-scale component generation due to macroparticle merging.

Given the finite dissipation ($\mu \to 0$), the shock fronts and, consequently, the macroparticles have a finite thickness. Taking into account that at $t \gg t_n$ the shock fronts have a universal structure (6.2.10), we have from (3.13) the expression for the gas density, namely

$$\rho(x, t) = \rho_0 \sum_j \eta_i \operatorname{sech}^2\left[\frac{x - x_j}{\delta_j}\right] \Big/ 2\delta_j,$$

and macroparticles diffuse to form clumps of finite density with thickness equal to

$$\delta_j = 4\mu t/\eta_j \sim 1/m_j,$$

varying inversely with particle mass.

Considering $\rho(x, t)$ as an ergodic process, one obtains for the macroparticle density probability distribution

$$W_m(\rho; t) = \lim_{L\to\infty} \frac{1}{L} \sum_j \left|\frac{\partial x}{\partial \rho}\right| = \frac{\rho_0 z}{2\rho^2} \lim_{L\to\infty} \frac{1}{L} \sum_j \frac{1}{\sqrt{\eta_i^2 - z^2}}$$

$$= \frac{\rho_0 z^2 n(t)}{2\rho} \int_z^\infty \frac{W_\eta(\eta; t)}{\sqrt{\eta^2 - z^2}} d\eta, \quad z^2 = \frac{8\mu t \rho}{\rho_0}.$$

Averaging this expression with the aid of the probability distribution $W_\eta(\eta; t)$ (3.22), we come to

$$W_m(\rho; t) = \exp(-\rho/\rho_0 R)/\rho R. \tag{3.27}$$

Here $R(t)$ is the effective Reynolds number, proportional to the ration of the characteristic distance between macroparticles $l(t)$ to their characteristic thickness $\rho(t) = 4\mu t/l(t)$,

$$R = \frac{2l}{\delta} = \frac{l^2}{2\mu t} \approx \frac{R_0}{\sqrt{\ln(t/t_n)}}, \quad R_0 = \frac{\sigma_v l_v}{\mu}. \tag{3.28}$$

It follows from (3.27) that the density probability distribution drops by a power law down to $\rho \sim \rho_0 R$, with an exponential decrease to follow. It is worth noting that although (3.27) has a non-integrable singularity, which relates to the fact that the clump tail overlapping effect was neglected when deducing (3.27), the average density of the macroparticle gas is equal, as is evident from (3.27), to the initial density,

$$\int_0^\infty W_m(\rho; t)\rho \, d\rho = \rho_0.$$

This means that under the approximation considered the density of the gas particles which have not stuck together to form macroparticles proves to be zero. A more detailed analysis of density in the regions between macro-particles can be carried out, supported by the formula

$$\rho(y, t) \approx \rho_0/(1 + v_0'(y)t), \tag{3.29}$$

which appears as a one-dimensional analogue of (1.20). Taking into account that the 'dark' regions initiated from the sectors where the initial action had a very deep minimum, the equality (3.29) can be readily rewritten as

$$\rho(y_k, t) \approx \rho_0/S_0''(y_k)t. \tag{3.30}$$

At $t \gg t_n$ 'dark' regions are corresponded to by an initial action with a characteristic value (4.12)

$$S_0 \approx -H_0, \; H_0 = \sigma_s\sqrt{\ln(t/t_n)} \gg \sigma_s, \quad \Delta S \ll \sigma_s,$$

and hence the density probability distribution in the 'dark' regions equals, according to (3.30),

$$W_d(\rho; t) = \frac{\rho_0}{t\rho^2} W_y(\rho_0/\rho t | -H_0),$$

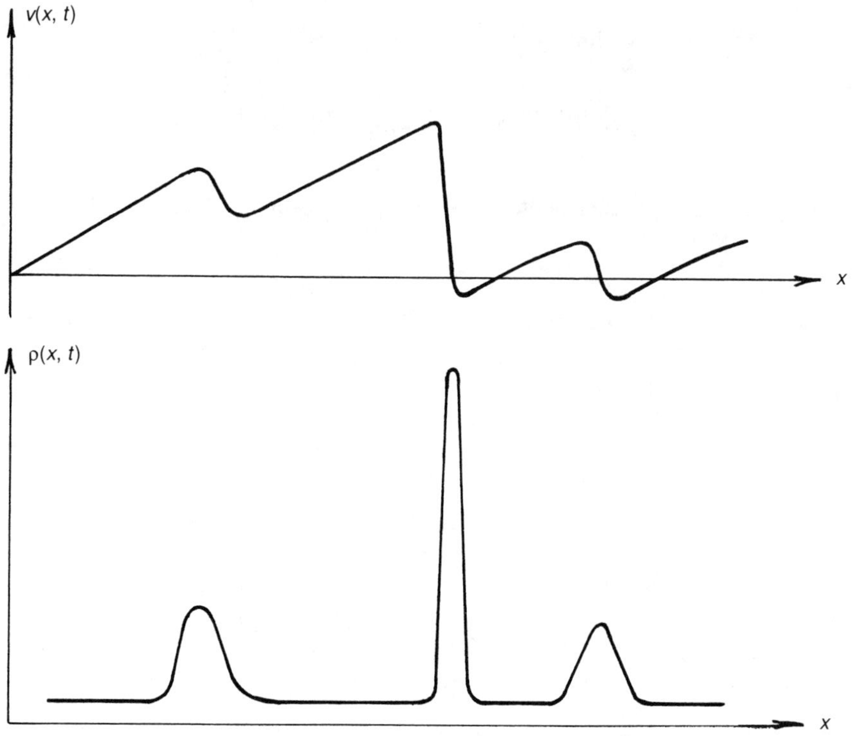

Fig. 6.23 Realisation of velocity and density fields and the density probability distribution type, at large Reynolds numbers

being the gas characteristic density in the 'dark' regions, which is much less than the initial density ρ_0, and

$$\Delta \approx \sqrt{\frac{1 - l_v^2/l_s^2}{\ln(t/t_n)}}$$

presents the relative width of the probability distribution (3.31), being small at $t \gg t_n$: $\Delta \ll 1$.

Combining (3.27) and (3.31) one obtains the total probability distribution of the density, allowing for both the macroparticle density and matter in the 'dark' regions,

$$W_\rho(\rho; t) = \begin{cases} W_m(\rho; t) + W_d(\rho; t), & \rho > \rho_d, \\ W_d(\rho; t), & \rho < \rho_d. \end{cases} \tag{3.32}$$

Normalisation of this distribution at $t \gg t_n$ and $R \gg 1$ is close to unity. Normalisation somewhat above unity can be attributed to the fact that (3.31) was deduced under an implicit assumption that the dark regions occupy the whole of the x-axis. Figure 6.23 depicts the velocity and density field realisations, as well as the characteristic shape of the probability distribution at $R \gg 1$.

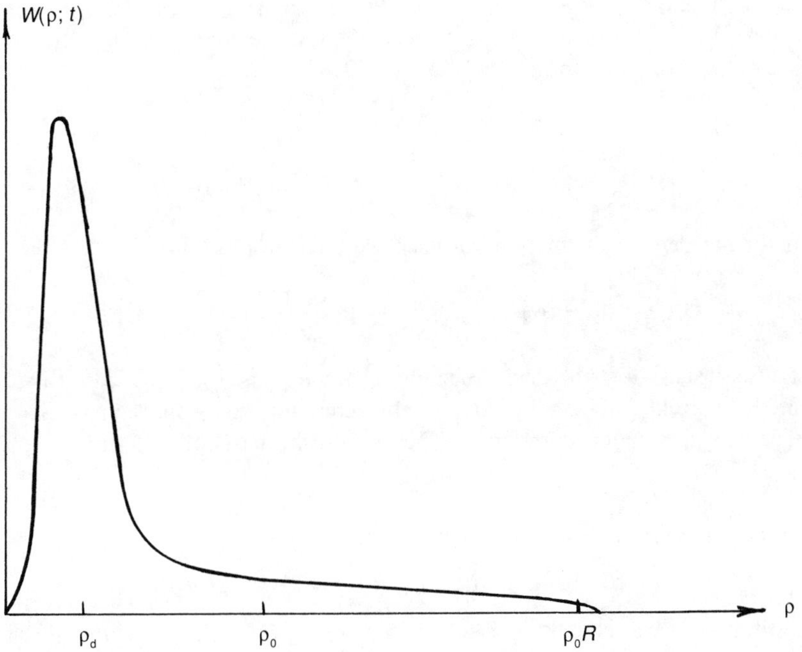

where $W_y(s''|s)$ is the conditional probability distribution of the initial action second derivative provided that the action itself is equal to S. Given Gaussian statistics for S, the latter expression leads to the following equation for the density probability distribution:

$$W_{\mathrm{d}}(\rho; t) = \frac{\rho_{\mathrm{d}}}{\sqrt{2\pi\rho^2\Delta}} \exp\left[-\frac{1}{2\Delta^2}\left(\frac{\rho_{\mathrm{d}}}{\rho} - 1\right)^2\right]. \qquad (3.31)$$

Here

$$\rho_{\mathrm{d}} \approx \frac{\rho_0 t_{\mathrm{n}}}{t\sqrt{\ln(t/t_{\mathrm{n}})}},$$

3. Consider now density fluctuations in a gas with pressure when $c_0 \neq 0$. Apart from the nonlinear overturning time $t_{\mathrm{n}} = l_v/\sigma_v$, one more temporal scale $t_l = l_v/c_0$ is observed here, offering the running-away time of the inhomogeneities with linear sound velocity c_0 over a scale l_v. In this case the density fluctuation evolution appears dramatically different when considered in a rare medium, with $\sigma_v \gg c_0 (t_{\mathrm{n}} \ll t_l)$, and in a dense medium with $\sigma_v \ll c_0 (t_l \ll t_{\mathrm{n}})$.

First we calculate the correlation function and the density fluctuation variance. It is obvious from (3.12) that they are formulated using the correlation function of the field $u(x, t)$ satisfying the BE (3.7),

$$\langle u(x, t)u(x + z, t)\rangle = \sigma^2(t)R(z; t), \qquad (3.33)$$

Here $\sigma^2(t) = \langle u^2 \rangle$ is the variance, and $R(z; t)$ is the field $u(x, t)$ correlation faactor. As concerns the density fluctuation correlation function $\tilde{\rho}(x, t) = \rho(x, t) - \rho_0$, we have form (3.12) and (3.33)

$$B_{\tilde{\rho}}(z; t) = \langle \tilde{\rho}(x, t)\, \tilde{\rho}(x + z, t) \rangle$$

$$= \rho_0^2 \frac{\sigma^2}{4c_0^2} [2R(z; t) - R(z - 2c_0 t; t) - R(z + 2c_0 t; t)], \tag{3.34}$$

and for the density fluctuation variance we have from (3.34)

$$\sigma_{\tilde{\rho}}^2(t) = \langle (\rho - \rho_0)^2 \rangle = B_{\tilde{\rho}}(0; t) = \rho_0^2 \frac{\sigma^2(t)}{2c_0^2} \left[1 - R(2c_0 t; t)\right]. \tag{3.35}$$

Let us discuss first the dense medium situation ($t_l \ll t_n$). Provided $c_0 t$ is less than the correlation function $R(z; t)$ characteristic scale, one cam write for the density fluctuation correlation function, on the basis of (3.34),

$$B_{\tilde{\rho}}(z; t) \approx -\rho_0^2 \sigma^2(t)\, R''_{zz}(z; t). \tag{3.36}$$

At $t \ll t_l$ nonlinear distortions can be ignored and

$$\sigma^2(t) \approx \sigma_v^2, \quad R(z; t) \approx R_0(z) \approx 1 - z^2/2l_v^2.$$

Substituting these equalities into (3.35) or (3.36) yields, for the density fluctuation variance,

$$\sigma_\rho^2(t) = \rho_0^2 \sigma_v^2 t^2 / l_v^2 = \rho_0^2 (t/t_n)^2, \tag{3.37}$$

and thus, the density fluctuation growth rate does not depend on the sound velocity. At $t_l \ll t \ll t_n$, when nonlinear distortions remain insignificant as yet, while the opposing waves have already covered a distance exceeding l_v, we see that $R_0(2c_0 t) \approx 0$ and the density fluctuation variance enters the stationary regime

$$\sigma_\rho^2(t) = \rho_0^2 \sigma_v^2 / 2c_0^2. \tag{3.37$'$}$$

Here, due to the pressure force action, the density fluctuation magnitude becomes much less than ρ_0. At $t \gg t_n$ and $\nu \to 0$ the field $u(x, t)$ comes out as a sequence of sawtooth pulses (3.21), while its energy decays as $l^2(t)/t^2$, where $l(t)$ is the external turbulence scale (3.23). Therefore, (3.35) yields, concerning the density fluctuation variance,

$$\sigma_{\tilde{\rho}}^2(t) = \rho_0^2 \frac{\sigma^2(t)}{2c_0^2} \approx \rho_0^2 \frac{\sigma_v^2}{2c_0^2} \left(\frac{t_n}{t} \right). \tag{3.38}$$

The density realisations in this case have the form of a sequence of square pulses, with characteristic amplitudes $\rho_0 \sigma(t)/c_0$ and scale $l(t)$.

In all the three cases considered here the density probability distribution $W_\rho(\rho; t)$ is Gaussian with an average ρ_0 and variance specified in (3.37′)–(3.38). At $t \ll t_n$ this is associated with the fact that the initial Gaussian field has not yet been nonlinearly distorted. At $t \gg t_n \gg t_l$ the Gaussian nature of $W_\rho(\rho; t)$ is ensured by the fact that the fields $u(x \pm c_0 t, t)$ appearing in (3.12) are statistically independent, and have Gaussian distribution (5.4.16).

4. Now let us proceed to analyse the density fluctuations in a rare medium ($t_n \ll t_l$). At $t \ll t_n$ the nonlinear distortions are negligibly small, and (3.36) holds true. At $t < t_n$, due to discontinuity formation in the field $u(x, t)$, the correlation function $R(z; t)$ analyticity is destroyed. If $n(t)$ is the average discontinuity number per unit length (5.1.4), then (5.1.19) yields for $R(z; t)$:

$$R(z; t) \approx 1 - n(t)\,|z| + \ldots, \quad n(t) \approx \exp\left(-\frac{t_n^2}{2t^2}\right)/l_v. \tag{3.39}$$

As a result, at $t > t_n$ the density fluctuation growth law (3.37) is replaced with

$$\sigma_\rho^2(t) = \rho_0^2 \frac{\sigma_v}{c_0} \frac{t}{t_n} \exp\left(-\frac{t_n^2}{2t^2}\right).$$

At $t \gg t_n$ the field $u(x, t)$ is observed to be a sequence of sawtooth pulses (3.21), and its statistical properties are self-preserving, depending only on a single scale $l(t)$ (3.23). In particular, as to variance $\sigma^2(t)$ and correlation function $R(z; t)$ we have

$$\sigma^2(t) = l^2(t)/t^2, \; R(z; t) = \tilde{R}(z/l(t)), \tag{3.40}$$

and for $\tilde{R}(z/l)$ at $z \ll l$ the equality (5.4.42) is valid,

$$\tilde{R}(z/l) \approx 1 - 2\,|z|/\sqrt{\pi}l.$$

The density fluctuation variance behaviour depends, in this instance, on the ration of the external scale $l(t)$ to the length of the opposing wave linear run, $2c_0 t$. For $2c_0 t \ll l(t)$, which is met when $t \ll t_l\, \sigma_v/c_0$, we have from (3.35)

$$\sigma_\rho^2(t) = \frac{2\rho_0^2 \sigma^2(t)\, t}{\sqrt{\pi}\, c_0 l(t)} \approx \rho_0^2 \frac{\sigma_v}{c_0} \sqrt{\frac{t_n}{t}}. \tag{3.41}$$

Moreover, in this case the density fluctuations prove rather high: $\sigma_\rho \gg c_0$. At $2c_0 t \gg l(t)$ ($t \gg t_l \sigma_v/c_0$) the 'running-away' waves become statistically independent, $R(2c_0 t; t) \approx 0$, and

$$\sigma_\rho^2(t) = \frac{\rho_0^2\, \sigma^2(t)}{2c_0^2} \approx \rho_0^2 \frac{\sigma_v}{c_0} \frac{t_l}{t}. \tag{3.42}$$

When $t > t_l \sigma_v/c_0$ the standard deviation of the density fluctuations becomes less than ρ_0.

Let us now address the density realisation forms and the density probability distribution for $t \gg t_n$ but $2c_0 t \ll l(t)$ ($t_n \ll t \ll t_l \sigma_v/c_0$) when nonlinear effects profoundly influence the statistical properties of the density fluctuations.

In this case it follows from (3.12) and (3.21) that the density realisations have the form of the square pulse sequence

$$\rho(x, t) = \sum_k \frac{\rho_0 \eta_k}{2c_0 t} \prod\left(\frac{x - x_k}{2c_0 t}\right),$$
$$\prod(x) = \begin{cases} 1, |x| \le 0.5, \\ 0, |x| > 0.5. \end{cases} \tag{3.43}$$

Here a kth pulse is interpreted as a macroparticle with mass $m_k = \rho_0 \eta_k$ and thickness $2c_0 t$. This only differs from a medium without pressure ($c_0 \to 0$) in that the particles have a finite thickness $2c_0 t$ defined by the linear sound velocity, rather than zero size.

In order to assess the density probability distributions corresponding to macroparticles with realisations of type (3.43), we make use of the equality

$$W_{\mathrm{m}}(\rho; t) = \lim_{L \to \infty} \frac{1}{L} \sum_k \left|\frac{\Delta x_k}{\Delta \rho}\right| = \frac{4}{\rho_0} c_0^2 t^2 n(t)\, W_\eta\left(\frac{2c_0 \rho t}{\rho_0}; t\right).$$

Substituting here the interval length $\eta_k - W_\eta(\eta; t)$ distribution (3.22), we get eventually

$$W_{\mathrm{m}}(\delta; t) = \frac{\rho}{2\sqrt{\pi \rho_0^2 \alpha^3}} \exp\left(-\frac{\rho^2}{4\rho_0^2 \alpha^2}\right). \tag{3.44}$$

Here

$$\alpha(t) = l(t)/2c_0 t \approx \frac{\sigma_v}{c_0} \sqrt{\frac{t_{\mathrm{n}}}{t}},$$

is the ratio of the macroparticle characteristic density to the gas average density. Within the time interval under consideration $\alpha \gg 1$, and the standard deviation of the density fluctuations proves far in excess of the average value ρ_0. It turns out form (3.44) that the average macroparticle gas density is equal to ρ_0, and the density fluctuation variance

$$\sigma_\rho^2(t) \approx \rho_0^2 \alpha = \rho_0^2 \frac{\sigma_v}{c_0} \sqrt{\frac{t_{\mathrm{n}}}{t}} \gg \rho_0^2.$$

The integral of the probability distribution (3.44) is

$$\int_0^\infty W_{\mathrm{m}}(\rho; t)\, d\rho = \frac{1}{\sqrt{\pi \alpha}} = \frac{2c_0 t}{\sqrt{\pi}\, t(t)} \ll 1,$$

and is equal to the relative part of the x-axis occupied by macroparticles. We see from (3.17) that the density probability distribution in dark regions (reduced density domains) has the form

$$W_d(\rho; t) = \delta(\rho)\tilde{P}(2c_0t; t), \tag{3.45}$$

where $\tilde{P}(s; t)$ is the probability for the field $u(x, t)$ to have no discontinuities within the interval of length s. At $t \gg t_n$ this probability is self-preserving, $\tilde{P}(s; t) = P(s/l)$ and is specified in (5.4.35). Utilising the asymptotic behaviour $P(z) = 1 - z/\sqrt{\pi}$ (5.4.41), one can see that the total density probability distribution

$$W_\rho(\rho; t) = W_d(\rho; t) + W_m(\rho; t) \tag{3.46}$$

is normalised to unity.

A more detailed study of the density distribution inside the dark regions between macroparticles can be performed, as was the case for gas without pressure, on the basis of (3.29). In so doing, the probability distribution of the dark region density is described by (3.45) where one should insert the distribution (3.31) instead of the delta function.

As time evolves, macroparticles occupy more of the x-axis, their density becoming lower to reach a state when at $t \approx t_l\sigma_v/c_0$, density fluctuations become of order ρ_0. At $t \gg t_l\sigma_v/c_0$, the density realisations also appear as a sequence of square pulses, although with an amplitude $\rho_0\sigma(t)/c_0 \ll \rho_0$, whereas the density probability distribution is Gaussian with average ρ_0 and variance (3.42). A characteristic type of density realisation and the probability distributions at $t \gg t_n$ is depicted, in the two cases $2c_0t \ll l(t)$ and $2c_0t \gg l(t)$, in Fig. 6.24. Observe that with respect to relatively weak acoustic perturbations in real gas, the oppositely travelling waves hardly interact at all, so that velocity and density fields can be expressed via superposition of two solutions (3.11) and (3.12) of BE subject to appropriate initial conditions [47]. At $\gamma \neq 3$ the mutual influence of the opposing waves leads to an additional phase modulation of the BE solutions. The wave amplitudes being small ($\sigma \ll c_0$), this modulation can be neglected when analysing the field statistical properties. This is, in particular, confirmed by numerical calculations [47].

5. We have shown in the present chapter that BE, when used to simulate the nonlinear stage of gravitational instability, predicts the formation of the matter distribution cellular structure. Such structures were encountered when numerically calculating the evolution of the gravitationally interacting particle two-dimensional gas (see Section 6.1). In addition to qualitative prediction, however, the Burgers model offers quantitative growth laws as to the cell scales (1.32) and (1.33), the structure element characteristic mass (1.36), the matter density in dark regions (2.31) and so on as well. One might check the model against a realistic situation by comparing its results with those of numerical analysis with respect to a gas of gravitationally interacting particles. Unfortunately, to compare the statistical properties one needs to perform

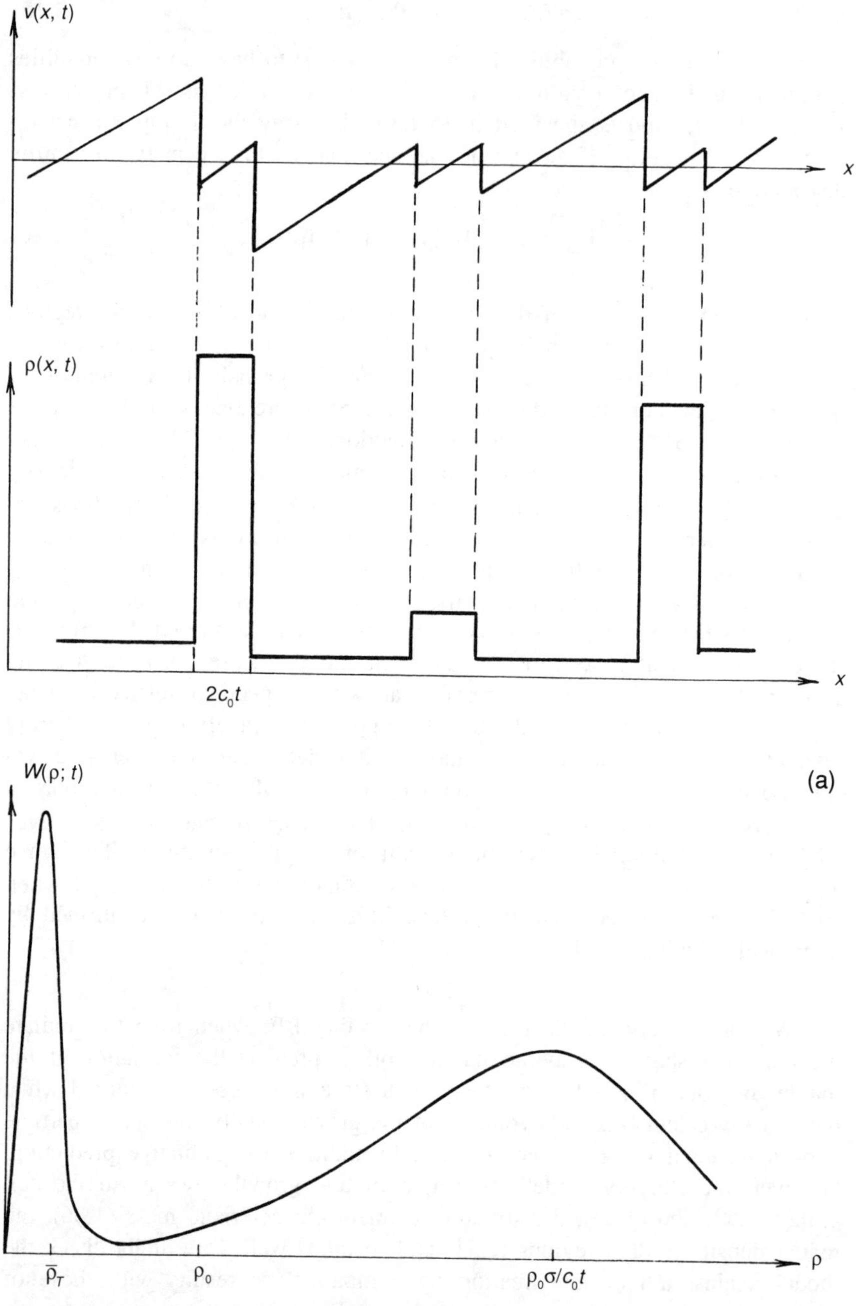

Fig. 6.24 Velocity and density realisations and the density probability distributions in (a) low-density gas; and (b) high-density gas

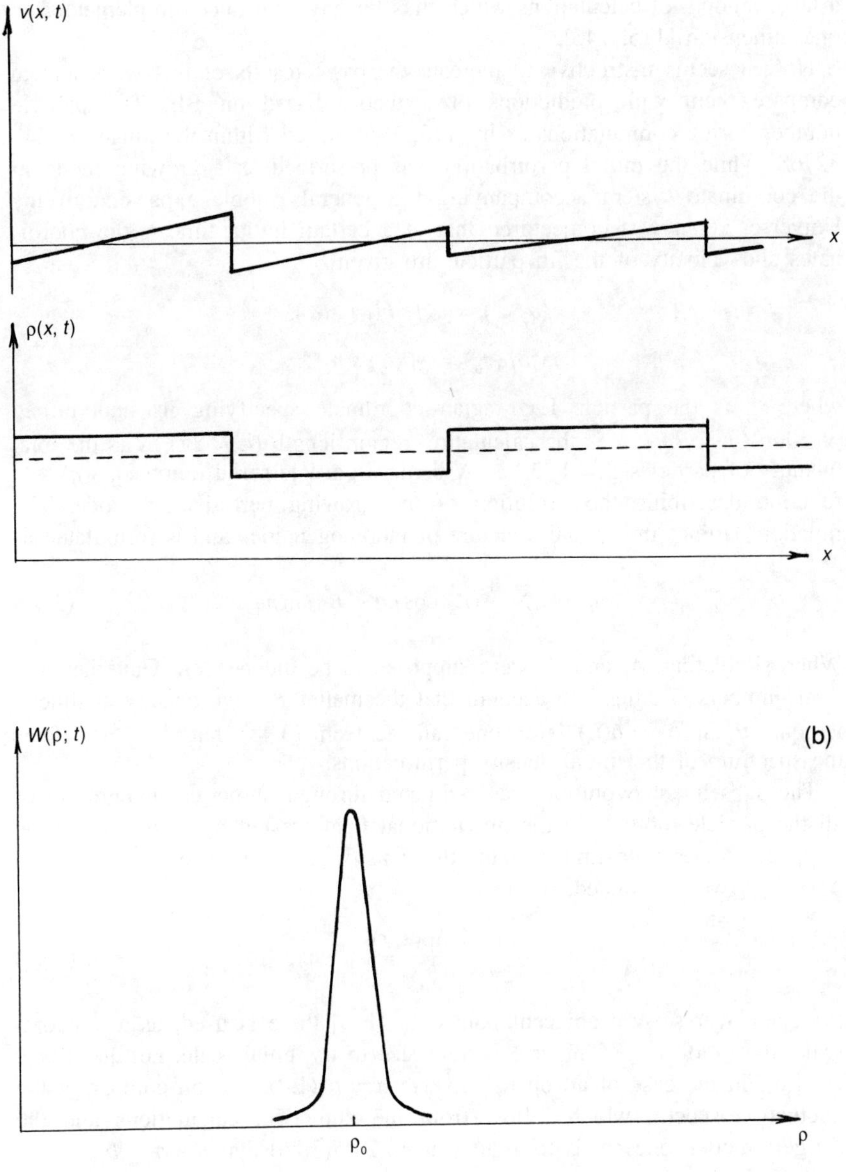
v(x, t)
x
ρ(x, t)
x
W(ρ; t)
(b)
ρ₀
ρ

a large amount of calculations which thus far have only been implemented in one dimension [115, 146].

Now it seems instructive to indicate the basic results of this work and to compare them with predictions of a theory based on BE. The particle number during computations as in [115, 146] varied within the range 16384–32768, while the initial perturbation was prescribed as a growing mode in the coordinate system accompanying the general Hubble expansion of the Universe, i.e. it was conjectured that at a certain initial time t_0 the coordinates and velocity of the ith particle are given by

$$\begin{aligned} x_i(q_i, t_0) &= q_i - b(t_0)\, s(q_i). \\ v_i(q_i, t_0) &= -\dot{b}(t_0)\, s(q_i), \end{aligned} \tag{3.47}$$

where q_i is the particle Lagrangian coordinate specifying its unperturbed position ($q_i = iL/N$, L is the calculation region length ($L = 2\pi$), N is the total quantity of particles, $i = 1, 2, \ldots, N$ being a particle serial number), $s(q)$ is a function describing the evolution of the growing perturbation mode. The function outlines the spatial structure of inhomogeneities and is formulated as

$$s(q) = \sum_{n=n_1}^{n_2} \frac{1}{n}(A_n \cos nq + B_n \sin nq). \tag{3.48}$$

When simulating, A_n and B_n were supposed to be independent Gaussian random numbers. Taking into account that the matter relative density at time t_0 is equal to $\Delta\rho/\rho_0 = b(t_0)\, \mathrm{d}s/\mathrm{d}q$ one can see from (3.48) that $A_n^2 + B_n^2$ defines the spectrum of the initial density perturbations.

The subsequent evolution was estimated through numerical integration of all the particle motions in the gravitational field created by their distribution in space. At separate time instants the density perturbation energy spectrum $\delta_n^2 = |\delta_n|^2$ was calculated, where

$$\delta_n = \frac{1}{2\pi} \int_0^{2\pi} \frac{\Delta\rho(x, t)}{\rho_0} \mathrm{e}^{ikx}\, \mathrm{d}x, \tag{3.49}$$

averaged over several adjacent points. In [115] there is used, as a temporal scale, the ratio of the Universe current size to its initial scale, i.e. quantity a in (1.6). In the case of an open Universe $a = b$ (1.8), and on comparing the fluctuation spectra which follow from the numerical calculations and the Burgers model, one needs to assume in (3.24)–(3.26) that $t \sim a = b$.

In [115] initial conditions of three types were imposed, in conformity with different density spectrum behaviour in the large-scale region. Let us show here the calculated results with respect to one of the types, when initial inhomogeneities are prescribed only within the range of the wave numbers from $n_1 = 513$ to $n_2 = 1024$, where $\langle A_n^2 + B_n^2 \rangle = \text{const}$. In the Burgers model (3.24) this corresponds to the following velocity fluctuation spectrum:

$$G_v(k;0) = \begin{cases} A_*^2/k^2, & k \in [k_1, k_2], \\ 0, & k \notin [k_1, k_2] \end{cases} \qquad (3.50)$$

$$(k_2 = 2k_1).$$

Within the framework of the linear theory of gravitational instability, when the density fluctuation spectrum is associated with the initial one through a linear relation, the perturbations with $k < k_1$ and $k > k_2$ are absent. For the (3.47)-type perturbation, however, when nonlinear effects are already substantial at time t_0, inasmuch as at $t = t_0$ the fluctuation value $\Delta\rho/\rho_0 \approx 0.3$, even in the initial density fluctuation spectrum both long-wave ($k < k_1$) and short-wave ($k > k_2$) components are present. Therefore, at all future times universal long-wave spectrum asymptotics $\delta_k^2 \sim k^4$ are observed (see Fig. 6.25). The spatial density distribution is presented in Fig. 6.26, where density perturbations $\Delta\rho/\rho_0$ are depicted at three time instants. the spatial scale has been chosen in such a way that the internal structure of the separate pancakes is not resolved, which makes them appear as density peaks. The regions between the latter, as for the Burgers model (Fig. 6.22), contain almost no matter.

The relative growth of harmonic amplitudes in time was also the subject of [115]. Experimental results, then, can be described as follows. At the initial stage, with $a < 3$, the harmonics present in the initial perturbation, i.e. at $n \in [n_1, n_2]$, increase as

$$\delta_n^2 \approx a^2 \quad (n \in [n_1, n_2]), \qquad (3.51)$$

while when $a > 3$ their growth has practically ceased. The long-wave part of the spectrum has a universal asymptotic behaviour $\delta_n^2 \sim n^4$, although the growth laws prove different at the initial (quasi-linear) and nonlinear stages, namely

$$\delta_n^2(a) = n^4 \begin{cases} a^4, a < 3, \\ a^{\frac{5}{2}}, a > 3. \end{cases} \qquad (3.52)$$

As harmonics attain a certain magnitude, their growth ends. On the short-wave side, the spectrum takes on a universal form $\delta_n^2 \sim n^{-1}$, such that from some time on it is stabilised at a constant level. Therefore the self-preserving spectrum of the density fluctuations is developed in the numerical experiments,

$$G_\rho(k) = \begin{cases} Ak_*^{-5}k^4, & k < k_*, \\ Ak^{-1}, & k > k_*, \end{cases} \qquad (3.53)$$

(where $k_* \sim a^{-\frac{1}{2}}$, $A = \text{const}$) with a comparatively narrow transient region at $k \approx k_*$ between the two universal laws (Fig. 6.25).

Now it is worthwhile to compare the results of numerical calculation of

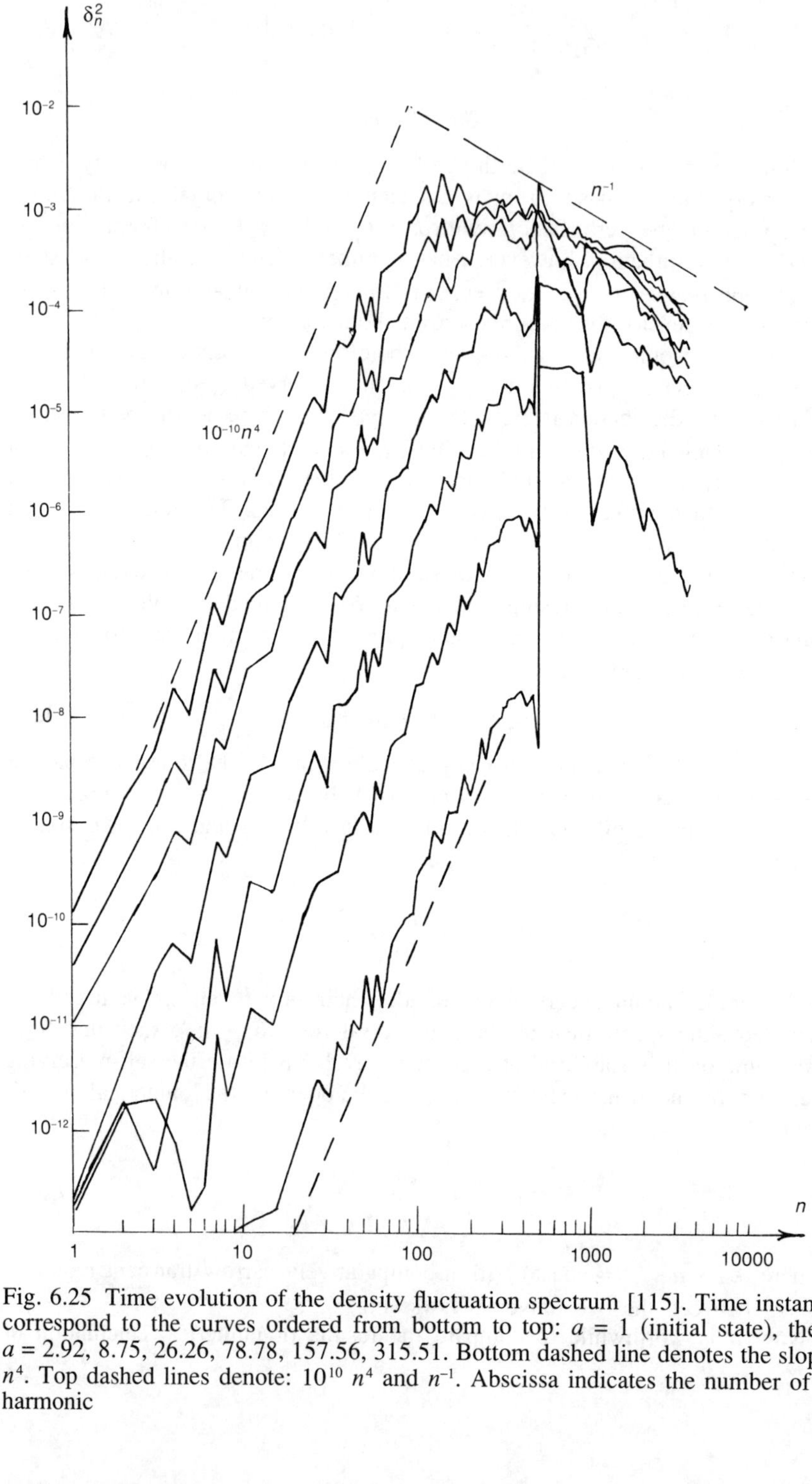

Fig. 6.25 Time evolution of the density fluctuation spectrum [115]. Time instants correspond to the curves ordered from bottom to top: $a = 1$ (initial state), then $a = 2.92$, 8.75, 26.26, 78.78, 157.56, 315.51. Bottom dashed line denotes the slope n^4. Top dashed lines denote: $10^{10}\, n^4$ and n^{-1}. Abscissa indicates the number of a harmonic

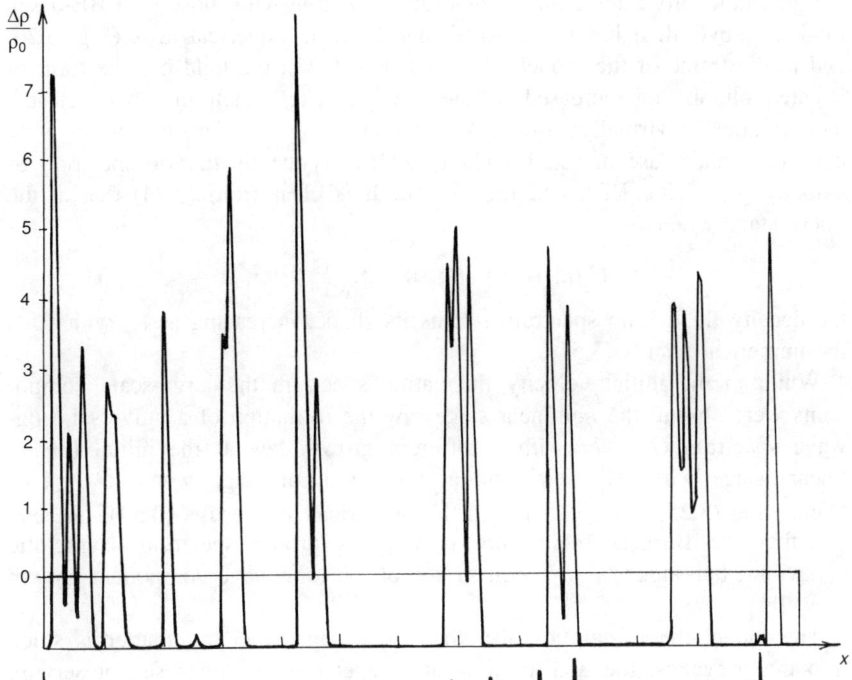

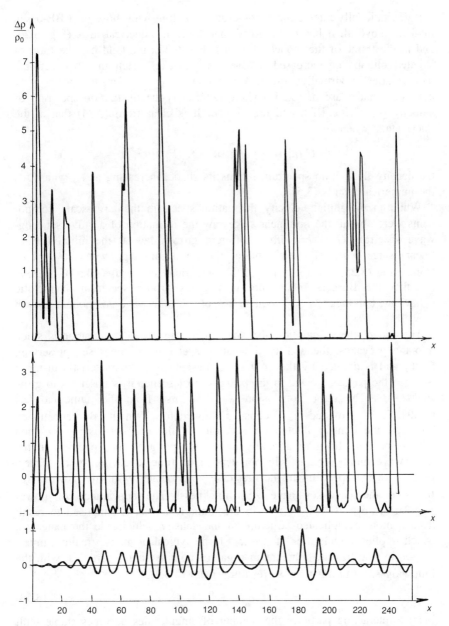

Fig. 6.26 Realisation of the density fluctuations $\Delta\rho/\rho_0$ at three time instants ordered from bottom to top: a = 1 (initial state), 26.26, and 315.51

the gravitationally interacting particle gas evolution with those of a BE-based model. Above all it has to be noted that, both in numerical tests (Fig. 6.26) and in the frame of the model equations (Fig. 6.22), the field has the form of isolated clumps of increased density, or pancakes, such that between the latter matter is virtually absent. When comparing the density spectra it is helpful to make use of (3.24), where $G_v(k;\ t)$ is the fluctuation spectrum of velocity (satisfying BE), and time $t \sim a$. It is clear from (3.24) that at the linear stage when

$$G_v(k;\ t) \approx G_v(k;\ 0) \approx G_\rho(k;\ 0)/k^2,$$

the density fluctuation spectrum retains its shape, increasing as t^2, which fits the numerical results (3.51).

While in the initial velocity fluctuation spectrum the large-scale components were absent, the nonlinear stage saw the formation of a universal long-wave spectrum $G_v \sim k^2$ with a different growth law at the initial (quasi-linear) stage when $G_v \sim k^2t^2$ and at the nonlinear stage when $G_v \sim k^2t^{\frac{1}{2}}$. Comparing (3.25) and (3.25′) with the numerical test results (3.52), one can see that the Burgers model not only gives proper spectrum asymptotic behaviour, but suggests an adequate law of the long-wave component growth as well.

The model theory implies also that, at a stage when all matter is stuck into the pancakes, the density fluctuation spectrum becomes self-preserving (3.26) and is defined by the only scale present, i.e. a characteristic distance between the pancakes. Due to the pancake coalescence it is observed to grow as $l(t) \sim t^{\frac{1}{2}}$. The same conclusion we have also from the numerical test results (3.53), provided $lk_* \approx$ const is allowed for. If in the Burgers model, however, the density fluctuation spectrum is flat, $G_\rho \sim t^{\frac{1}{2}}$ as $k \to \infty$, then with respect to the gravitationally interacting particles the spectrum in the large wave number range will not depend on time, decaying as k^{-1}. Emergence of a flat spectrum in the first case is associated with the fact that for $\mu \to 0$ the pancakes have zero thickness. The appearance of power singularities $\delta_k^2 \sim k^{-1}$ as $k \to \infty$ in the numerical tests is supposed to result from a more complicated structure of the density field inside the pancakes, which implies singularities of $\rho \sim x^{-\frac{1}{2}}$ form, typical of many-stream regimes. These singularities are inherent in a gas of noninteracting particles, with the following spectrum at $t \to \infty$, resulting from (4.3.7),

$$\delta_k^2 \sim k^{-1}t^{-1}. \tag{3.54}$$

For noninteracting particles the number of singularities becomes stable with time, and their number per unit length is inversely proportional to l_0, i.e. to the initial field scale. The decrease of the spectral density relates to the fact that the singularity field dies away as $\rho \sim (xt)^{-\frac{1}{2}}$ with time. The density spectrum of the gravitationally interacting particle gas enters a constant, time-independent level, which is connected with the fact that the number of singularities *grows*, due to the additional formation of oscillatory many-stream regions inside each pancake, rather than being stabilised.

As distinct from [115], the main emphasis in [146] is placed on analysis of the pancake mass probability distribution, such that singling out of clouds of grouped particles was carried out with the help of one-dimensional cluster-analysis. One calculation model only allowed for the gravitational particle interaction, while another suggested the incorporation in the evolution equations of sticking-together of the particles on their intersection, in order to simulate the inelastic collisions of gas-stellar complexes. It has been noted that while the first model provided an acceptable agreement (both qualitative and quantitative) with the theoretical distribution of the pancake masses (3.23) proposed from the Burgers model, in the second model the sticking effect 'boosted' the evolution ensuring, hence, a very good agreement between the theoretical and numerical distributions. It seems noteworthy, however, that a discrepancy exists between the theoretical and experimental probability distributions of the pancake masses, in a region corresponding to low-mass pancakes which do not stick together, for a long time, to form more massive clumps. As it seems to us, this discrepancy can be attributed to the absence, in the cluster-analysis, of a clear-cut definition of a massive pancake (usually having a complicated internal structure).

Appendix Properties of delta functions and their statistical averages

The following properties of delta functions of different arguments and their integration are based on the unit (Heaviside) function definition

$$\mathbf{1}(x) = \begin{cases} 0, & x < 0, \\ \frac{1}{2}, & x = 0, \\ 1, & x > 0, \end{cases} = \tfrac{1}{2}(1 + \operatorname{sgn} x).$$

It is also supposed that functions of type $y(x)$, $\varphi(z)$ and so on involved in the following formulae do not have singularities in a region wherein the delta function is singular.

$$\frac{\mathrm{d}}{\mathrm{d}y}\mathbf{1}(y - a) = \delta(y - a) = \delta(a - y). \tag{A.1}$$

$$\frac{\mathrm{d}}{\mathrm{d}y}\mathbf{1}(a - y) = -\delta(y - a). \tag{A.2}$$

$$\frac{\partial}{\partial x}\mathbf{1}(z - y(x)) = -\delta(z - y(x))\, y'(x). \tag{A.3}$$

$$\mathbf{1}(y - v) = \int_{-\infty}^{y} \delta(v - u)\, \mathrm{d}u, \quad \int_{-\infty}^{\infty} \delta(v - u)\, \mathrm{d}u = 1. \tag{A.4}$$

$$\int_{-\infty}^{\infty} \varphi(z)\, \delta\, (z - a)\, \mathrm{d}z = \varphi\, (a). \tag{A.5}$$

$$\int_{-\infty}^{\infty} \varphi(z)\, \delta^{(n)}\, (z - a)\, \mathrm{d}z = (-1)^n\, \varphi^{(n)}\, (a). \tag{A.6}$$

$$\int_{-\infty}^{\infty} \varphi(z)\, \delta^{(n)}\, (a - z)\, \mathrm{d}z = \varphi^{(n)}\, (a). \tag{A.7}$$

$$\delta(f(x)) = \sum_i \frac{1}{|f'(x_i)|}\, \delta(x - x_i), \tag{A.8}$$

where x_i are the real roots of equation $f(x) = 0$. Summation is performed over all the roots.

$$\delta(y - z) f(y) = \delta(y - z) f(z). \tag{A.9}$$

From formula (A.8) it follows that

$$\int_{-\infty}^{\infty} \varphi(x)\, \delta(f(x))\, dx = \sum_i \frac{\varphi(x_i)}{|f'(x_i)|}. \tag{A.10}$$

If $y = f(x)$ and $x = \varphi(y)$ are mutually single-valued functions, then

$$\delta(y - f(x)) = \delta(x - \varphi(y))\, |\varphi'(y)| = \frac{\delta(x - \varphi(y))}{|f'(x)|}. \tag{A.11}$$

If the function inverse to $y = f(x)$ is multi-valued and has branches

$$x = \begin{cases} \varphi_1(y), \\ \varphi_2(y), \\ \ldots, \end{cases}$$

then

$$\delta(y - f(x)) = \sum_i \delta(x - \varphi_i(y))\, |\varphi_i'(y)|, \tag{A.12}$$

where summation is over all branches.

Formula (A.12) implies that

$$\int_{-\infty}^{\infty} \delta(y - f(x))\, \Phi(x)\, dx = \sum_i \Phi(\varphi_i(y))\, |\varphi_i'(y)|. \tag{A.13}$$

If $y'(x) > 0$, then

$$\delta(x - x(y)) = \delta(y - y(x))\, y'(x) = -\frac{d}{dx}\mathbf{1}(y - y(x)). \tag{A.14}$$

Multiplying the first formula of (A.11) by $|f'(x)|$ and making use of property (A.9) one can readily obtain

$$|f'(x)|\, \delta(y - f(x)) = \delta(x - \varphi(y)). \tag{A.15}$$

Performing a similar procedure with respect to formula (A.12), i.e. for the case when the inverse function $x = \varphi(y)$ is multi-valued, we get

$$|f'(x)|\, \delta(y - f(x)) = \sum_i \delta(x - \varphi_i(y)). \tag{A.16}$$

$$\int_{-\infty}^{\infty} e^{iuz}\, du = 2\pi\, \delta(z). \tag{A.17}$$

Now we give the definition of two integrals which are often used in the book.

$$\Phi(z) = \sqrt{\frac{2}{\pi}} \int_0^z \exp\left(-\frac{x^2}{2}\right) \mathrm{d}x. \tag{A.18}$$

$$g(z) = \int_{-\infty}^{z} \exp\left(-\frac{x^2}{2}\right) \mathrm{d}x = \sqrt{\frac{\pi}{2}} \, (\Phi(z) + 1). \tag{A.19}$$

In the formulae given below, angle brackets $\langle \ldots \rangle_{\xi(\alpha)}$ denote statistical averaging over a random value ξ depending on parameter α. This can be understood as the parameter dependence of the probability density of these random value.

$$\langle \delta(x - \xi) \rangle_{\xi(\alpha)} = \langle \delta(\xi - x) \rangle_{\xi(\alpha)} = W_\xi(x; \alpha). \tag{A.20}$$

$$\langle f(\xi) \rangle_{\xi(\alpha)} = \int_{-\infty}^{\infty} f(u) \, W_\xi(u; \alpha) \, \mathrm{d}u. \tag{A.21}$$

$$\langle 1(x - \xi) \rangle_{\xi(\alpha)} = F_\xi(x; \alpha) = \int_{-\infty}^{x} W_\xi(u; \alpha) \, \mathrm{d}u. \tag{A.22}$$

$$\langle 1(\xi - x) \rangle_{\xi(\alpha)} = 1 - F_\xi(x; \alpha) = \int_{x}^{\infty} W_\xi(u; \alpha) \, \mathrm{d}u. \tag{A.23}$$

$$\left\langle \frac{\mathrm{d}\eta}{\mathrm{d}\alpha} \varphi(\eta) \right\rangle_{\eta(\alpha)} = \frac{\mathrm{d}}{\mathrm{d}\alpha} \left\langle \int_{-\infty}^{\eta} \varphi(u) \, \mathrm{d}u \right\rangle_{\eta(\alpha)}. \tag{A.24}$$

$$\left\langle \frac{\mathrm{d}\eta}{\mathrm{d}\alpha} \delta(v - \eta) \right\rangle_{\eta(\alpha)} = -\int_{-\infty}^{v} \frac{\partial}{\partial \alpha} W_\eta(u; \alpha) \, \mathrm{d}u. \tag{A.25}$$

Supplement 1 Nonlinear gravitational instability of random density waves in the expanding Universe

Adrian L. Melott[1]
and
Sergei F. Shandarin[2]

Abstract

We present the results of a series of very high-resolution simulations of gravitational clustering in two dimensions [1]. These 2D simulations are equivalent to a cross-section of an $\Omega = 1$ Friedmann Universe which is homogeneous perpendicular to the plane of the calculation. Our simulations have equal resolution in space and in mass, achieved by having both large numbers of particles and mesh cells when solving the Poisson equation. The evolution of structure at late nonlinear stages includes filament-like structures suggesting that filaments and pancakes (in 3D) are generic structures.

Gravitational instability and formation of the large-scale structure in the Universe

Gravitational instability plays a very important role in the Universe, resulting in the growth of primordial density perturbations and formation of the observable structure (e.g [2]). Primeval perturbations presumably arose as vacuum fluctuations during the very early inflationary stage when the Universe was expanding exponentially (e.g. [3]). At present, the scale factor characterising the Hubble (i.e. isotropic and homogeneous) expansion of the Universe is $a(t) \propto t^{\frac{2}{3}}$ (here t is time), assuming that the mean density is close to the closure value when the kinetic energy of the expansion is exactly balanced by the gravitational potential energy. The density perturbations had a long and dramatic history before they became galaxies, clusters of galaxies, superclusters, or voids. The evolution of density perturbations and the

[1] Department of Physics and Astronomy
University of Kansas
[2] Institute for Physical Problems, Moscow, USSR*;
Department of Physics and Astronomy, University of Kansas
*Permanent Address

formation of the structure in the Universe is one of the most important problems in modern cosmology.

There are several rival theories to explain this process. They can be classified by the type of primordial perturbations. Some theories are based on the assumption that primeval fluctuations were random fields of the Gaussian type. Others (e.g. the model of cosmic strings) assume that initial perturbations were non-Gaussian. Here we discuss some problems arising in models based on the assumption of Gaussian fluctuations.

One of the most difficult problems arising in understanding such models is the analysis of the evolution of perturbations at the nonlinear stage, when the typical amplitudes of density inhomogeneities become larger than mean density in the universe: $\delta\rho/\rho \geq 1$. The most straightforward way to solve the problem is to do 3D numerical simulations. Usually in simulations of this type the medium is assumed to consist of collisionless particles, in agreement with the popular hypothesis that most of the mass in the Universe is in the form of weakly interacting particles such as massive neutrinos, photinos, or axions. However, 3D *N*-body simulations take much storage and computer time, since the number of particles needs to be large. A practical solution to this problem is either to perform detailed simulations of an isolated object or else to make rough simulations in a rather large volume containing many objects. The former suffers from unrealistic assumptions about the environment while the latter has poor resolution of the internal structure of individual objects.

Numerical model

Our numerical simulations were performed in co-moving coordinates with a two dimensional PM (particle mesh) code [4]. (Co-moving coordinates assume that the general expansion of the Universe is excluded by simple transformation $\mathbf{r}_{\text{comoving}} = \mathbf{r}_{\text{phys.}}/a(t)$, where $a(t)$ is the scale factor.) The code solves for the force by Fourier transform. In the models shown, we have $512^2 = 262,144$ particles on an equivalent mesh. Our models are equivalent to a cross-section of an Einstein–de Sitter Universe homogeneous perpendicular to the plane and doubly periodic in the plane.

Initial conditions were of nine types. All models had random smooth Gaussian perturbations. All had pure power-law spectra $\langle \delta_k^2 \rangle \propto k^n$ where $n = 2$ (Q series), $n = 0$ (J series), or $n = -2$ (N series). In each case, three alternative cut-offs were imposed on the initial spectrum: waves longer than $L/4$, $L/32$, or $L/256$ were kept where L is the side of the square. All models began with $\frac{\delta\rho}{\rho} = 0.25$ on the scale of cells, and were evolved and saved at expansion factors $a_n = a_0 2^{n-1}$ or extrapolating the linear theory up to $(\frac{\delta\rho}{\rho})_{\text{linear theory}}$ of 32, as $(\frac{\delta\rho}{\rho})_{\text{linear theory}} \propto a$. Thus, file N4F5 corresponds to a power law $n = -2$, cut off at $L/4$, and stopped at an expansion factor of 16. Amplitudes and phases of waves are identical for models with identical cut-offs.

Our resolution is *L*/512. It is important to note that we have not only correct gravity to that *length* scale, but also we have enough particles to resolve *mass* to that scale. The method we use is by far the fastest when mass is resolved as well as length. When mass is not resolved as well as length, 2-body scattering may mean that the code is an incorrect description of a Vlasov-type system as a dark matter Universe. Further discussion of some numerical issues can be found in references [5] and [6].

The 2D simulations suffer the disadvantage of not being fully generic, and their results cannot be directly compared with observational structure in our 3D Universe. On the other hand, they have definite advantages in spatial resolution and in visualising the process. The results of our simulations with truncated initial power spectra (in particular the *L*/4 series) most nearly describe the dynamics of collisionless particles in the hot dark matter (HDM) scenario. This theory is presently unpopular, but cannot be considered totally excluded. However the *L*/32 and especially *L*/256 series are useful for understanding gravitational processes in the cold dark matter (CDM) scenario. In any case, these simulation highlight new effects in a collisionless self-gravitating medium at advanced nonlinear stages. We have also found it very useful to make different simulations (e.g. with different spectra) keeping the amplitude and phases the same.

In this paper, we present a qualitative discussion of the phenomena found in our simulations. Pattern recognition is not a simple question. No statistics exist which can adequately describe all the patterns and structures one can easily see in the pictures. Since developing statistics of this type is a separate complicated problem, it seems useful to show pictures which provide a great deal of information. We hope our pictures will encourage the development of new quantitative statistics as well as intuition in the very complicated field of nonlinear dynamics in a self-gravitating medium.

Internal structure of mass concentrations

The *L*/4 series simulations address the problem of the evolution of internal structure in the densest regions. This is very important for the theory of galaxy formation in the HDM model, including variants with unstable particles. In addition, it is a very interesting theoretical question related to the structures of generic singularities in a self-gravitating medium [7].

Two pictures (Figs S1.1a and b) show the time evolution of density perturbations in the J4 model. One can easily see the complicated internal structures arising at late nonlinear stages. However, even the photographic plates used here cannot show all the detail which exists in the computation.

Careful inspection of previous stages (not shown) gives evidence that the first objects to form at the beginning of the nonlinear stage were pancakes but located close to the places where clumps would later form. It is well known [7–9] that pancakes form from Lagrangian regions containing maxima

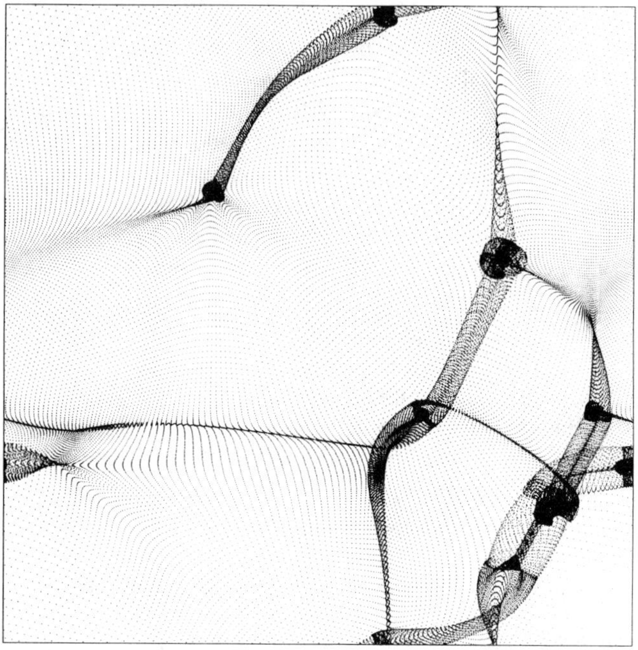

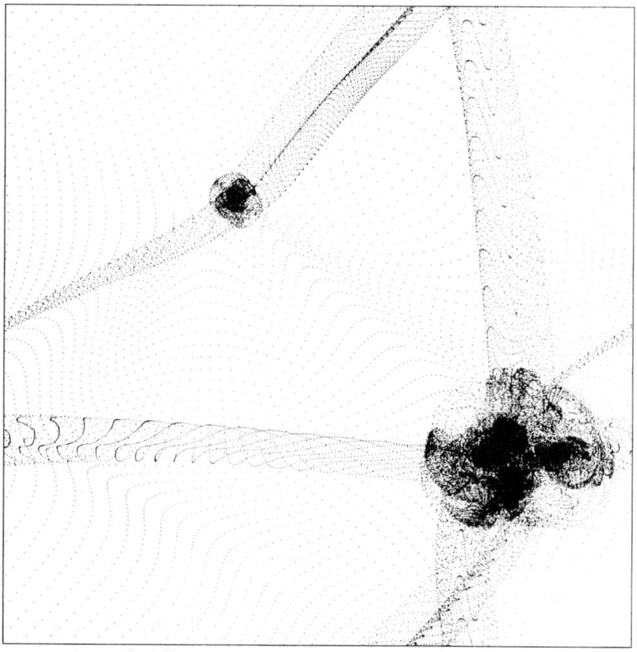

Fig. S1.1 (a) J4F4, (b) J4F6 (see text for notation)

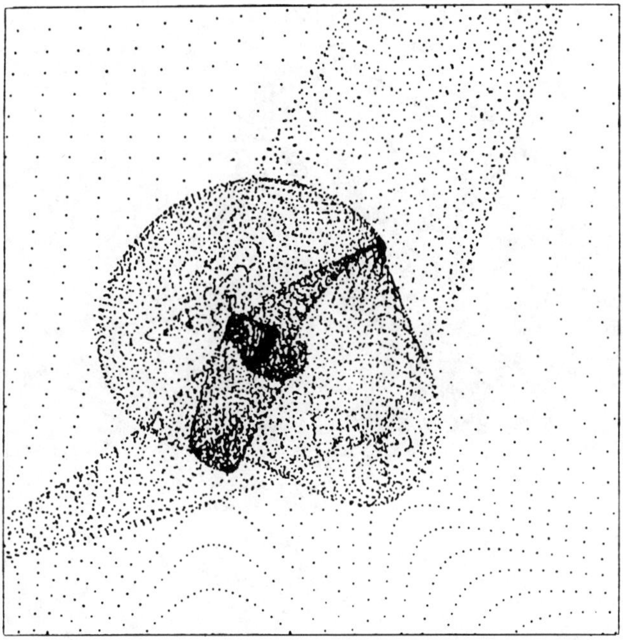

Fig. S1.2 J4F4 Centre 171, 363 (in cell units). The image is magnified by a factor 16 compared to Fig. S1.1a

of the largest eigenvalue of the deformation tensor. It is determined by one-dimensional contraction along the direction determined by the eigenvector related to this eigenvalue. The formation of clumps is connected with contraction along the other principal direction and probably begins near the Lagrangian sites of maxima of the second eigenvalue, as predicted by the Zel'dovich [8] approximation. In generic fields, maxima of the largest eigenvalue rarely coincide with maxima of the smallest one, so formation of the clumps does not always happen precisely at the sites of the pancake formation. However, the Lagrangian positions of the maxima of the second eigenvalue probably correlate with the positions of the maxima of the largest eigenvalue. Thus, soon after formation the oldest part of the pancake changes its shape and becomes a clump of mass embedded in a highly asymmetrical halo. Usually these parts of pancakes become nodes of the cellular structure.

But even pictures with high resolution (Figs S1.1a and b) cannot show the real structures that arise in the densest regions. Some of these structures are shown in Figs S1.2 and S1.3. These pictures show small pieces cut out of Fig. S1.1a. Figure S1.2 is a magnified view of the part that can be seen in Fig. S1.1a lying close to the main diagonal at about a quarter of the distance form the upper left-hand corner. Figure S1.3 shows the structure in the lower right-hand corner of Fig. S1.1a.

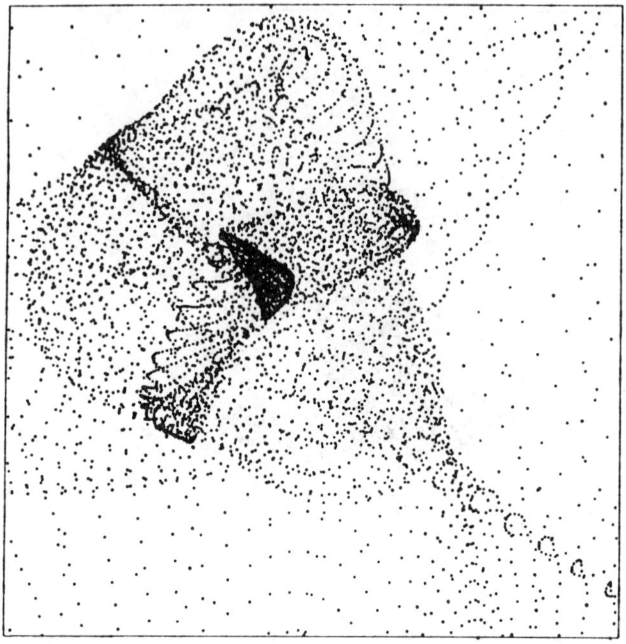

Fig. S1.3 J4F4 Centre 416, 64. The image is magnified by a factor of 32 compared to Fig. S1.1a

The first impression of these pictures is that they are complicated constructions of generic caustic patterns arising in geometrical optics or in noninteracting media [7]. In order to make more definite conclusions, additional analysis is needed, especially in the clusters. Some structures are so complicated that one cannot be sure that totally new structures are absent. Also cannot be sure that our resolution is enough to show all structures in the densest regions.

One common feature of these structures attracts attention: they look like a sequence of very elongated structures embedded one into another. It seems that they formed as a result of folding of the phase hypersurface in two orthogonal directions similar to what is known in one-dimensional collisionless self-gravitating systems [10, 11]. This may explain the very strong coherence in the orientations of the structures with different scales. It is tempting to relate this phenomenon with the coherence of the Local Supercluster and much larger structures found by Tully [12], and possibly also the 'sandwich-like' structure found by de Vaucouleurs [13].

Further evolution shows that the first clumps acquire mass by non-spherical accretion from dense bridges connecting the clumps.

Comparing the structures arising from different initial spectra shows that the larger the value of the spectral index n, the more 'spiky' the structure.

Global structure

We have also studied models of types 32 and 256 (the latter with initial power right up to the Nyquist frequency). Such simulations address the question of long nonlinear evolution, which must inevitably arise in the CDM model. The large number of particles is important because at late nonlinear stages most of the mass becomes concentrated in large clumps. Thus, weak filaments as well as small clumps can easily be lost in simulations with low mass resolution. We believe we can demonstrate that pancake-like structures can be expected in models with spectra flatter than $n = 0$ ($n \leq 0$ in 2D, which is equivalent to $n \leq -1$ in 3D) even in hierarchical clustering. Figures S1.4 and S1.5 show models with $n = -2$ and $n = 0$ power spectra, with the same phases.

We have used simple power-law spectra for initial conditions with $n = 0$ and $n = 2$ because they are similar to $n = -1$ and $n = -3$ in 3D, in the following sense: integrals of the spectrum determining statistical properties of random fields behave in the same way. The linear spectrum of CDM after decoupling gradually changes from $\langle\delta_k^2\rangle \propto k^{-3}$ to $\langle\delta_k^2\rangle \propto k^{-1}$ or k^0 as the scale increases from globular clusters through galaxy clusters and superclusters. We believe that study of simple power-law spectra can give rough but useful ideas about features of dynamics with other spectra, because they clearly show the dependence of morphology on spectral index.

One of the interesting features of the N32 simulation is that at the beginning of the nonlinear stage (Fig. S1.4a), small pancakes having the size of the short-wave cut-off form a pattern that definitely reflects the long-wave part of the spectrum which one can easily see in Fig. S1.4b.

However, the most remarkable feature of the J32 simulation (which is similar to $n = -1$ in 3D) is that it demonstrates the existence of cellular structure for rather long time. The scale factor between Figs S1.5a and S1.6b has increased 16 times. But the cellular structure is still obvious. At this stage the typical sizes of the cells are noticeably large than in Fig. S1.5a and are not related to the initial cut-off. Obviously, structures of clumps in Fig. S1.5b coincide with filaments in Fig. S1.4b, and are related to long-wave amplitudes. Also the filaments appear to be oriented parallel to nearly linear sequences of massive clumps. This can justify the presence of filaments in CDM simulations [14, 15].

Summary

Summarising our results, we wish to stress the main points. Assuming that initial density fluctuations represent a random Gaussian field one can formulate general features of the density distribution at the nonlinear stage of gravitational instability as follows.

If the initial spectrum of density fluctuations has a cut-off of small-scale

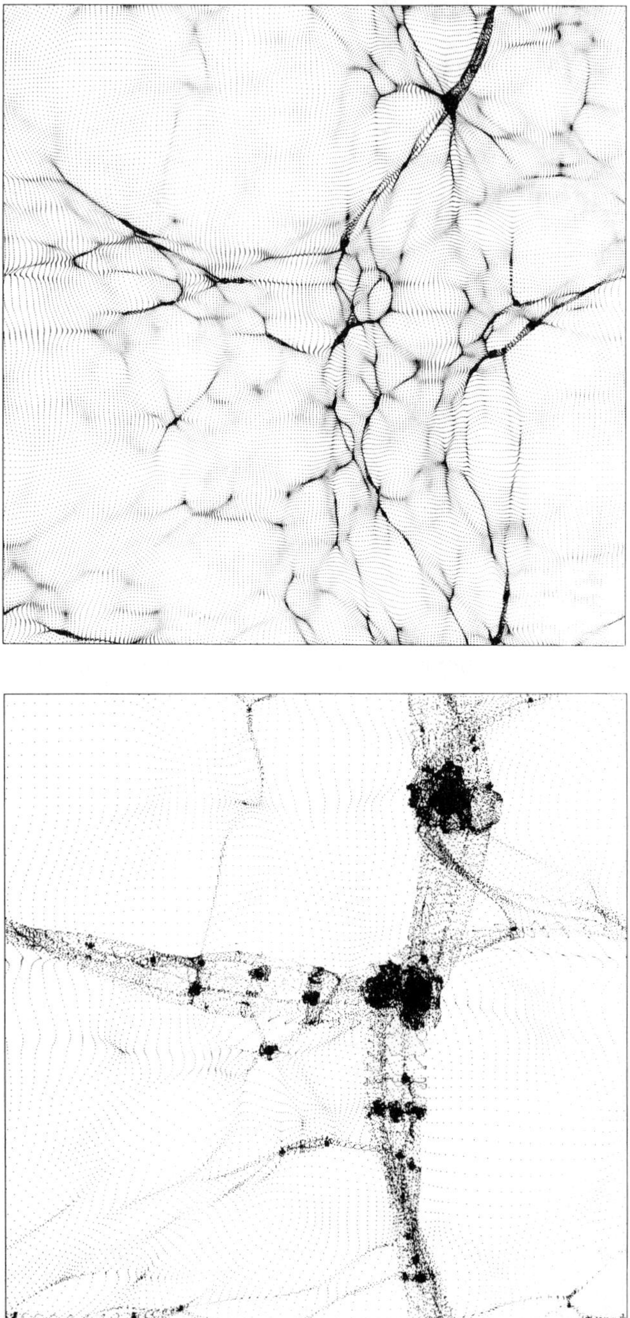

Fig. S1.4 (a) N32F3, (b) N32F5

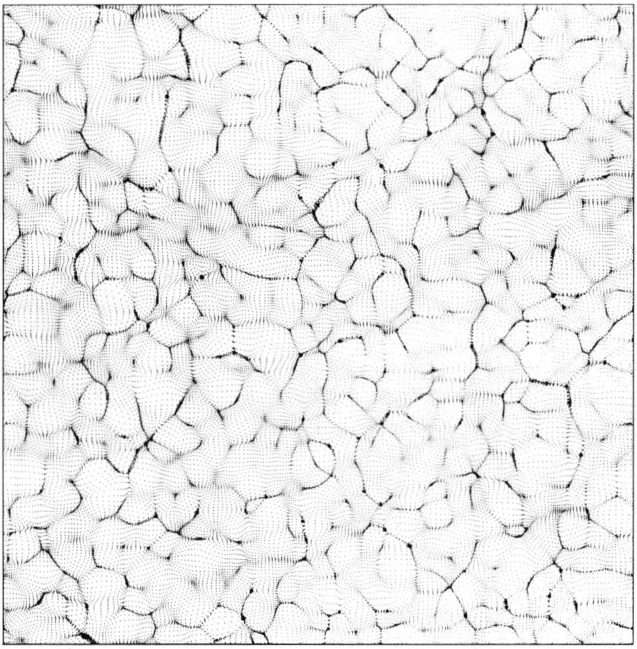

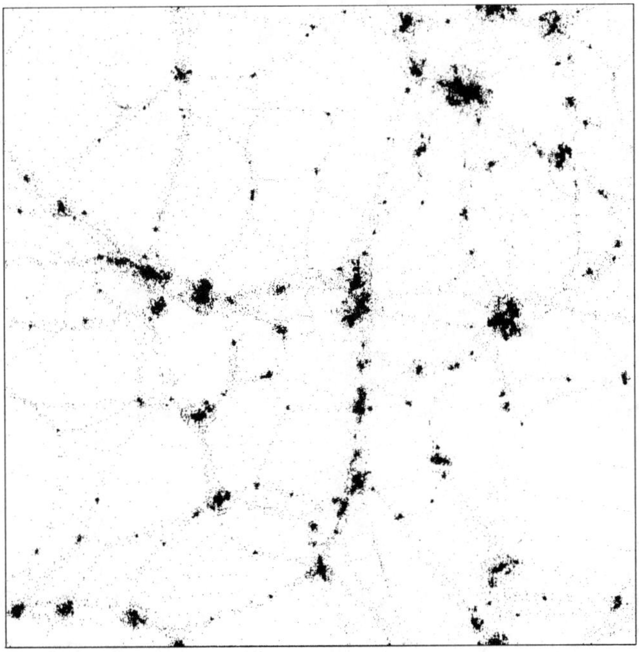

Fig. S1.5 (a) J32F3, (b) J32F7

perturbations at some scale k_{cutoff} then the nonlinear stage begins with formation of pancakes of size $\lambda_{cutoff} = 2\pi/k_{cutoff}$. Very quickly the pancakes merge into a cellular structure (Figs S1.1a, S1.4a and S1.5a): large under-dense regions are surrounded by thin walls of high density.

In the course of time the cellular structure evolves: some cells (usually smaller ones) shrink, others expand resulting in the structure with larger (in co-moving coordinates) cells. It is very important that during this process the walls of the structure remain thin compared to the sizes of under-dense regions (Figs S1.1b, S1.4b and S1.5b). This observation can be explained as a result of the action of gravity which prevents the walls from expanding quickly.

Neglecting the details of the internal structure of dense regions one can develop an analytical model for the global evolution of the structure which combines the Zel'dovich approximation for nonlinear gravitational instability and the Burgers model for potential turbulence [16] which is in good agreement with the results of N-body simulations [17–19].

The detailed study of the internal structures of the dense regions has shown that they can be strongly aligned with the overall distributions of mass.

In terms of the fraction of mass associated with various types of structure the evolution proceeds roughly as follows. After the formation of the cellular structure, mass leaves the interiors of the cells for the walls and later leaves the walls for clumps. The clumps themselves merge with each other resulting in the growth of typical mass of clumps. Finally most of the mass ends up in the clumps.

We acknowledge NASA Grants NAGW-1288, NAGW-1793, for financial support and the National Center for Supercomputing Applications for CRAY-2 time.

References to Supplement 1

[1] Melott, A. L. & Shandarin, S. F. *Astrophys. J.*, **343** (1989), 26.
[2] Peebles, P. J. E. *The Large-Scale Structure of the Universe*. Princeton, NJ: Princeton University Press, 1980.
[3] Brandenberger, R, *J. Phys.*, **G15** (1990), 1.
[4] Hockney, R. W. & Eastwood, J. W. *Computer Simulation Using Particles*. New York: McGraw-Hill, 1981.
[5] Peebles, P. J. E., Melott, A. L., Holmes, M. R. & Jiang, L. R. Preprint, 1988.
[6] Melott, A. L. *Comments on Astrophysics*, in press.
[7] Arnold, V. I., Shandarin, S. F. & Zel'dovich, Ya. B. *Geophys. Ap. Fluid Dynamics*, **20** (1982), 111.
[8] Zel'dovich, Ya. B. *Astron. Ap.*, **5** (1970), 84.
[9] Shandarin, S. F. & Zel'dovich, Ya. B. *Rev. Mod. Phys.*, **61** (1989), 185.
[10] Doroshkevich, A. G., Kotok, E. V., Novikov, I. D., Polyudov, A. N., Shandarin, S. F. & Sigov, Yu. S. *Mon. Not. R. Astr. Soc.*, **192** (1980), 321.
[11] Melott, A. L. *Astrphys. J.*, **264** (1983), 59.

[12] Tully, B. *Astrophys. J.*, **323** (1987), 1.
[13] de Vaucouleurs, G. *Bull. Astr. Soc. India*, **9** (1981), 1.
[14] Melott, A. L., Einasto, J., Saar, E., Suisalu, I., Klypin, A. A. & Shandarin, S. F. *Phys. Rev. Lett.*, **51** (1983), 935.
[15] White, S. D. M., Davis, M., Efstathiou, G. & Frenk, C. S. *Nature*, **330** (1987), 451.
[16] Gurbatov, S. N., Saichev, A. I. & Shandarin, S. F. *Mon. Not. R. Astr. Soc.*, **236** (1989), 385.
[17] Kofman, L., Pogosyan, D. & Shandarin, S. *Mon. Not. R. Astr. Soc.*, **242** (1990), 200.
[18] Nusser, A. & Dekel, A. Preprint, 1989.
[19] Weinberg, D. H. & Gunn, J. E. *Astrophys. J.*, in press.

Supplement 2 Singularities and bifurcations of potential flows

V. I. Arnold, Yu. M. Baryshnikov
and
I. A. Bogayevsky

1 Singularities of caustics

Singularities of potential flows in collisionless media are described by the theory of caustics (see [1], [2]). The typical singularities of caustics in the 3-space – the swallowtail, the pyramid and the purse – are depicted in Fig. S2.1.

A caustic may be described in terms of classical mechanics as a set of critical values of the projection of a Lagrange submanifold from the phase space to the configuration space, $(p, q) \to q$ (see Fig. S2.2). A Lagrange submanifold is a submanifold of the phase space, of dimension one-half the dimension of the phase space, along which the integral $\int p\mathrm{d}q$ does not depend (locally) on the integration path. The graph of a potential velocity field $p = \partial S/\partial q$ is a Lagrange submanifold. A general Lagrange submanifold may be viewed as a graph of the multi-valued potential velocity field. The Kolmogorov tori of integrable or almost integrable Hamiltonian systems are also Lagrange submanifolds.

The theory of singularities and of bifurcations of caustics is the basis of the theory of the large-scale structure of the Universe, due to Ya. B. Zel'dovich [3]. The evolution of a Lagrange submanifold under any Hamiltonian phase flow leaves it a Lagrange submanifold (because of the conservation of the Poincaré integral invariant. The slowest particles are overrun by the fastest ones, and the caustic appears in the course of the time, even in the case when the initial Lagrange submanifold was the graph of a potential velocity field and hence had no caustic (Fig. S2.3). The density of the particles has a singularity of order $1/\sqrt{r}$ at distance r from the caustics.

After the formation of the caustics, most of the particles are concentrated in the neighbourhood of the surfaces of the caustics. The density is even larger at the singular lines of the caustics, and maximal at the most singular points where these lines intersect each other.

Thus a complicated large-scale cellular structure emerges with large voids

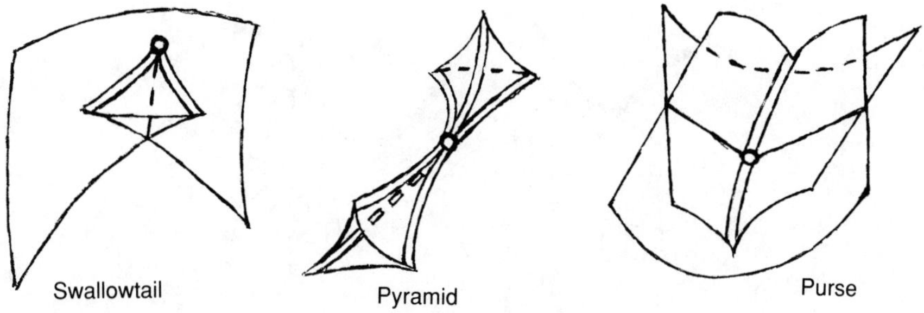

Fig. S2.1

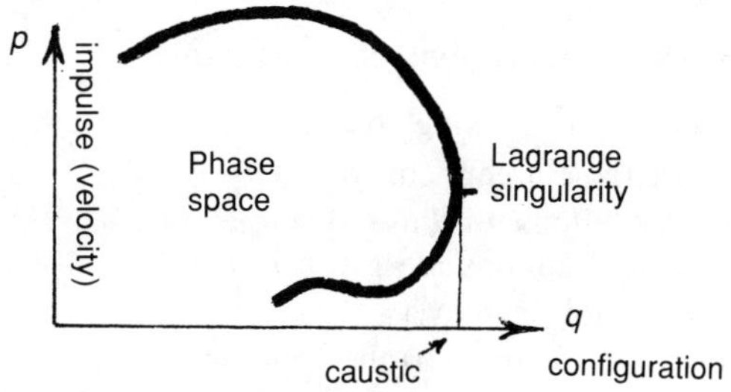

Fig. S2.2

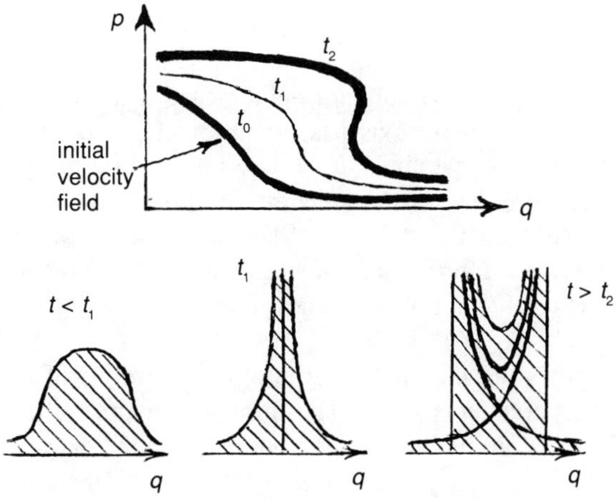

Fig. S2.3

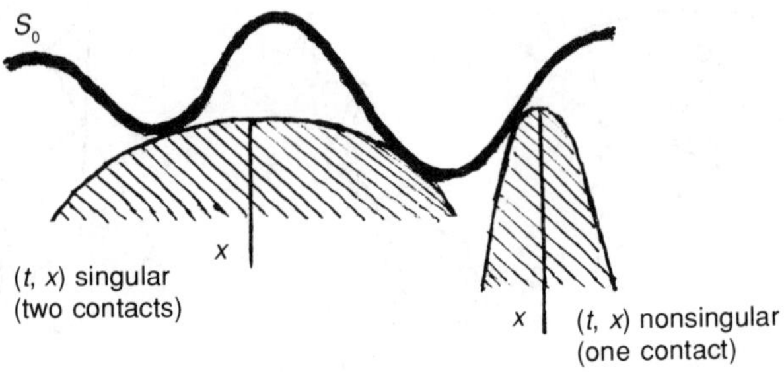

Fig. S2.4

inside the cells and a net of surfaces and lines of concentration of matter. This unusual structure is consistent with the astronomical observations.

2 Singularities and bifurcations of shock waves

The simplest model of sticking particle flow is described by potential solutions $\mathbf{v} = \nabla S$ of the Burgers equation $\mathbf{v}_t + (\mathbf{v}.\ \nabla)\ \mathbf{v} = \mu\Delta\mathbf{v}$ with vanishing viscosity μ. (The solutions of this equation and their statistical properties under random initial conditions are given in Chapter 6 of the present book – Gurbatov, Malakhov and Saichev.)

It is well known that for $\mu \to 0$ the solution acquires singularities, namely the shock waves. The Florin formula $S = -2\nu \ln \Phi$ [4] reducing the Burgers equation to the heat equation (and usually called the Hopf–Cole transformation) implies the description of the shock wave in space–time as via the set of non-smoothnesses of the function

$$F(x, t) = \min_y f(y, x, t), \quad f(y, x, t) = S_0(y) + \frac{(x - y)^2}{2t}.$$

Geometrically (Fig. S2.4) this formula means that the graph of $S_0(\cdot) - f(\cdot, x, t)$ is a parabola whose axis has the abscissa x and whose width grows with t. One pushes the parabola up till it touches the graph of the initial function $S_0(\cdot)$. The ordinate of the touching parabola vertex is equal to $F(x, t)$. It depends smoothly on x and t, if the parabola touches the graph at exactly one point. If there are more points of tangency the function F is, generically, non-smooth. Such values of x and t form the shock wave hypersurface in the space–time.

The function F is the minimal function of the family f of real functions on y, depending on parameters (x, t). The minima function of any smooth family of smooth functions is continuous, but generically, is not smooth. There exist only five typical singularities of the minimal function, depending on three parameters (this case corresponds to the space–time of dimension $3 = 2 + 1$, that is, to the shock waves propagating in a 2-plane). The sets of discontinuity of the derivatives of these minimal functions are represented in Fig. S2.5. The same picture describes the typical singularities of a momentary shock wave in the 3-space. There exist eight typical singularities of the minimal functions in the case of four parameters (Table 1, second column). The singularities of the sets of non-smoothness of the minimal functions of typical smooth families of smooth functions depending on four parameters are those presented in the table (the proof is given in [2], sect. 17.2).

The shock wave travelling in x-space experiences perestroikas at some discrete time instants. (The word 'perestroika' was always used in Russian mathematics as an equivalent of bifurcation and metamorphosis. Now we no longer need to translate it into English.) To investigate these perestroikas one should intersect the shock wave hypersurface in the space–time with the isochrones t = const. The shock wave hypersurface in space–time is the set of non-smoothness of the minimal function F. If S_0 is a generic initial function, the minimal function F has only the typical singularities. To investigate the generic perestroikas in the 2-space or in the 3-space we consider the typical singularities of the minimal functions and generic time functions, described in Theorem 1.

Theorem 1 (see [5], Chapter 2, section 3; the proof is given in [6]). A generic smooth function in the 3-space or in the 4-space, containing a typical shock wave hypersurface is reducible in some neighbourhood of a fixed point to the normal form given in Table 1 (columns 3, 4) by a continuous change of variables, preserving the shock wave hypersurface.

Thus all the perestroikas of a shock wave hypersurface of dimension 2 or 3 generated by the typical initial function are presented in Figs S2.5 and S2.6. S. N. Gurbatov and A. I. Saichev discovered in 1984 that only some of these perestroikas are observable [7].

In some cases, such as the vanishing of a triangle formed by the shock waves, the perestroika is possible only in one direction: the triangle may not appear. In other cases both directions are admissible; in yet other cases, none. Figs S2.5 and S2.6 the admissible perestroika directions are indicated by the arrows.

The admissible directions of the shock wave perestroikas depend on the convexity properties of the Hamilton function defining the Hamilton–Jacobi equation, restricted to the space of impulses (for the case of the Burgers equation this Hamilton function is $P^2/2$).

Theorem 2 (see [5], Chapter 2, section 3; the proof is given in [6]). A perestroika direction is admissible if and only if the local shock wave, born after the perestroika is contractible, is in a neighbourhood of the perestroika point.

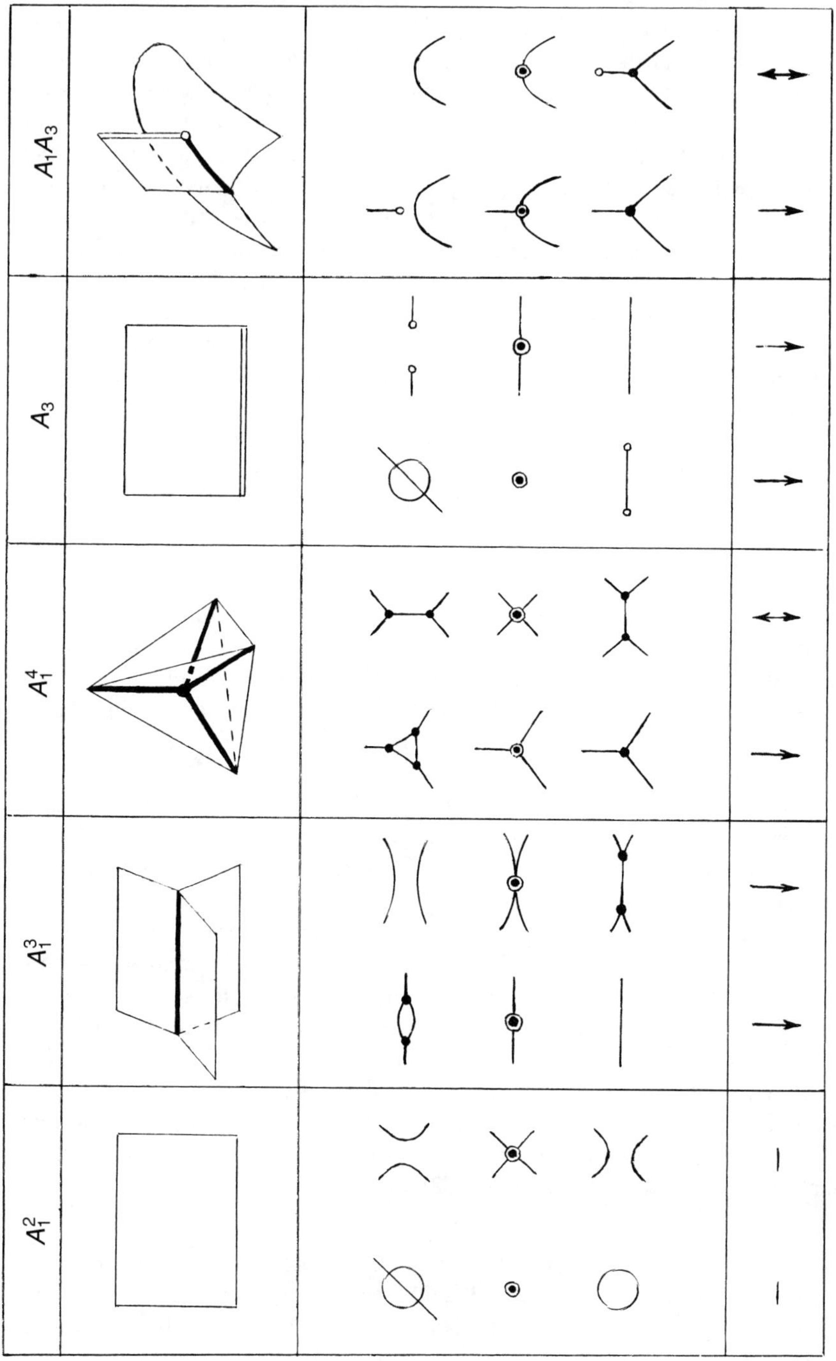

Fig. S2.5

Table 1.

Name	Minima function	Time function in the 3-space	Time function in the 4-space
A_1^2	$\min\{a, b\};\ a + b = 0$	$(a - b) \pm c^2 \pm d^2;\ c$	$(a - b) \pm c^2 \pm d^2 \pm e^2;\ c$
A_1^3	$\min\{a, b, c\};\ a + b + c = 0$	$(a \pm b - c) \pm d^2;\ d$	$(a \pm b - c) \pm d^2 \pm e^2;\ d$
A_1^4	$\min\{a, \ldots, d\};\ a + \ldots + d = 0$	$a \pm b \pm c - d$	$(a \pm b \pm c - d) \pm e^2;\ e$
A_1^5	$\min\{a, \ldots, e\};\ a + \ldots + e = 0$	—	$a \pm b \pm c \pm d - e$
A_3	$\min_x(x^4 + ax^2 + bx)$	$\pm a \pm c^2;\ c$	$\pm a \pm c^2 \pm d^2;$
A_1A_3	$\min\left\{\min_x(x^4 + ax^2 + bx), c\right\}$	$\pm a \pm c$	$\pm a \pm c \pm d^2;\ d$
$A_1^2A_3$	$\min\left\{\min_x(x^4 + ax^2 + bx), c, d\right\}$	—	$\pm a \pm c \pm d$
A_5	$\min_x(x^6 + ax^4 + bx^3 + cx^2 + dx)$	—	$\pm a$

Yu. M. Baryshnikov's efforts to explain Bogayevsky's veto led him to the following statements.

Let M be a smooth manifold, N be a parameter space, $t \in \mathbb{R}$ be the time variable. Let us consider the following smooth functions: $f\colon M \to \mathbb{R}$ which describe an 'initial data', and $K\colon M \times N \times \mathbb{R} \to \mathbb{R}$ an action. Put $h(x, y, t) = K(x, y, t) - f(x)$. We say that the point x is the non-degenerate minimum of $h(\cdot, y, t)$ is the global minimum of that function and if x is a non-degenerate critical point of h. The set $F = \{(y, t)\colon h(\cdot, y, t)$ does not possess a non-degenerate minimum$\}$ $N \times \mathbb{R}$ is called the (large) shock wave hypersurface; its sections $F_t = F \cap \{t\}$ are called the instant fronts.

For every $t \in \mathbb{R}$ the mapping θ_t of the complement of the instant front into M is defined, which associates the (unique) point x of the global minimum of $h(\cdot, y, t)$ with the parameter value y. The mapping θ_t is smooth: that is the consequence of the non-degeneracy of x.

We shall say that the action K is non-degenerate if θ_t maps $N \setminus F_t$ onto its image diffeomorphically (for any initial data).

The image of θ_t we shall call the support set $S_t \subset M$ for the action K and the initial data f.

The action K is called Huygensian if the support set for any initial data $f = K(\cdot, y, t)$ and for any y, t is the whole space M for every $t' < t$.

Theorem 3. For a Huygensian action the support set S_t does not increase, i.e. $S_{t'} < S_t$ when $t' < t$.

Corollary. Let the action be Huygensian and non-degenerate, and $t = 0$ be an isolated point of the perestroika of the instant front. Then the homotopy

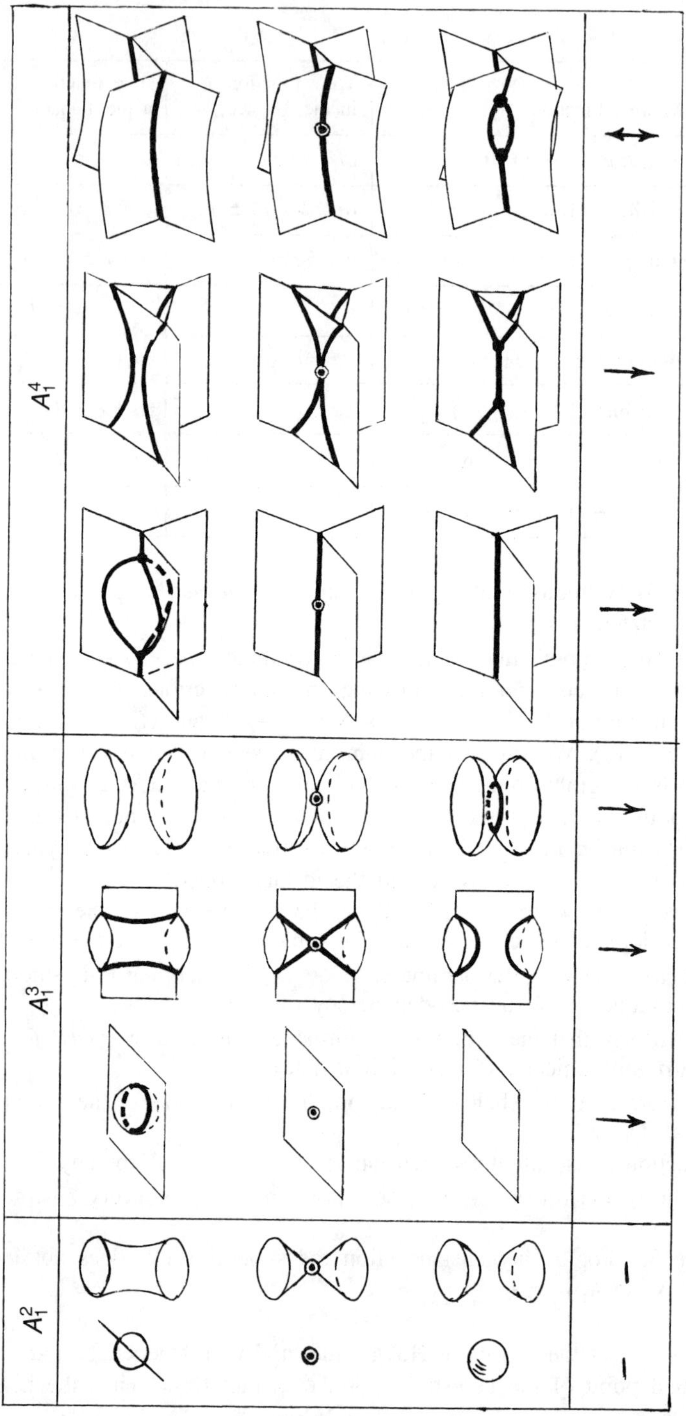

Fig. S2.6

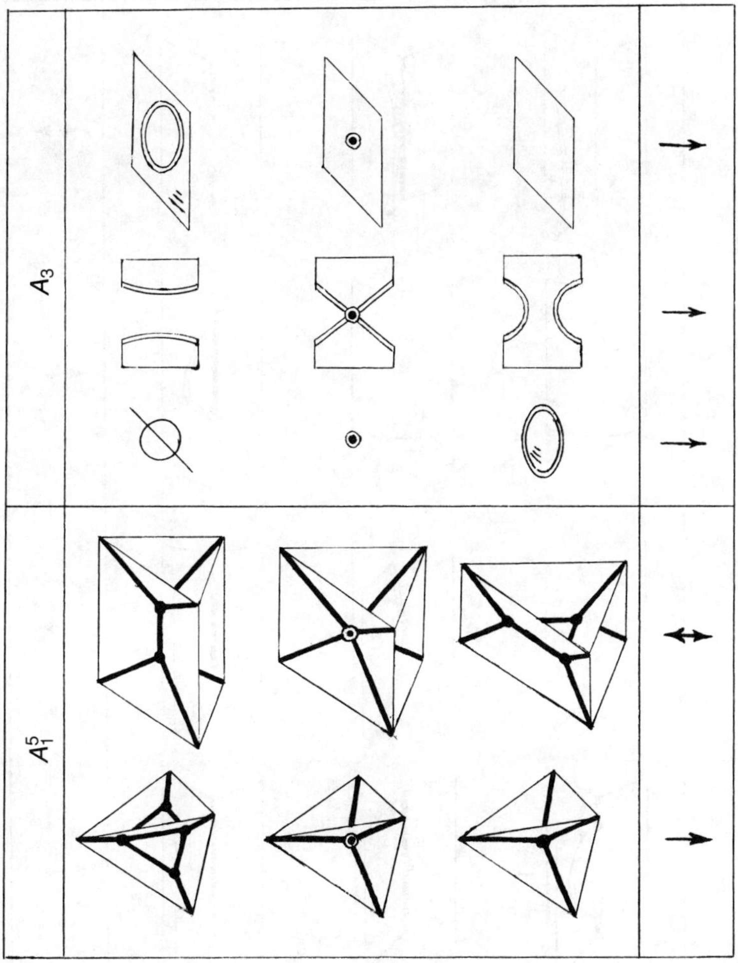
A_3
A_1^5

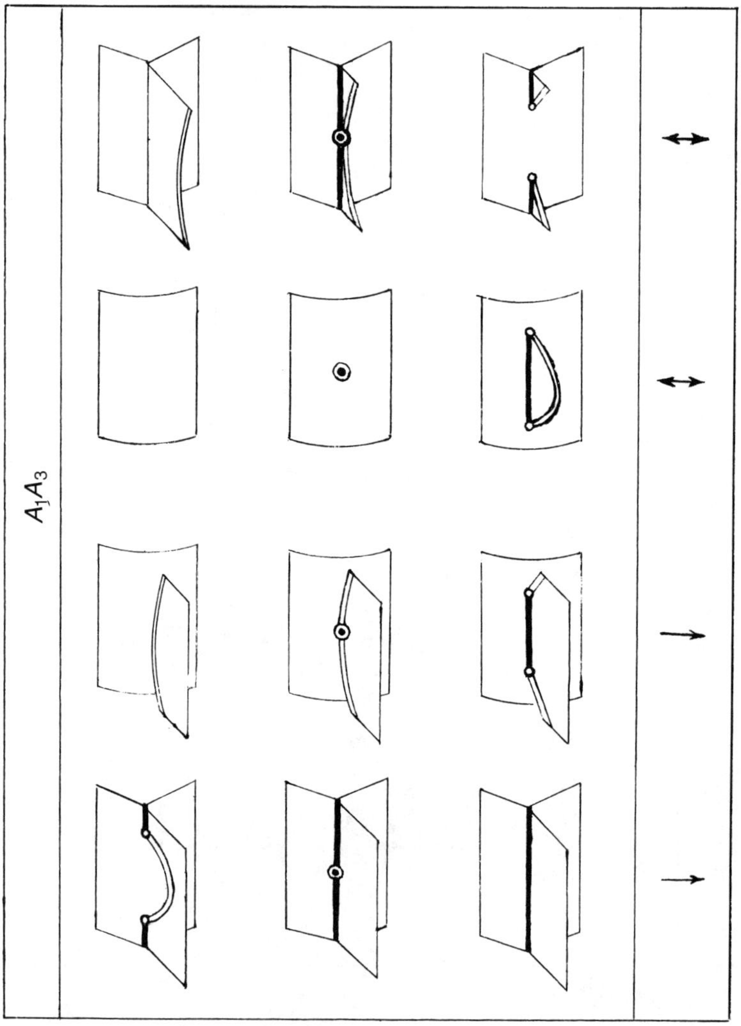
A_1A_3

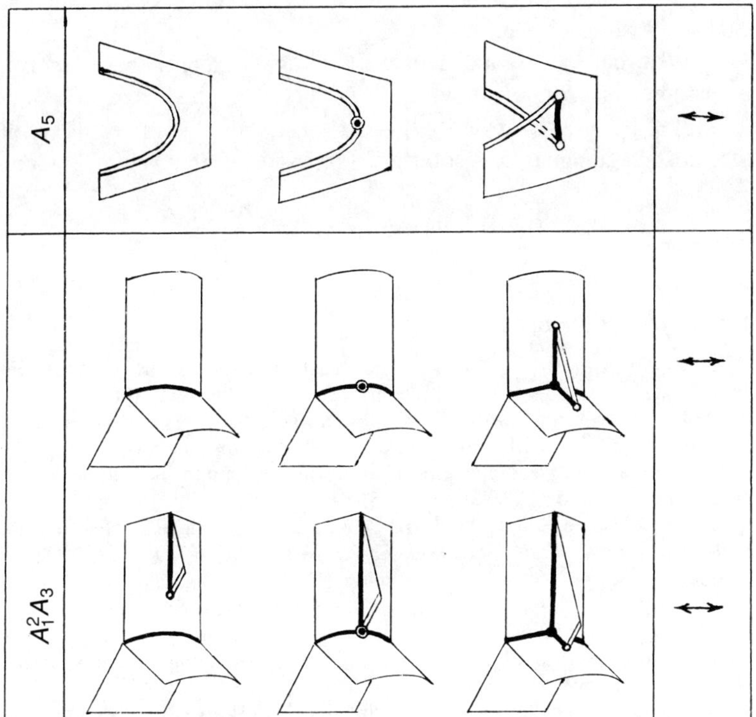
A_5
$A_1^2A_3$

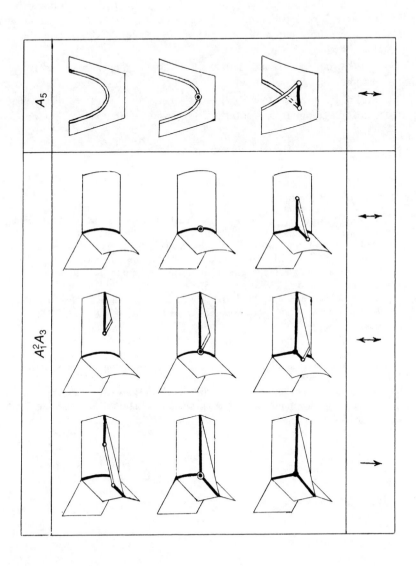

type of the complement of the instant front when $t = 0$ is equal to the homotopy type of the complement of the instant front when $t > 0$.

The theorem shows that the complement is diffeomorphic to the set of lesser values for a function on M (which *a priori* is not-ever continuous). But for the generic action and initial data in the small dimensions of space–time ($\leq 3 + 1$) that function $w(x) = \sup\{t: xS_t\}$ is continuous and topologically equivalent to a Morse function. So, Bogayevsky's veto follows the known fact that the evolutions of the set of lesser values for a Morse function are obtained via pasting of a disk.

The theorem can be expanded to the actions K which depend upon additional parameters, for example, being curves going from x to y. If $K(x, y, \lambda, t) = \int_0^t z(\lambda, \dot{\lambda}, t)\, d\tau$; $\lambda: [0, t] \to M$; then K always is Huygensian provided that Lagrangian Z is 'optical'. For instance, if $z = |\dot{\lambda}|^2/2$ then

$$K(x, y, t) = \inf_{\lambda} K(x, y, \lambda, t) = \frac{|x - y|^2}{2t}.$$

References to Supplement 2

[1] Arnold, V. I., *Catastrophe Theory.* Translated from Russian by, G. Wassermann, based on a translation by R. K. Thomas, Springer, 2nd rev., 1986.

[2] Arnold, V. I., Varchenko, A. N. & Gusein-Zade, S. M. *Singularities of Differentiable Maps. Vol. I* Moscow: Nauka, 1982; Boston: Birkhäuser, 1985.

[3] Zel'dovich, Ya. B. Decomposition of uniform matter into pieces due to gravitation. *Astrophysics* **6**, 2 (1970), 319–35.

[4] Florin, V. A. Some Simple Nonlinear problems of consolidation of watersaturated soils. *Isvestia Ak. Nauk SSSR, Otd. Tekhn. Nauk*, No. 9, (1948), 1389–97.

[5] Arnold, V. I. (ed.) *Dynamical Systems VIII. Encyclopaedia of Mathematical Sciences, Vol. 39.* Moscow: VINITI, 1989; Engl. Evans. Springer, 1990.

[6] Bogayevsky, I. A. Perestroikas of singularities of minima functions and bifurcations of shock waves of Burgers equation with vanishing viscosity. *Algebra i Analiz* no. 4 (1989).

[7] Gurbatov, S. N., Saichev, A. I. & Shandarin, S. F., Large scale structures of the Universe in model nonlinear diffusion equation. Preprint 152, IPM im. M. V. Keldysh, Moscow, 1984.

References to the main text

[1] Whitham, G. B. *Linear and Nonlinear Waves*. New York: Wiley, 1974. Russ. transl. Moscow: Mir, 1977.
[2] Kadomtsev, B. V. & Karpman, V. I. *Usp. Fiz. Nauk* **103** (1971), 193. *Sov. Phys. Usp.* **14** (1972), 40.
[3] Karpman, V. I. *Nonlinear Waves in Dispersive Media*. Moscow: Nauka, 1973.
[4] Rozhdestvensky, B. L. & Yanenko, N. N. *Systems of Quasilinear Equations*. Moscow: Nauka, 1978.
[5] Courant, R. & Hilbert, D. *Methods of Mathematical Physics*, Vol. II. New York: Interscience, 1962.
[6] Dafermos, C. M. In: *Nonlinear Waves*, S. Leibovich & A. R. Seebass (eds). Ithaca, NY and London: Cornell University Press, 1974.
[7] Burgers, J. M. *The Nonlinear Diffusion Equation*. Dordrecht: Reidel, 1974.
[8] Bateman, H. *Monthly Weather Rev.* **43** (1915), 163–70.
[9] Burgers, J. M. *Trans. Roy. Neth. Acad. Sci.* (Amsterdam), **17** (1939), 1–53.
[10] Burgers, J. M. *Proc. Roy. Neth. Acad. Sci.* (Amsterdam), **43** (1940), 2–12.
[11] Akhmanov, S. A. & Chirkin, A. S. *Statistical Phenomena in Non-linear Optics*, Moscow: Nauka, 1975.
[12] Akhmanov, S. A., D'yakov, Yu. E. & Chirkin, A. S. *Introduction to Statistical Radiophysics and Optics*. Moscow: Nauka, 1981.
[13] Kadomtsev, B. V. & Kontorovich, V. M. *Radiophys. Quant. Electr.* **17** (1974), 511.
[14] Zakharov, V. E. *Radiophys. Quant. Electr.* **17** (1974), 431–53.
[15] Mirabel, A. P. & Monin, A. S. Two-dimensional turbulence. *Usp. Mekh.* **2**, 3 (1979), 47–95.
[16] Kraichnan, R. H. & Montgomery, D. *Rep. Prog. Phys.* **43** (1980), 547–619.
[17] Gorev, V. V., Kingsep, A. S. & Rudakov, L. I. *Radiophys. Quant. Electr.* **19**, 5–6 (1976), 486–507.
[18] Monin, A. S. & Yaglom, A. M. *Statistical Fluid Mechanics*. Moscow: Nauka; Part 1, 1965; Part 2, 1967; Cambridge, Mass.: MIT Press, 1975.
[19] Malakhov, A. N. *Cumulant Analysis of Random Non-Gaussian Processes and their Transformations*. Moscow Sov. Radio, 1978.
[20] Ogura, Y. *J. Fluid. Mech.* **16** (1963), 38–41.
[21] Tatsumi, T. *Adv. Appl. Mech.* **20** (1980), 39–133.
[22] Tanaka, H. *J. Phys. Soc. Japan* **34** (1973), 373–83.
[23] Mizushima, J. & Sagami, *A. Phys. Fluids* **23**, 12 (1980), 2559–61.

[24] Kuwabara, S. Statistical hydrodynamics for Burgers turbulence, *Mem. Fac. Engin. Nagoya University* **30**, 2 (1978), 245–88.
[25] Kraichnan, R. H. *Phys. Fluids* **11**, 3 (1968), 265–77.
[26] Walton, J. J. *Phys. Fluids* **13**, 6 (1970), 1634–5.
[27] Mond, M. & Knorr, G. *Phys. Fluids* **23**, 7 (1980), 1306–10.
[28] Brissaud, A. *et al. Ann. Geophys.* **29** (1973), 539–45.
[29] Frisch, U., Lesieur M. & Brissaud A. *J. Fluid. Mech.* **65** (1974), 145–52.
[30] Bardos, C., Penel, P., Frisch, U. & Sulem, P. L. *Arch. Ration. Mech. and Anal.* **71**, 3 (1979), 237–56.
[31] Meecham, W. C. & Siegel, A. *Phys. Fluids.* **7**, 8 (1964), 1178–90.
[32] Orszag, S. A. & Bissonette, L. R. *Phys. Fluids* **10**, 12 (1967), 2603–13.
[33] Meecham, W. C., Jyer, P. & Clever, W. C. *Phys. Fluids* **18**, 12 (1975), 1610–22.
[34] Khristov, Kh. *Teoretichna i prilozhna mekhanika* (Sofia) **11** (1980), 59.
[35] Lee, J. *J. Fluid. Mech.* **101**, 2 (1980), 349–76.
[36] Grigor'yev, Yu. N. Diagram method of Prigozhin–Balesky with respect to the simplest models of hydrodynamical turbulence. Preprint no 3. Inst. teoret, i prikl. mekh. SO AN SSSR. Novosibirsk, 1979.
[37] Kollman, W. *J. Statist. Phys.* **14**, 3 (1976), 291–303.
[38] Goodarr, A. *Appl. Sci. Res.* **32**, 2 (1976), 207–15.
[39] Moiseev, S. S., Tur, A. V. & Yanovsky, V. V., *Radiophys. Quant. Electr.* **20**, 7 (1977), 1032–9.
[40] Khokhlov, R. V. *Radiotekh. Elektron.* **6** (1961), 917.
[41] Ostrovsky, L. A. *Radiophys. Quant. Electr.* **19**, 5–6 (1976), 661–90.
[42] Gaponov, A. V., Ostrovsky, L. A. & Freidman, G. I. *Radiophys. Quant. Electr.* **10**, 9 (1967), 772–93.
[43] Khokhlov, R. V. *Usp. Fiz. Nauk* **87** (1965), 17. *Sov. Phys. Usp.* **8** (1965), 642.
[44] Kataev, I. G. *Udarnye electromagnitnye volny*, Sov. Radio, Moscow, 1963. Engl. transl. *Electromagnetic Shock Waves*, London: Iliffe, 1966.
[45] Engel' brecht, J. K. *Nonlinear Wave Processes of Deformation in Solids*. London: Pitman, 1983.
[46] Tatsumi, T. & Tokunaga, H. *J. Fluid. Mech.* **65** (1974), 581–601.
[47] Tokunaga, H. *J. Phys. Soc. Japan* **41**, 1 (1976), 328–37.
[48] Kaner, V. V. Rudenko, O. V. & Khokhlov, R. V. *Akust. Zh.* **23** (1977), 756. *Sov. Phys. Acoustics* **23** (1977), 432.
[49] Nimmo, J. J. & Crighton, D. G. *Phil. Trans. R. Soc. Lond. A* **320** (1986), 1–35.
[50] Bekki, N. *J. Phys. Soc. Japan* **52**, 5 (1983), 1505–8.
[51] Zarembo, L. K. & Krasil'nikov, V. A. *Vvedenie v nelineinuyu akustiku* (*Introduction to Nonlinear Acoustics*), Moscow: Nauka, 1975.
[52] Rudenko, O. V. & Soluyan, S. I. *Teoreticheskie osnovy nelineinoi akustiki* (*Theoretical Foundations of Nonlinear Acoustics*). Moscow: Nauka, 1975; Engl. transl. New York: Plenum Press, 1977.
[53] Vinogradova, M. B., Rudenko, O. V. & Sukhorukov, A. P. *Theory of Waves*. Moscow: Nauka, 1979.
[54] Zabolotskaya, E. A. & Khokhlov, R. V. *Akust. Zhurnal*, **15**, 1 (1969), 40.
[55] Kuznetsov, V. P. *Akust. Zhurnal* **6**, 4 (1970), 548.
[56] Bahvalov, N. S., Zhileikin, Ya. M. & Zabolotskaya, E. A. *Nonlinear Theory of Sound Beams*. Moscow: Nauka, 1982.
[57] Novikov, B. K., Rudenko, O. V. & Timoshenko, V. I. *Nonlinear Underwater Acoustics*. Leningrad Sudostrojenie, 1981; English translation, ed. M. F. Hamitton, New York: American Institute of Physics, 1987.

[58] Rudenko, O. V. *Usp. Fiz. Nauk*, **149**, 3 (1986), 413–47.
[59] Sutin, A. M. In: *Nelineinaya akustika* (*Nonlinear Acoustics*), V. A. Zverev & L. A. Ostrovsky (eds), Gorky: IPF AN SSSR, 1980.
[60] Pelinovskii, E. N., Petukhov, Yu. A. & Fridman, V. E. *Izv. Vyssh. Uchebn. Zaved., Ser. Fiz. Atm. Okeana* **15** (1979), 436.
[61] Rytov, S. M., Kravtsov, Yu. A., & Tatarsky, V. I. *Introduction to Statistical Radiophysics. Part II.* Moscow: Nauka, 1978.
[62] Kravtsov, Yu. A. & Orlov, Yu. I. *Geometricheskaya optika neodnorodnykh sred* (*Geometrical Optics of Inhomogeneous Media*) Moscow: Nauka, 1980.
[63] Ostrovsky, L. A. & Stepanov, N. S. *Radiophys. Quant. Electr.* **14**, 4 (1971), 489–529.
[64] Hopf, E. *Comm. Pure Appl. Math.* **3** (1950), 201–3.
[65] Cole, J. D. *Quart. Appl. Math.* **9** (1951), 225–36.
[66] Benton, E. R. & Platzman, G. N. *Quart. Appl. Math.* (1972). N 7. 195–212.
[67] Sobolev, S. L. *Equations of Mathematical Physics.* Moscow: Nauka, 1966.
[68] Fedoryuk, M. V. *Method of Saddle-point.* Moscow: Nauka, 1977.
[69] Tatsumi, T. & Kida, S. *J. Fluid. Mech.* **55**, 4 (1972) 659–75.
[70] Landau, L. D. & Lifshitz, E. M. *Mechanics.* Moscow: Nauka, 1965.
[71] Lyamshev, L. M. & Naugol'nykh, K. A. *Akust. Zhurn.*, **27**, 5 (1981), 641–68.
[72] Vasil'yeva, O. A., Karabutov, A. A., Lapshin, E. A. & Rudenko, O. V. *Interaction of One-dimensional Waves in Nondispersive Media*, Moscow: MGU, 1983.
[73] Gossard, E. E. & Khuk, U. K. *Waves in Atmosphere.* Moscow: Mir, 1978.
[74] Yakushkin, I. G. *Radiophys. Quant. Electr.* **24**, 1 (1981), 59–67.
[75] Gurbatov, S. N. *Radiophys Quant. Electr.* **26**, 3 (1983), 283–94.
[76] Nakazawa, H. *Progr. Theoret. Phys.* **64**, 5 (1980), 1551–64.
[77] Klyatskin, V. I. *Stochastic Equations and Waves in Randomly-Inhomogeneous Media.* Moscow: Nauka, 1980.
[78] Rytov, S. M. *Introduction to Statistical Radiophysics. Part I.* Moscow: Nauka, 1976.
[79] Malakhov, A. N. *Fluktuatsii v avtokolebatel'nykh sistemakh (Fluctuations in Self-Oscillating Systems).* Moscow: Nauka, 1968.
[80] Doroshkevitch, A. G. *Astrofizika* **6** (1970), 581.
[81] Tikhonov, V. I. *Yybrosy slutchainykh protsessov* (*Overshoots of Random Processes*). Moscow: Nauka, 1970.
[82] Klyatskin, V. I. *Method of Submersion in a Theory of Wave Propagation.* Moscow: Nauka, 1986.
[83] Zel'dovich, Ya. B. & Myshkis, A. D. *Elementy mathematicheskoi fiziki* (*Elements of Mathematical Physics*) Moscow: Nauka, 1973.
[84] Gurbatov, S. N. *Radiophys. Quant. Electr.* **20**, 1 (1977), 73–6.
[85] Kadomtsev, B. V. & Petviashvili, V. I., *Dokl. Akad. Nauk SSSR* **208** (1973), 794. *Sov. Phys. Dokl.* **18** (1973), 115.
[86] Kuznetsov, V. P. *Akust. Zh.* **16** (1970), 155. *Sov. Phys. Acoustics* **16** (1970) 129.
[87] Rudenko, O. V. & Chirkin, A. S. (a) *Dokl. Akad. Nauk SSSR* **225** (1975), 520. *Sov. Phys. Dokl.* **20** (1975), 748. (b) *Akust. Zh.* **20** (1974), 297. *Sov. Phys. Acoustics* **20** (1974), 181.
[88] Gurbatov, S. N. & Malakhov, A. N. *Sov. Phys. Acoust.* **23** (1977), 325. *Akust. Zh.* **23** (1977) 569.
[89] Gurbatov, S. N. & Shepelevich, L. G., *Radiophys. Quant. Electr.* **21**, 11 (1978), 1131–8.
[90] Gurbatov, S. N., Saichev, A. I. & Yakushkin, I. G. *Usp. Fiz. Nauk* **141** (1983), October, 221–55. *Sov. Phys. Usp.* **26**, 10 (1983), Oct. 857–76.

[91] Tatsumi, T. & Kida, S. *J. Fluid. Mech.* **55**, 4 (1972), 659–75.
[92] Kida, S. *J. Fluid. Mech.* **93**, 2 (1979), 337–77.
[93] Cramer, H. & Leadbetter, M. R. *Stationary and Related Stochastic Processes*, New York: Wiley, (1967), Russ. transl., Moscow: Mir, 1969.
[94] Tikhonov, V. I. *Vybrosy sluchainykh protsessov (Exceedances of Random Processes)*, Moscow: Nauka, 1970.
[95] Gurbatov, S. N. & Saichev, A. I., *Zh. Eksp. Teor. Fiz.* **80** (1981), 689. *Sov. Phys. JETP* **53** (1981), 347
[96] Pelinovsky, E. N., *Radiophys. Quant. Electr.* **19**, 3 (1976), 262–70.
[97] Lipatov, S. K. & Saichev, A. I. *Radiophys. Quant. Electr.* **21**, 12 (1978), 1252–6.
[98] Gurbatov, S. N. *Radiophys. Quant. Electr.* 10 (1984), **27**, 1248–55.
[99] Kuznetsov, N. N. & Rozhdestvensky, B. L. *ZhVMMF* **1**, 2 (1961), 217–23.
[100] Longeir, M. & Einisto. J. (eds) *Large-scale Structure of the Universe.* Moscow: Mir, 1981.
[101] Zel'dovich, Ya. B., Einasto, J. & Shandarin, S. F. Giant voids in the Universe. *Nature.* **300**, 5891 (1982), 407–13.
[102] Peebles, P. J. E. *The Large Scale Structure of the Universe.* Princeton, NJ: Princeton University Press, 1983.
[103] Shandarin, S. F., Doroshkevich, A. G. & Zel'dovich, Ya. B. *Usp. Fiz. Nauk* **139** (1983), 83. *Sov. Phys. Usp.* **26** (1983) 46.
[104] Arnold, V. I., Varchenko, A. N. & Gusein-Zade, S. M. *Singularities of Differential Mappings.* Moscow: Nauka, 1982.
[105] Arnold, V. I. *Usp. Fiz. Nauk.* 1983. **141**, vyp. 4, 569–90.
[106] Zel'dovich, Ya. B. *Pis'ma v Astron. Zh.* (*Letters to Astron. Journal*) **8** (1982), 195–7.
[107] Zel'dovich, Ya. B. & Shandarin, S. F. *Pis'ma v Astron. Zh.* (*Letters to Astronom. Journal*) **8**, 3 (1982), 131–5.
[108] Zel'dovich, Ya. B., Mamaev, A. V. & Shandarin, S. F., *Usp. Fiz. Nauk* **139**, 1 (1983), 153–63. *Sov. Phys. Usp.* **26** (1983), 77.
[109] Gurbatov, S. N. & Saichev, A. I. *Radiophys. Quant. Electr.* **27**, 4 (1984), 456–68.
[110] Gurbatov, S. N. & Saichev, A. I. & Shandarin, S. F. Large-scale structure of the universe within the framework of model equation for nonlinear diffusion. Preprint no 152. IPM AN SSSR im. Keldysha, Moscow, 1984; *DAN SSSR*, **285** (1985), 323; *Sov. Phys. Dokl.* **30** (1985), 321.
[111] Kofman, L., Pogosyan, D. & Shandarin, S. *Mon. Not. R. Ast. Soc.* **242** (1989), 200.
[112] Melott, A. L. & Shandarin, S. F. *Astrophys. J.* **343** (1989), 1.
[113] Morfey, C. L. & Howell, G. P. *AIAA Journal* **19**. 8 (1981), 986–92.
[114] Howell, G. P. & Morfey, C. L. *J. Sound and Vibration* **114**, 2 (1987), 189–201.
[115] Falkovich, G. E. & Shafarenko, A. V. *Physica* **270** (1987), 399–411.
[116] Falkovich, G. E. & Shafarenko, A. V. Preprint no 344 Inst. Avt. Electrometr. SO AN SSSR, 1987.
[117] Gurbatov, S. N. & Saichev, A. I. *Radiophys. Quant. Electr.* **19**, 8 (1976), 807–11.
[118] Saichev, A. I. *Prikl. matem. mekh.* **41** (1977) 1107.
[119] Gurbatov, S. N. & Demin, I. Yu. *Akust. Zh.* **28**, 5 (1982), 634–40. *Sov. Phys. Acoust.* **28**, 5 (1982), 375.
[120] Silk, G. *The Big Bang.* San Francisco: W. H. Freeman, 1980; Moscow: Mir 1982.

[121] Lyubimov, B. Ya. *Dokl. Akad. Nauk SSSR* **184** (1969), 1069. *Sov. Phys. Dokl.* **14** (1969), 97.
[122] Malakhov, A. N. & Saichev, A. I., *Radiophys. Quant. Electr.* **19**, 10 (1976), 1092–4.
[123] Malakhov, A. N. & Saichev, A. I. *Radiophys. Quant. Electr.* **19** (1976), 1368.
[124] Rudenko, O. V. & Chirkin,. A. S., *Dokl. Akad. Nauk SSSR* **214** (1974), 1045. *Sov. Phys. Dokl.* **19** (1974), 64.
[125] Saichev, A. I. *Radiophys. Quant. Electr.* **17**, 7 (1974), 1025–34.
[126] Rudenko, O. V. & Chirkin, A. S. *Zh. Eksp. Teor. Fiz.* **67** (1974), 1903. *Sov. Phys. JETP* **40** (1974), 945.
[127] Gurbatov, S. N. *Akust. Zh.* **26** (1980), 551; *Sov. Phys. Acoustics* **26** (1980), 302; *Akust. Zh.* **27** (1981), 862; *Sov. Phys. Acoustics* **27** (1981), 475.
[128] Yakushkin, I. G. *Zh. Eksp. Teor. Fiz.* **81** (1981), 967. *Sov. Phys. JETP* **54** (1981), 513.
[129] Akhiezer, A. P. & Lyubarsky, G. Ya. *Dokl. Acad. Nauk SSSR*, **80** (1951), 193.
[130] Gurbatov, S. N., Demin, I. Yu. & Pronchatov-Rubtsov, N. V. *Zh. Eksp. Teor. Fiz.* **91**, 11 (1986), 1352. *Sov. Phys. JETP* **64**, 4 (1986), 797.
[131] Rudenko, O. V. & Khokhlova, V. A. *Akust. Zh.* **34**, 3 (1988), 500–506.
[132] Gurbatov, S. N., Demin, J. Yu. & Saichev, A. I. *Zh. Eksp. Teor. Fiz.* **87** (1984), 497. *Sov. Phys. JETP*, **60**, 2 (1984), 284.
[133] Qian, J. *Phys. Fluids* **27**, 8 (1984), 1957–65.
[134] Yin-Chih & Reed, *Phys. Fluids* **28**, 7 (1985), 2088–99.
[135] Mizushima, J. *Phys. Fluids* **28**, 5 (1985), 1294–8.
[136] Pestorius, F. M. & Blackstock, D. T. *Proc. 1973, Symp. Finite Amplitude Waves Effects in Fluid. Copenhagen*, 1974, pp. 22–4.
[137] Bjørnø, L. & Gurbatov, S. N. *Akust. Zh.* **31**, 3 (1985), 179.
[138] Gurbatov, S. N., Dubkov, A. A. & Malakhov, A. N., *Zh. Eksp. Teor. Fiz.* **72** (1977), 456. *Sov. Phys. JETP* **45** (1977), 239.
[139] Yakushkin, I. G. *Radiophys. Quant. Electr.* **25**, 1 (1982), 33–9.
[140] Oort, J. H. *Ann. Rev. Astron. Astrophys.* **21** (1983), 373.
[141] Doroshkevich, A. G., Kotok, E. V., Novikov, I. D., Polyudov, A. N., Shandarin, S. F. & Sigov, Yu. S. *Mon. Not. R. Ast. Soc.* **19** (1980), 321.
[142] Zel'dovich, Ya. B. *Astrofizika* **6** (1970), 373.
[143] Rozhansky, L. V. & Shandarin, S. F. Large-scale structure of the Universe. A three-dimensional model. Preprint no 79 IPM AN SSSR im. Keldysha, 1988.
[144] Gurbatov, S. N., Saichev, A. I. & Shandarin, S. F. *Mon. Not. R. Ast. Soc.* **236** (1989), 385.
[145] Kravtsov, Yu. A. & Orlov, Yu. A. *Usp. Fiz. Nauk*, **141**, 4 (1983), 591–628.
[146] Doroshkevich, A. G. & Kotok, E. V. Preprint no 137, IPM AN SSSR im. Keldysha, 1988.
[147] Press, W. H. & Schechter, P. *Astrophys. J.*, **187** (1974), 425.
[148] Doroshkevich, A. G. & Zel'dovich, Ya. B. *Astrophys. Space Sci.* **35** (1975), 55.
[149] Barenblatt, G. I. *Radiophys. Quant. Electr.* **19**, 5–6 (1976), 903–31.
[150] Barenblatt, G. I. *Similarity, Self-similarity, and Intermediate Asymptotics*. English Translation, Consultants Bureau (Plenum Press), 1979.
[151] Gurbatov, S. N. & Saichev, A. I. *Radiophys. Quant. Electr.* **31**, 12 (1988), 1043–54.
[152] Shevtchik, V. N. *Fundamentals of Super-high Frequency Electronics*. Moscow: Sov. Radio, 1959.
[153] Rabinovich, M. I. & Trubetskov, D. I. *Introduction to the Theory of Oscillations and Waves*. Moscow: Nauka, 1984; English Transl., Dordrecht; Kluwer, 1989.

[154] Molodtsov, S. N. & Saichev, A. I. *Radiophys. Quant. Electr.* **20**, 5 (1977), 496–502.
[155] Malakhov, A. N., Molodtsov, A. I. & Saichev, A. I. *Radiophys. Quant. Electr.* **20** 2 (1977) 169–76.
[156] Krupnik, A. V., Molodtsov, S. N. & Saichev, A. I. *Radiophys. Quant. Electr.* **22**, 12 (1979), 1472–7.
[157] Salpeter, E. E. *Astrophys. J.* **147**, 2 (1967), 433.
[158] Arnold, V. I. Catastrophe theory. Moscow: Znanie, 1981 (ser. 'Matematika i kibernetika', no. 9), *Usp. Fiz. Nauk* **4**, (1983), 569–90; English Transl., Berlin: Springer, 1984.
[159] Bogayevsky, I. A. *Algebra i analiz.* **1**, 4 (1989), 1.
[160] Ostrovsky, L. A. & Naugol'nykh, K. A Nonlinear Acoustic Wave Processes in fluids, Cambridge: Cambridge University Press, 1991.
[161] Moffatt, H. K. Simple Topological Aspects of Turbulent Velocity Dynamics, *Proc. IUTAM Symp. on Turbulence and Chaotic Phenomena in Fluids*, Elsevier, 1984, p. 223.
[162] Hentschel, H. G. E. & Procaccia, I. *Phys. Rev.* **A29** (1984), 1461.
[163] Ball, R. C. & Kingdon, R. D. The fractal dimension of clouds, Preprint, Cavendish Laboratory, University of Cambridge, 1986.
[164] Voss, R. F. *Fractals in Nature, The Science of Fractal Images* (ed. H. O. Peitgen & D. Saupe). Berlin, Heidelberg: Springer, 1988.
[165] Vassilicos, J. C. On the geometry of lines in two-dimensional turbulence; In: *Advances in Turbulence 2* (ed. N. H. Fernholz & H. E. Fiedler). Berlin, Heidelberg: Springer, 1989.
[166] Vassilicos, J. C. Fractal dimensions and spectra in turbulence, *Proc. Royal. Soc. London* (submitted in 1990).
[167] Weinberg, D. H. & Gunn, J. E. Large scale structure and the adhesion approximation, *Mon. Not. Roy. Astr. Soc.* (submitted in 1990).
[168] Weinberg, D. H. PhD thesis, Princeton University, 1989.
[169] Melott, A. Preprint, University of Pittsburgh, 1982.

Index